Applied Optimal Designs

Applied Optimal Designs

Edited by

Martijn P. F. Berger

Department of Methodology and Statistics, University of Maastricht, The Netherlands

Weng Kee Wong

Department of Biostatistics, UCLA, Los Angeles, USA

John Wiley & Sons, Ltd

Telephone (+44) 1243 779777

Email (for orders and customer service enquiries): cs-books@wiley.co.uk
Visit our Home Page on www.wiley.com

Other Wiley Editorial Offices

John Wiley & Sons Inc., 111 River Street, Hoboken, NJ 07030, USA

Jossey-Bass, 989 Market Street, San Francisco, CA 94103-1741, USA

Wiley–VCH Verlag GmbH, Boschstr. 12, D-69469 Weinheim, Germany

John Wiley & Sons Australia Ltd, 33 Park Road, Milton, Queensland 4064, Australia

John Wiley & Sons (Asia) Pte Ltd, 2 Clementi Loop #02-01, Jin Xing Distripark, Singapore 129809

John Wiley & Sons Canada Ltd, 22 Worcester Road, Etobicoke, Ontario, Canada M9W 1L1

Library of Congress Cataloging-in-Publication Data

Applied optimal designs/edited by Martijn P. F. Berger, Weng Kee Wong.
p. cm.
Includes bibliographical references and index.
ISBN 0-470-85697-1 (alk. paper)
1. Optimal designs (Statistics) 2. Experimental design. I. Berger, Martijn P. F. II. Wong, Weng Kee.

QA279.A67 2005
519.5′7–dc22 2004058017

British Library Cataloguing in Publication Data

A catalogue record for this book is available from the British Library

ISBN 0-470-85697-1

Typeset in 10/12pt Times by Thomson Press (India) Limited, New Delhi
Printed and bound in Great Britain by TJ International Ltd., Padstow, Cornwall
This book is printed on acid-free paper responsibly manufactured from sustainable forestry in which at least two trees are planted for each one used for paper production.

Contents

List of Contributors

Mekibib Altaye
Center for Epidemiology and Biostatistics
Cincinnati Children's Hospital
and
The University of Cincinnati College of Medicine
Cincinnati, Ohio
USA

Michaela Brocke
Westfälische Wilhelms-Universität Münster
Psychologisches Institut IV
Fliednerstr. 21
D-48149 Münster
Germany

Steven Buyske
Rutgers University
Department of Statistics
110 Frelinghuysen Rd
Pitscataway
NJ 08854-8019
USA

Holger Dette
Ruhr-Universität Bochum
Fakultät und Institut für Mathematik
44780 Bochum
Germany

Allan Donner
Department of Epidemiology and Biostatistics
Faculty of Medicine and Dentistry
University of Western Ontario
and
Robarts Clinical Trials
Robarts Research Institute
London Ontario
Canada

Valerii Fedorov
GlaxoSmithKline
1250 So. Collegeville Road
PO Box 5089, UP 4315
Collegeville
PA 19426-0989
USA

Peter Goos
Department of Mathematics, Statistics & Actuarial Sciences
Faculty of Applied Economics
University of Antwerp
Prinsstraat 13
2000 Antwerpen
Belgium

Ulrike Graßhoff
Otto-von-Guericke-Universität Magdeburg
Insitut für Mathematische Stochastik
Postfach 4120
D-39016 Magdeburg
Germany

Heiko Großmann
Westfälische Wilhems-Universität
Münster
Psychologisches Institut IV
Fliednerstr. 21
D-48149 Münster
Germany

Heinz Holling
Westfälische Wilhelms-Universität
Münster
Psychologisches Institut IV
Fliednerstr. 21
D-48149 Münster
Germany

Peter W. James
University of Newcastle
School of Mathematics and Statistics
Newcastle Upon Tyne
NE1 7RU
Newcastle
UK

Douglas J. Jones
Rutgers Business School
111 Washington Avenue
Newark
NJ 07102
USA

Sergei Leonov
GlaxoSmithKline
1250 So. Collegeville Road
PO Box 5089, UP 4315
Collegeville
PA 19426-0989
USA

John N. S. Matthews
University of Newcastle
School of Mathematics and Statistics
Newcastle Upon Tyne
NE1 7RU
Newcastle
UK

James McPhee
UCLA
Department of Civil and Environmental Engineering
5732B Boelter Hall
Los Angeles
CA 90095-1593
USA

Viatcheslav B. Melas
St. Petersburg State University
Department of Mathematics
St. Petersburg
Russia

Mikhail S. Nediak
Queen's School of Business
Goodes Hall
Queen's University
143 Union St.
Kingston, Ontario, Canada
K7L 3N6

Marie Reilly
Karolinska Institutet
Department of Medical Epidemiology and Biostatistics
PO Box 281
SE-17177
Stockholm
Sweden

Agus Salim
National Centre for Epidemiology and Population Health
The Australian National University
Canberra
Australia

Rainer Schwabe
Otto-von-Guericke-Universität
Magdeburg
Insitut für Mathematische Stochastik
Postfach 4120
D-39016 Magdeburg
Germany

Nikolay Strigul
Princeton University
Department of Ecology and
Evolutionary Biology
Princeton, NJ 08540
USA

Lieven Tack
Katholieke Universiteit Leuven
Department of Applied
Economics
Leuven
Belgium

Martina Vandebroek
Katholieke Universiteit Leuven
Department of Applied Economics
Leuven
Belgium

William W-G. Yeh
UCLA
Department of Civil and Engineering
5732B Boelter Hall
Los Angeles
CA 90095-1593
USA

Editors' Foreword

There are constantly new and continuing applications of optimal design ideas in different fields. An impetus behind this driving force is the ever-increasing cost of running experiments or field projects. A well-designed study cannot be overemphasized because a carefully designed study can provide accurate statistical inference with minimum cost. Optimum design of experiments is therefore an important subfield in statistics. This book is a collection of papers on applications of optimal designs to real problems in selected fields. Some chapters include an overview of applications of optimal design in specific fields. Because optimal design ideas are widely used in many disciplines and researchers have different backgrounds, we have tried to make this book accessible to our readers by minimizing the technical discussion. Our purpose here is to expose researchers to applications of optimal design in various fields and hope that in so doing we will stimulate further work in optimal experimental designs. In the next few paragraphs, we provide a sample of applications of optimal design theory in different fields.

Optimal design theory has been frequently applied to engineering (Gianchandani and Crary, 1998; Crary *et al.*, 2000; Crary 2002), chemical engineering (Atkinson and Bogacka, 1997), and calibration problems (Cook and Nachtsheim, 1982). Optimal design theory has also been applied to the design of electronic products. For example, Clyde *et al.* (1995) used Bayesian optimal design strategies for constructing heart defibrillators. In bioengineering, Lutchen and Saidel (1982) derived an optimal design for nonlinear pulmonary models that described mechanical and gas concentration dynamics during a tracer gas washout. Nathanson and Saidel (1985) also constructed an optimal design for a ferrokinetics experiment. Beginning in the late 1990s, applications of optimal designs are being increasingly used in food engineering (Cunha *et al.*, 1997, 1998; Cunha and Oliverira, 2000).

Another field with many applications of optimal design ideas is the broad area of biomedical and pharmaceutical research. Applications of optimal designs can be found in toxicology (Gaylor *et al.*, 1984; Krewski *et al.*, 1986; Van Mullekom and Myers, 2001; Wang, 2002), rhythmometry (Kitsos *et al.*, 1988), bioavailability studies for compartmental models (Atkinson *et al.*, 1993), pharmacokinetic studies (Landaw, 1984; Retout *et al.*, 2002; Green and Duffull, 2003), cancer research

(Hoel and Jennrich, 1979), drug, neurotransmitter and hormone receptor assays (Bezeau and Endrenyi, 1986; Dunn, 1988; Minkin, 1993; Lopez-Fidalgo and Wong, 2002; Imhof *et al.*, 2002, 2004). A recent application of optimal design theory is in the study of viral dynamics in AIDS trials (Han and Chaloner, 2003). Optimal designs for clinical trials are described in Atkinson (1982, 1999), Zen and DasGupta (1998), Mats *et al.* (1998) and Haines *et al.* (2003). In a related set-up, Zhu and Wong (2000, 2001) discussed optimal patient allocation schemes in group randomized trials. Recently, optimal design strategies are increasingly being used in event-related fMRI-experiments in brain mapping studies, see Dale (1999) and the references therein.

Optimal design theory is also widely used in improving the design of tests in education. There are two types of designs here: calibration or sampling designs and test designs. Optimal sampling designs have been developed for efficient item parameter estimation (Berger, 1994; Jones and Jin, 1994; Buyske, 1998; Berger, *et al.*, 2000; Lima Passos and Berger, 2004), and optimal test designs have been studied for efficient latent trait estimation (Berger and Mathijssen, 1997; Van der Linden, 1998). Optimal design issues have also been applied to computer adaptive testing (CAT) (Van der Linden and Glas, 2000).

Another two areas where optimal design ideas are used are in the field of environmental research and epidemiology. Good designs for studying spatial sampling in air pollution monitoring and contamination problems were proposed by Fedorov (1994, 1996) and Abt *et al.* (1999) respectively; see also Mueller and Zimmerman (1999) where they constructed efficient designs for variogram estimation. Applications of optimal design theory can also be found in environmental water-related problems. Zhou *et al.* (2003) provided optimal designs to estimate the smallest detectable trace limit in a water contamination problem. In epidemiology, optimal designs were used to estimate the prevalence of multiple rare traits (Hughes-Oliver and Rosenberger, 2000) or in estimating different types of risks (Dette, 2004).

In the above papers, a common approach to constructing optimal designs is to treat them as continuous designs. These designs are treated as probability measures on a known design space and the design points and the proportion of observations to be taken at each design point are determined. The total number of observations of the experiment is assumed to be predetermined either by cost or practical considerations, and the implemented design then takes the appropriate number of observations at each point prescribed by the continuous design. There is no guarantee that observations at each point will be an integer; in practice, simple rounding to an integer will suffice. Optimal rounding schemes are given in Pukelsheim and Rieder (1992).

Continuous designs, sometimes also called approximate designs, are the main focus in this book. Such optimal designs were proposed by Kiefer in the late 1950s and his research in this area is voluminously documented in Kiefer (1985). Monographs on optimal design theory for continuous designs include Silvey (1980), Atkinson and Donev (1992) and Pukelsheim (1993), among others. Wong and Lachenbruch (1996) gave a tutorial on application of optimal design theory to design a dose response study. More complicated design strategies are described in Cook and Wong (1994) and Cook and Fedorov (1995). Wong (2000) gave an overview of recent developments in optimal design strategies.

In the simplest case, the set-up for application of optimal design theory to find an optimal design for a statistical model is as follows. Suppose that we can adequately describe the relationship between the mean response and a predictor variable x by $\eta(x, \theta)$. Here x takes on values in a user-selected design space $\boldsymbol{X}$, η is assumed known and θ is a vector of unknown parameters. The space $\boldsymbol{X}$ is usually an interval if there is only a single independent variable x in the study; otherwise $\boldsymbol{X}$ is a multi-dimensional Euclidean space. The responses or observations are assumed to be independent normally distributed variables and the error variance of each observation is assumed to be constant. If the design ξ has trials at m distinct points on the design space $\boldsymbol{X}$, the design is written as

$$\xi = \left\{ \begin{matrix} x_1 & x_2 & \dots & x_m \\ w_1 & w_2 & \dots & w_m \end{matrix} \right\},$$

where the first line represents the m distinct values of the independent variable x and the second line represents the associated weights w_i, such that $0 < w_i < 1$ for all i's and $\int_X \xi(\mathrm{d}x) = 1$. Apart from a multiplicative constant, the expected Fisher's information of the design ξ is given by

$$M(\xi, \theta) = \int_X f(x, \theta) f(x, \theta)^T \xi(\mathrm{d}x),$$

where $f(x, \theta)$ is the derivative of $\eta(x, \theta)$ with respect to θ. In our set-up, the objective of our study, like many of the objectives in this book, is a convex function of the expected information matrix. This formulation ensures that the optimal designs and their properties can be readily found and studied using tools from convex analysis.

The optimal design ξ^* is the one that minimizes a user-selected objective function Φ over all designs on the design space $\boldsymbol{X}$. In general, the optimal design problem can be described as a constrained non-linear mathematical programming problem, i.e.

$$\text{minimize } \Phi\{M(\xi, \theta)\},$$

where the minimization is taken over all designs on $\boldsymbol{X}$. Sometimes, the minimization is taken over a restricted set of designs on $\boldsymbol{X}$. For example, if it is expensive to take observations at a new location or administer a drug at a new dose, one may be interested in designs with only a small number of points. Typically, when this happens, the minimization is over all designs supported at only k-points and k is the length of the vector θ. Such optimal designs are called k-point optimal designs and they can be described analytically (Dette and Wong, 1998) even when there is no closed form description for the optimal designs found from the unrestricted search on X.

One of the most frequently used objective functions is D-optimality, defined by the functional $\Phi\{M(\xi,\theta)\} = -\ln|M(\xi,\theta)|$. This is a convex function over the space of all designs ξ on space X (Silvey, 1980). A natural interpretation of a D-optimal design is that it minimizes the generalized variance of the estimated θ, or equivalently, a D-optimal design has the minimal volume of the confidence ellipsoid of θ, the vector of all the model parameters. A nice property of D-optimal designs is that for quantitative variables x_i, they do not depend on the scale of the variables. This is an advantage that may not be shared by other design criteria.

Other alphabetic optimality criteria used in practice are A-optimality and E-optimality criteria. An A-optimal design minimizes the sum of the variances of the parameter estimates, i.e. minimizes the objective functional $\Phi\{M(\xi,\theta)\} = \text{trace}\left\{M(\xi,\theta)^{-1}\right\}$. In terms of the confidence ellipsoid, the A-optimality criterion minimizes the sum of the squares of the lengths of the axes of the confidence ellipsoid. The E-optimality criterion minimizes the least well-estimated contrast of the parameters. In other words, an E-optimal design minimizes the squared length of the major axis of the confidence ellipsoid. Other popular design criteria are D_s-optimality and I-optimality. The former criterion minimizes the volume of the confidence ellipsoid of a user-selected subset of the parameters, while I-optimality averages the predictive variance of the design over a given region using a user-selected weighting measure. In particular, c-optimality, which is a special case of I-optimality, is often used to estimate a given function of the model parameters. For instance, Wu (1988) used c-optimality to construct efficient designs for estimating a single percentile in different quantal response curves. Silvey (1980), Atkinson and Donev (1992) and Pukelsheim (1993) provide further discussion of these criteria and their properties.

Following convention, we measure the efficiency of any design by the ratio, or some function thereof, of the objective functions evaluated at the design relative to the optimal design. In practice, the efficiency is scaled between 0 and 1 and is reported as a percentage. Designs with high efficiency are sought in practice. A design with 50% efficiency means the design requires 50% more resources than what would have been required if an optimal design had been used, without loss of accuracy in the statistical inference.

There are computer algorithms for generating many of the optimal designs described here. A starting design is required to initiate the algorithm. At each iteration, a design is generated and eventually the designs converge to the optimal design. Details of the algorithms, convergence and computational issues are discussed in the design monographs. The verification of the optimality of a design over all designs on X is usually accomplished graphically using an equivalence theorem, again widely discussed in the design monographs. The directional derivative of the convex functional is plotted versus the values of $\boldsymbol{X}$ and the equivalence theorem tells us that the design is optimal if the graph satisfies certain properties required for an optimal design. This plot can be easily constructed and visually inspected if X is an interval. Equivalence theorems also provide us with a useful lower bound on the efficiency of each of the generated designs and the lower bound can help the practitioner specify a stopping rule in the numerical algorithm (Dette and Wong, 1996).

The ten chapters in the book contain reviews and sample applications of optimal design theory to real problems. The application areas are broadly divided under the following headings (i) education, (ii) business marketing, (iii) epidemiology, (iv) microbiology and pharmaceutical research, (v) medical research, (vi) environmental science and (vii) manufacturing industry.

(i) Education

Large-scale standardized testings in educational institutions, the US military and multinational companies have been popular for the past 50 years. At the same time there is interest in testing large samples of pupils, workers and soldiers as efficiently as possible. Optimal design ideas were applied with the aim of reducing the costs of administering the traditional paper and pencil test. This has led to so-called tailored tests and, more recently, computerized adaptive tests (CAT). All these tests are now widely used at reduced cost, thanks in part to the successful application of optimal design theory.

In Chapter 1, Buyske reviews the development of optimal designs in educational testing. Two distinct design problems exist in testing. The first has to do with the design of a test. How can a test be composed with a minimum number of items to estimate the proficiency or attitude of examinees as efficiently as possible? The second problem is a calibration problem. How can the item parameters be estimated as efficiently as possible? Buyske considers not only fixed-form tests, but also adaptive tests, with dichotomous and polytomous responses. Research on the application of optimal design theory to testing is ongoing and may very well lead to further developments in CAT and expansions to models that include multidimensional traits or non-parametric measurement models.

One of the promising developments in testing is the design of so-called testlets. Testlets are small tests consisting of a set of related items tied to a common stem. Jones and Nediak describe in Chapter 2 how the parameters of the items in such testlets can be estimated efficiently by formulating the design as a network-flow problem. They incorporate optimal design theory and study the feasibility of sequential estimation with D-optimal designs. This research is still in progress. Possible extensions include the employment of informative priors and other optimality criteria.

(ii) Business Marketing

A subfield in social sciences where optimal design theory can be applied is the measurement of preferences. Großmann and colleagues describe in Chapter 3 optimal designs for the measurement of preferences, and empirically test their relevance. Using a general linear model, the authors evaluate various consumers' preferences using paired comparisons. The problem of choosing the paired comparisons is an optimal design problem. Großmann *et al.* use a *DS*-optimality criterion (Sinha, 1970) to find optimal designs for paired comparison experiments and compare their performances with heuristic designs. The results indicate that *DS* optimal designs for paired comparison experiments provide good guidance for choosing an appropriate design for practitioners.

(iii) Epidemiology

A popular and efficient design in epidemiology is the case-control design. A balanced design with equal numbers of cases and controls in the various exposure strata is usually efficient when the cost of sampling is not taken into account (Cain and Breslow, 1988). When the cost of measurement is an important consideration, Reilly and Salim in Chapter 4 show how to derive optimal two-stage designs, where cheap measurements are obtained for a cross-sectional, cohort or case-control sample in the first stage, and more expensive measurements are obtained for a limited subgroup of subjects in the second stage. The authors also provide software for deriving optimal designs using the R, S-Plus and STATA statistical packages.

(iv) Microbiology and Pharmaceutical research

In pharmaceutical experiments, nonlinear models are often applied and the optimal design problem has received much attention. In Chapter 5, Fedorov and Leonov present an overview of optimal design methods and describe some new strategies for drug development. First the basic concepts are introduced and the optimal design problem is described for a general nonlinear regression model. Multi-response problems and models with a non-constant variance function are included. They also incorporate cost considerations in their designs and discuss the usefulness of adaptive designs in drug development.

In microbiology the regression models are often nonlinear and quite complex. This makes the design problem much more complicated. Dette *et al.* present an overview of these problems in Chapter 6. They explain optimal design theory for different exponential nonlinear models, including the Monod differential model. Because optimal designs for such models are usually locally optimal, Dette *et al.* also describe three sophisticated procedures to handle this problem, namely the sequential design procedure, the maximin design procedure and Bayesian designs. Their chapter clearly demonstrates the benefits of optimal design methodology in microbiology.

(v) Medical Research

Before mounting a large-scale clinical trial, sometimes pilot studies are carried out to ascertain whether outcomes can be accurately measured. For example, skin scores frequently serve as a primary outcome measure for Scleroderma patients even though skin scores are subjectively measured by the rheumatologists. Inter-rater agreement becomes an important issue and in such studies, the design problem concerns the optimal number of subjects and the optimal number of raters. Donner and Altaye discuss these design issues in Chapter 7 and show how statistical power is affected by dichotomization of continuous or polytomous outcomes, and budgetary constraints. This chapter demonstrates that precision of the estimate can be improved by judicious choice of the number of raters and subjects, or a binary or polytomous outcome.

The Bayesian approach to designing a study is gaining popularity. Matthews and James use a Bayesian paradigm in Chapter 8 and construct optimal designs for measurement of cerebral blood flow. The problem is particularly challenging because for patients with severe neurological traumas, such blood-flow measurement needs to be monitored at the bedside. The authors use a nonlinear model to describe the cerebral blood-flow and apply Bayesian procedures to this design problem. The optimal design is then used to assess efficiency of competing designs and to search for more practical designs.

(vi) Environmental Science

The pollution of groundwater is a major source of concern today. It may not be possible in the future to clean polluted groundwater at a reasonable cost and in a reasonable time. Knowledge about flow and mass transportation of groundwater is therefore of crucial importance. The flow and mass transport of groundwater can be modelled by partial differential equations, and optimal design theory can play a critical role in constructing monitoring networks that maximize plume characterization with a minimum of sampling costs. In Chapter 9, McPhee and Yeh review the application of experimental design theory in two areas of groundwater modelling, namely, to parameter estimation and to monitoring the network design for contaminant plume characterization.

(vii) Manufacturing Industry

Optimal designs have a long tradition in industrial experiments. These experiments have experimental factors, such as material, temperature or pressure, but also extraneous sources of variation or blocking factors, which are not subject to experimental manipulation. Examples of blocking factors are location, plots of land or time. Such experiments are usually referred to as blocked experiments, where the blocks are frequently considered as random factors. Goos *et al.* review the literature on the design of blocked experiments in Chapter 10. Factorial designs and response surface designs are discussed for experiments when blocks are considered fixed or random. Optimal ways to run a blocked experiment are discussed, including instances when the trend and cost of the experiment have to be incorporated into the study.

Acknowledgements

The editors are most grateful to all authors for their contribution to this volume and to all referees who helped with the review process. The referees provided valuable assistance in selecting and finalising papers appropriate for the volume.

References

Abt, M., Welch, W. J., and Sacks, J. (1999). Design and analysis for modeling and predicting spatial contamination. *Mathematical Geology*, **31**, 1–22.

Atkinson, A. C. (1982). Optimum biased coin designs for sequential clinical-trials with prognostic factors. *Biometrika*, **69**, 61–67.

Atkinson, A. C. (1999). Optimum biased-coin designs for sequential treatment allocation with covariate information. *Statistics in Medicine*, **18**, 1741–1752.

Atkinson, A. C. and Bogacka, B. (1997). Compound D- and Ds-optimum designs for determining the order of a chemical reaction. *Technometrics*, **39**, 347–356.

Atkinson, A. C. and Donev, A. N. (1992). *Optimum Experimental Design*. Clarendon Press, Oxford.

Atkinson, A. C., Chaloner, K., Herzberg, A. M. and Jurtiz, J. (1993). Optimum experimental designs for properties of a compartmental model. *Biometrics*, **49**, 325–337.

Berger, M. P. F. (1994). D-optimal sequential sampling designs for item response theory models. *Journal of Educational Statistics*, **19**, 43–56.

Berger, M. P. F. and Mathijssen, E. (1997). Optimal test designs for polytomously scored items. *British Journal of Mathematical and Statistical Psychology*, **50**, 127–141.

Berger, M. P. F., King, J. and Wong, W. K. (2000). Minimax designs for item response theory models. *Psychometrika*, **65**, 377–390.

Bezeau, M. and Endrenyi, L. (1986). Design of experiments for the precise estimation of dose-response parameters: the Hill equation. *Journal of Theoretical Biology*, **123**, 415–430.

Buyske, S. G. (1998). Optimal design for item calibration in computerized adaptive testing: the 2PL case. In *New Developments and applications in Experimental Design*, Flournoy, N., Rosenberger W. F. and Wong (eds), W. K. Institute of Mathematical Statistics, Hayward, Calif. Monograph Series, 34, 115–125.

Cain, K. C. and Breslow, N. E. (1988). Logistic regression analysis and efficient design for two-stage studies. *American Journal of Epidemiology*, **128**(6): 1198–1206.

Clyde, M., Muller, P. and Parmigiani, G. (1995). Optimal design for heart defibrillators. In *Bayesian Statistics in Science and Engineering: Case Studies II*, Gatsonis, C., Hodges, J. S., Kass, R. E., Singpurwalla, N. D. (eds), Springer-Verlag, Berlin/Heidelberg/New York, 278–292.

Cook, R. D. and Fedorov, V. V. (1995). Constrained optimization of experimental design. *Statistics*, **26**, 129–178.

Cook, R. D. and Nachtsheim, C. J. (1982). Model robust linear-optimal designs. *Technometrics*, **24**, 49–54.

Cook, R. D. and Wong, W. K. (1994). On the equivalence of constrained and compound optimal designs. *Journal of the American Statistician Association*, **89**, 687–692.

Crary, S. B. (2002). Design of experiments for metamodel generation. Special invited issue of the *Journal on Analog Integrated Circuits and Signal Processing*, **32**, 7–16.

Crary, S. B., Cousseau, P., Armstrong, D., Woodcock, D. M., Mok, E. H., Dubochet, O., Lerch, P. and Renaud, P. (2000). Optimal design of computer experiments for metamodel generation using I-OPT^{TM}. *Computer Modeling in Engineering and Sciences*, **1**, 127–140.

Cunha, L. M. and Oliverira, F. A. R. (2000). Optimal experimental design for estimating the kinetic parameters of processes described by the first-order Arrhenius model under linearly increasing temperature profiles. *Journal of Food Engineering*, **46**, 53–60.

Cunha, L. M., Oliverira, F. A. R., Brandao, T. R. S. and Oliveira, J. C. (1997). Optimal experimental design for estimating the kinetic parameters of the Bigelow model. *Journal of Food Engineering*, **33**, 111–128.

Cunha, L. M., Oliverira, F. A. R. and Oliveira, J. C. (1998). Optimal experimental design for estimating the kinetic parameters of processes described by the Weibull probability distribution function. *Journal of Food Engineering*, **37**, 175–191.

Dale, A.M. (1999). Optimal experimental design for event-related fMRI. *Human Brain Mapping*, **8**, 109–114.

Dette, H. (2004). On robust and efficient designs for risk estimation in epidemiologic studies. *Scandinavian Journal of Statistics*, **31**, 319–331.

Dette, H. and Wong, W. K. (1996). Bayesian optimal designs for models with partially specified heteroscedastic structure. *The Annals of Statistics*, **24**, 2108–2127.

Dette, H. and Wong, W. K. (1998). Bayesian D-optimal designs on a fixed number of design points for heteroscedastic polynomial models. *Biometrika*, **85**, 869–882.

Dunn, G. (1988). Optimal designs for drug, neurotransmitter and hormone receptor assays. *Statistics in Medicine*, **7**, 805–815.

Fedorov, V. V. (1994). Optimal experimental design: spatial sampling. *Calcutta Statistical Association Bulletin*, **44**, 17–21.

Fedorov, V. V. (1996). *Design of Spatial Experiments: Model Fitting and Prediction*. Oak Ridge National Laboratory Report, ORNL/TM-13152.

Gaylor, D. W., Chen, J. J. and Kodell, R. L. (1984) Experimental designs of bioassays due for screening and low dose extrapolation. *Risk Analysis*, **5**, 9–16.

Gianchandani, Y. B. and Crary, S. B. (1998). Parametric modeling of a microaccelerometer: comparing I- and D-optimal design of experiments for finite element analysis. *JMEMS*, 274–282.

Green, B. and Duffull, S. B. (2003). Prospective evaluation of a D-optimal designed population pharmacokinetic study. *Journal of Pharmacokinetics and Pharmacodynamics*, **30**, 145–161.

Haines, L.M., Perevozskaya, I. and Rosenburger, W.F. (2003). Bayesian optimal designs for Phase I clinical trials. *Biometrics*, **59**, 591–600.

Han, C. and Chaloner, K. (2003). D-and c-optimal designs for exponential regression models used in viral dynamics and other applications. *Journal of Statistical Planning Inference*, **115**, 585–601.

Hoel, P. G. and Jennrich. R. I. (1979). Optimal designs for dose response experiments in cancer research, *Biometrika*, **66**, 307–316.

Hughes-Oliver, J. M. and Rosenberger, W. F. (2000). Efficient estimation of the prevalence of multiple rare traits, *Biometrika*, **87**, 315–327.

Imhof, L., Song, D. and Wong, W. K. (2002). Optimal designs for experiments with possibly failing trials. *Statistica Sinica*, **12**, 1145–1155.

Imhof, L., Song, D. and Wong, W. K. (2004). Optimal design of experiments with anticipated pattern of missing observations. *Journal of Theoretical Biology*, **228**, 251–260.

Jones, D. H. and Jin, Z. (1994). Optimal sequential designs for on-line item estimation. *Psychometrika*, **59**, 59–75.

Kiefer, J. (1985). *Jack Carl Kiefer Collected Papers III: Design of Experiments*. Springer-Verlag New York Inc.

Kitsos, C. P., Titterington, D. M. and Torsney, B. (1988). An optimal design problem in rhythmometry. *Biometrics*, **44**, 657–671.

Krewski, D., Bickis, M., Kovar, J. and Arnold, D. L. (1986). Optimal experimental designs for low dose extrapolation I: The case of zero background. *Utilitas Mathematica*, **29**, 245–262.

Landaw, E. (1984). Optimal design for parameter estimation. In *Modeling Pharmacokinetic/ Pharmacodynamic Variability in Drug Therapy*, Rowland, M., Sheiner, L. B. and Steimer, J-L. (eds), Raven Press, New York, 51–64.

Lima Passos, V. and Berger, M.P.F. (2004). Maximin calibration designs for the nominal response model: an empirical evaluation. *Applied Psychological Measurement*, **28**, 72–87.

Lopez-Fidalgo, J. and Wong, W. K. (2002). Optimal designs for the Michaelis–Menten model. *Journal of Theoretical Biology*, **215**, 1–11.

Lutchen, K. R. and Saidel, G. M. (1982). Sensitivity analysis and experimental design techniques: application to nonlinear, dynamic lung models. *Computers and Biomedical Research*, **15**, 434–454.

Mats, V. A., Rosenberger, W. F. and Flournoy, N. (1998). Restricted optimality for Phase 1 clinical trials. In *New Developments and Applications in Experimental Design*, Flournoy, N., Rosenberger and Wong, W. K. (eds), Institute of Mathematical Statistics, Hayward, Calif. Lecture Notes Monograph Series Vol. 34, 50–61.

Minkin, S. (1993). Experimental design for clonogenic assays in chemotherapy. *Journal of the American Statistician Association*, **88**, 410–420.

Mueller, W. G. and Zimmerman, D. L. (1999). Optimal designs for variogram estimation. *Environmetrics*, **10**, 23–37.

Nathanson, M. H. and Saidel, G. M. (1985). Multiple-objective criteria for optimal experimental design: application to ferrokinetics. *Modeling Methodology Forum*, 378–386.

Pukelsheim, F. (1993). *Optimal Design of Experiments*. Wiley Series in Probabilty and Mathematical Statistics, John Wiley & Sons, Ltd, New York.

Pukelsheim, F. and Rieder, S. (1992). Efficient rounding of approximate designs. *Biometrika*, **79**, 763–770.

Retout, S., Mentre, F. and Bruno, R. (2002). Fisher information matrix for non-linear mixed-effects models: evaluation and application for optimal design of enoxaparin population pharmacokinetics. *Statistics in Medicine*, **21**, 2633–2639.

Silvey, S.D. (1980). *Optimal Design.* Chapman and Hall, London, New York.

Sinha, B. K. (1970). On the optimality of some designs. *Calcatta Statistical Association Bulletin*, **20**, 1–20.

Van der Linden, W.J. (1998). Optimal test assembly of psychological and educational tests. *Applied Psychological Measurement*, **22**, 195–211.

Van der Linden, W. J. and Glas, C. A. W. (2000). *Computerized Adaptive Testing: Theory and Practice*. Kluwer Academic Press, Pordrecht.

Van Mullekom, J. and Myers, R. (2001). *Optimal Experimental Designs for Poisson Impaired Reproduction*. Technical Report 01-1, Department of Statistics, Virginia Tech., Blackburg, Va.

Wang, Y. (2002). Optimal experimental designs for the Poisson regression model in toxicity studies. PhD thesis, Department of Statistics, Virginia Tech., Blackburg, Va.

Wong, W. K. (2000). Advances in constrained optimal design strategies. *Statistica Neerlandica*, **53**, 257–276.

Wong, W. K. and Lachenbruch, P. A. (1996). Designing studies for dose response. *Statistics in Medicine*, **15**, 343–360.

Wu, C. F. J. (1988). Optimal design for percentile estimation of a quantal response curve. In *Optimal Design and Analysis of Experiments*, Dodge, J., Fedorov, V. V. and Wynn, H. P. (eds), North-Holland, Amsterdam, 213–233.

Zen, M. M. and DasGupta, A. (1998). Bayesian design for clinical trials with a constraint on the total available dose. *Sankhya*, Series A, 492–506.

Zhou, X., Joseph, L., Wolfson, D. B. and Belisle, P. (2003). A Bayesian A-optimal and model robust design criterion. *Biometrics*, **59**, 1082–1088.

Zhu, W. and Wong, W. K. (2000). Optimum treatment allocation in comparative biomedical studies. *Statistics in Medicine*, **19**, 639–648.

Zhu, W. and Wong, W. K. (2001). Bayesian optimal designs for estimating a set of symmetric quantiles. *Statistics in Medicine*, **20**, 123–137.

1

Optimal Design in Educational Testing

Steven Buyske

Rutgers University, Department of Statistics, 110 Frelinghuysen Rd, Piscataway, NJ 08854-8019, USA

1.1 Introduction

Formal job testing of individuals goes back more than 3000 years, while formal written tests in education go back some 500 years. Although the earliest paper on optimal design in statistics appeared at about the same time as multiple choice tests appeared, at the beginning of the twentieth century, optimal design theory was first applied to issues arising in standardized testing 40 years ago.

Van der Linden and Hambleton (1997b) suggest thinking of a test as a collection of small experiments (that is, the questions, or items) for which the observations are the test-taker's responses. These observations allow one to infer a measurement of the test-taker's proficiency in the subject of the test. As with most experimental settings, the application of optimal design principles can offer great gains in efficiency, most obviously in shorter tests. Since the cost of producing items can easily exceed US$100 per item, more efficient testing can lead to substantial savings.

The theory underlying most of modern testing is known as item response theory (IRT). In contrast to traditional test theory, IRT considers individual test items, rather than the entire test, to be the fundamental unit. It assumes the existence of an unobserved, or latent, underlying trait for both the proficiency of the test-taker and

Applied Optimal Designs Edited by M.P.F. Berger and W.K. Wong
 ISBN: 0-470-85697-1 (HB)

for the difficulty of the individual item. The difference between the two, as well as other characteristics of the item, determine the probability that the test-taker will answer the item correctly.

1.1.1 Paper-and-pencil or computerized adaptive testing

Traditionally, standardized educational testing has been conducted in large-scale paper-and-pencil administrations of fixed-form tests. For example, in the United States some 3 million students take the SAT I and II tests on seven separate dates annually. These administrations feature a large number of students taking a small number of distinct, essentially equivalent, test forms. After the administration, both test-taker and item parameters are estimated simultaneously.

Although fixed-form tests can also be administered by computer, in recent years the leading alternative to paper-and-pencil testing has been computerized adaptive testing (CAT). In a CAT administration, a test-taker works at a computer. Because each item can be scored as quickly as the answer is recorded, the computer can adaptively select items to suit the examinee. The idea is that by avoiding items that are too hard or too easy for the examinee, a high-quality estimate of the examinee's proficiency can be made using as few as half as many items than in a fixed-form test. CAT administrations can be ongoing. In the United States some 350 000 students take the Graduate Record Examination over more than 200 possible test days annually. Because of the need for on-line proficiency estimation, the item parameters are estimated as part of earlier administrations, known as item calibration, and so CAT is heavily dependent on efficient prior estimation of item parameters. Such testing is not limited to an educational setting; the US military and companies such as Oracle and Microsoft use CAT. Wainer (2000) gives a complete introduction to the subject, while Sands *et al.* (1997) and Parshall *et al.* (2002) give details on the implementation of computer-based testing.

1.1.2 Dichotomous response

The simplest IRT models apply when the answer is dichotomous: either right or wrong. By far the most common model for this situation are the 1-, 2- and 3-parameter logistic models (1-PL, 2-PL and 3-PL). The number of parameters refers to the parameters needed to describe each item. In the 3-PL model, the probability that a test-taker with proficiency θ correctly answers an item with parameters (a, b, c) is

$$P(\theta \mid a, b, c) = c + \frac{1-c}{1+e^{-a(\theta-b)}}, \tag{1.1}$$

where $a \in (0, \infty)$, $b \in (-\infty, \infty)$, and $c \in [0, 1)$. Typical ranges in practice might be $a \in [0.3, 3]$, $b \in [-3, 3]$ and $c \in [0, 0.5]$. The c parameter is often known as the 'pseudo-guessing' parameter. In a plot of the function P, c represents the height of the left asymptote and may be thought of as the probability of a correct response for a test-taker with very low ability. Surprisingly, the estimated value of c is often less than the reciprocal of the number of possible answers. This phenomenon is usually ascribed to the existence of what are known as 'attractive distractors', or wrong answers that seem especially attractive to a guesser. It may also be an artefact of the parameter estimation process. The b parameter is the measure of item difficulty. When θ is much larger than b, the test-taker has a high probability of answering correctly, while when θ is much lower than b, the probability of a correct response is near c. Note that θ and b are on the same scale. Finally, the a parameter is known as the discrimination parameter. It equals $(1-c)/4$ times the slope of the tangent to the graph of P when $\theta = b$. The higher the value of a the more the item discriminates between test-takers with θ a bit below b and test-takers with θ a bit above b, in the sense of the probability of those test-takers getting the item right. Figure 1.1 shows graphs of P for three 3-PL items. These graphs are known as trace lines or item characteristic curves in the IRT literature.

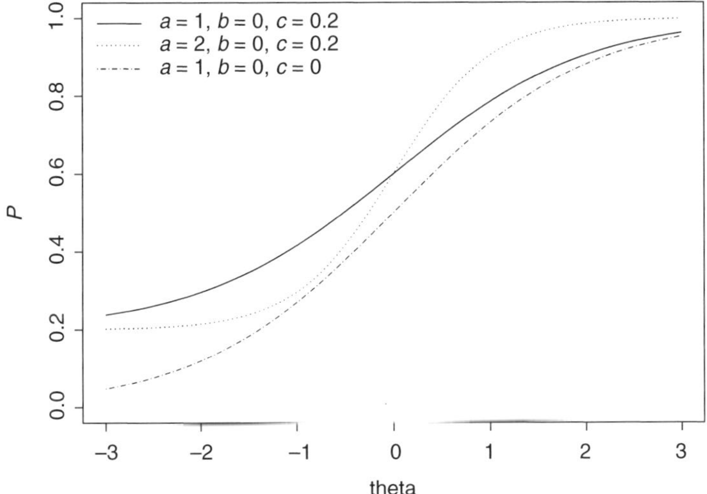

Figure 1.1 Item characteristic curves for three dichotomous items

The 2-PL model is simply the 3-PL model with the pseudo-guessing parameter c fixed at zero. For the 1-PL model, c is fixed at zero and a is also fixed. The well-known Rasch model is equivalent to the 1-PL model. In all three cases, a scaling factor of $D = 1.7$ is often included in front of the a term to approximate the normal ogive. A brief, historically based introduction to dichotomous IRT models can be found in Thissen and Orlando (2001). Lord (1980) and Hambleton and Swaminathan (1985) have book-length treatments.

1.1.3 Polytomous response

There has been substantial recent interest in modelling all of the individual possible responses to an item, rather than just whether the response is correct or not (see, for example, van der Linden and Hambleton, 1997a; Thissen *et al.*, 2001). There are a variety of models for polytomous response. We consider the nominal response model, first put forward by Bock (1972), as it includes several of the others as special cases (see Thissen and Steinberg, 1986). In the statistics literature it is also sometimes known as the multinomial logistic model. The nominal response model can be thought of as a combination of 2-PL models, one for each possible response to an item. If there are $m + 1$ possible responses, then the probability of a test-taker with proficiency θ choosing response u is

$$P_{\mathrm{u}}(\theta \mid \mathbf{a}, \mathbf{b}) = \frac{\exp(a_{\mathrm{u}}\theta + b_{\mathrm{u}})}{\sum_{v=0}^{m} \exp(a_v\theta + b_v)}. \tag{1.2}$$

Because this model is underidentified, a constraint of either $\sum_v a_v = \sum_v b_v = 0$ or $a_0 = b_0 = 0$ is usually added. The vector pair $(\boldsymbol{a}, \boldsymbol{b}) = (\{a_1, \ldots, a_m\}', \{b_1, \ldots, b_m\}')$ thus characterizes the psychometric properties of the item. Figure 1.2 shows graphs of P_u for a single item with three possible responses.

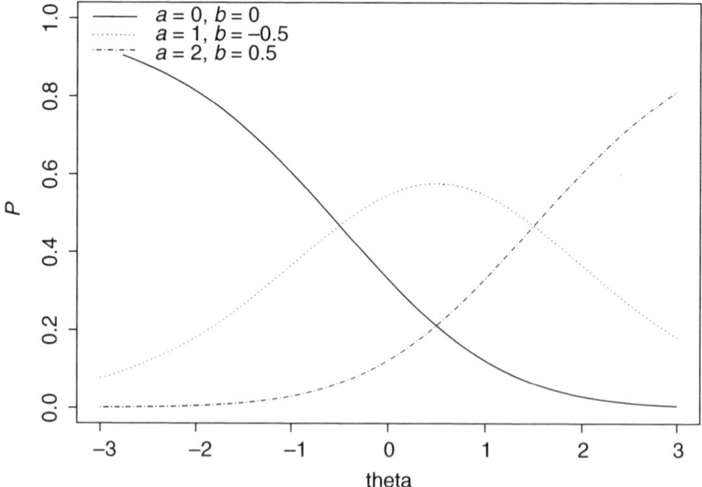

Figure 1.2 Item characteristic curves for three responses to one item, using the nominal response model

1.1.4 Information functions

For future reference it will be useful to write the Fisher information functions for proficiency and for the item parameters. The information for θ from a dichotomous item with parameters (a, b, c) is

$$I_{a,b,c}(\theta) = \frac{P'(\theta)^2}{P(\theta)Q(\theta)} = \frac{a^2(P(\theta) - c)^2 Q(\theta)}{(1 - c)^2 P(\theta)}, \tag{1.3}$$

where $P(\theta)$ is the probability of a correct response, as in Equation (1.1), and $Q(\theta) = 1 - P(\theta)$ is the probability of an incorrect response. Note that when $c = 0$, $I_{a,b}(\theta) = a^2 P(\theta)Q(\theta)$, which is maximized when $P(\theta) = Q(\theta) = 0.5$, or when $b = \theta$. Figure 1.3 shows information functions for the three items displayed in Figure 1.1.

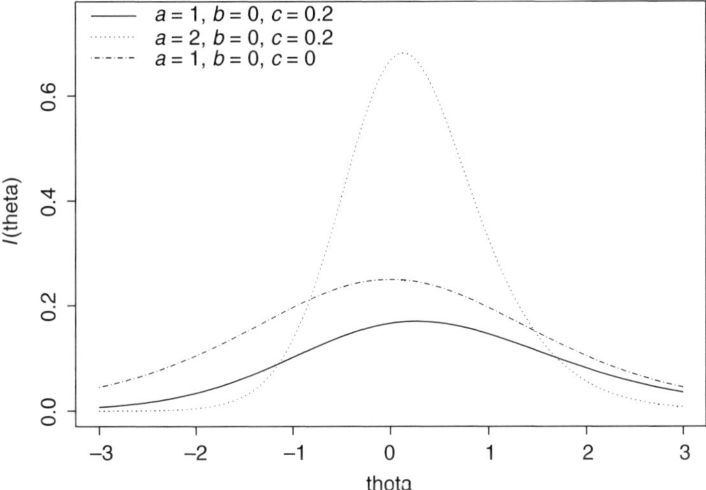

Figure 1.3 Information functions for θ for three dichotomous items. When $c \neq 0$ the function is asymmetric

The information matrix for an item with parameters (a, b, c) from a test-taker with proficiency θ is, writing $\tilde{P}$ for $P - c$,

$$I_\theta(a, b, c) = \frac{Q(\theta)}{(1 - c)^2 P(\theta)} \begin{pmatrix} (\theta - b)^2 \tilde{P}^2(\theta) & -a(\theta - b)\tilde{P}^2(\theta) & (\theta - b)\tilde{P}(\theta) \\ & a^2 \tilde{P}^2(\theta) & -a\tilde{P}(\theta) \\ & & 1 \end{pmatrix}. \tag{1.4}$$

For the nominal response model, the information for θ from an $(m+1)$-choice item with parameters $(\boldsymbol{a}, \boldsymbol{b})$ is

$$I_{\boldsymbol{a},\boldsymbol{b}}(\theta) = \sum_{i,\, j=0}^{m} a_i a_j (\delta_{ij} - P_i) P_j, \tag{1.5}$$

where P_i is the probability of responding with choice i, and $\delta_{ij} = 1$ when $i = j$ and is zero otherwise. Notice that in the case of a dichotomous $(m = 1)$ item, this function reduces to the 2-PL information function for θ above. The information matrix for the item from a test-taker with proficiency θ has the form

$$\begin{pmatrix} A & C \\ C^t & B \end{pmatrix} \tag{1.6}$$

where A, B and C are $m \times m$ matrix with entries of the form

$$A_{ij} = (\theta - b_i)(\theta - b_j)(\delta_{ij} - P_i)P_j, \tag{1.7}$$

$$B_{ij} = a_i a_j (\delta_{ij} - P_i) P_j, \tag{1.8}$$

$$C_{ij} = -(\theta - b_i) a_j (\delta_{ij} - P_i) P_j. \tag{1.9}$$

Notice again that in the case of a dichotomous item, this function reduces to the 2-PL information function for (a, b) above. The information function for θ for the polytomous item in Figure 1.2 is shown in Figure 1.4.

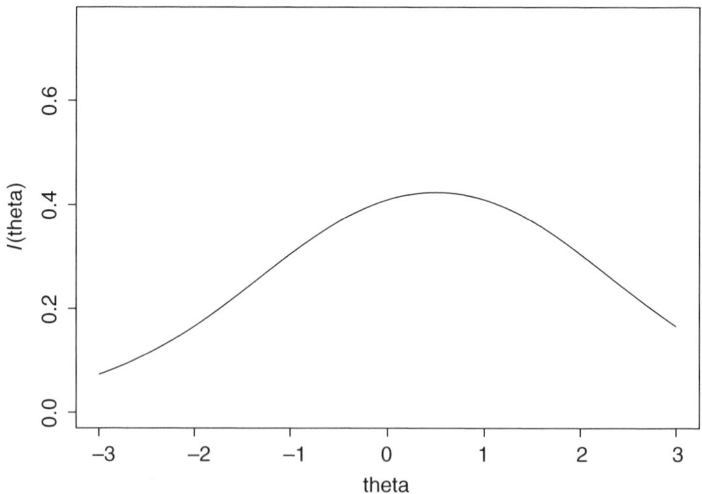

Figure 1.4 Item information for θ for a polytomous item

1.1.5 Design problems

Design problems in educational testing are of two types. The first type, known as test design, has to do with the optimal selection of items for proficiency estimation. The second type, sometimes called the sampling design problem, has to do with the sampling of test-takers for optimal estimation of the item parameters.

There are two important aspects to optimal design in educational testing. The first is that the model is nonlinear. In a linear optimal design problem, the question of how to sample the levels of the independent variables to estimate the model parameters efficiently depends on a function of Fisher's information matrix for the model parameters. In the nonlinear case the Fisher information matrix will also depend in an essential way on the model parameters. This dependence has two consequences. First, the design will be optimal only locally, in the sense of only for the specific model parameter values. Other values will lead to different designs. Second, as the model parameters are themselves unknown, it is not a simple matter to find even a locally optimal design. The most common approach is to work sequentially. As the parameter estimates change, the design changes with them. An alternative is to use Bayesian methods. A reasonable prior for the parameters can be integrated away, eliminating the design dependence on the specific values of the parameters.

The second important aspect is that the underlying model for IRT is a latent variable model. Whether one is creating a design for estimating proficiencies or item parameters, one cannot exactly specify the design levels. They are latent variables, and cannot be directly observed. They can, however, be estimated, and one can work with those estimates. Even with the estimates, however, one cannot specify an observation at a specific value. In particular, a design may call for an item to be tested on an examinee with a particular value of θ, but one should not expect that such an examinee will be available. Instead, one must work with random samples.

1.2 Test Design

Birnbaum (1968) first proposed the modern conceptual procedure for test construction. First, one determines the purpose of the test. Second, the purpose of the test determines the target for the test information function (TIF). The function TIF(θ) is the expected Fisher information from the test for an examinee with proficiency θ. For an aptitude test the ideal TIF might be uniform across a wide range of proficiencies, while for a certification test the ideal TIF might be high near the certification boundary and lower away from the boundary. The third step in test assembly is the selection of items so that the actual TIF, the sum of the item information functions, comes close to the target, subject to a variety of constraints discussed below. A thorough review of test design is given in van der Linden (1998), which introduces a special issue of *Applied Psychological Measurement* devoted to test construction issues.

1.2.1 Fixed-form test design

Fixed-form test design problems can be classified as unconstrained or constrained. Unconstrained problems consider optimal designs in the purely psychometric IRT setting, where there are no external constraints as to the items selected. Constrained problems incorporate these constraints, which may force a balance among items of a certain type, limit the total word length of items, or restrict the use of too similar items. Solutions to constrained problems have cast the problem as a 0–1 linear programming problem, as a problem in network flow programming, or used less formal heuristic approaches. Readers not familiar with optimization techniques may wish to consult texts such as Nocedal and Wright (1999) and Ahuja *et al.* (1993).

Berger (1998) reviews the use of optimal design theory to design tests. Suppose that instead of starting with an ideal test information function one simply wanted to efficiently estimate the θ's for a population of test-takers. Begin with a vector $\boldsymbol{\theta} = (\theta_1, \ldots, \theta_k)'$ of target proficiency values and their weights $\boldsymbol{W}$ corresponding to a discrete proficiency distribution. The set of possible item vector pairs $(\boldsymbol{a}, \boldsymbol{b})$ or triplets $(\boldsymbol{a}, \boldsymbol{b}, \boldsymbol{c})$ forms the design space. The optimal test design problem can then be stated as a problem of maximizing a function of the Fisher information for the vector $\boldsymbol{\theta}$ with weights $\boldsymbol{W}$ over the design space $\{(\boldsymbol{a}, \boldsymbol{b})\}$ or $\{(\boldsymbol{a}, \boldsymbol{b}, \boldsymbol{c})\}$. This design space may be constrained by practical issues. D-optimality seems a natural design criterion, or one could use the maximin criterion, which maximizes the minimum information over θ_i (van der Linden and Boekkooi-Timminga, 1989). Berger (1998) reports that for very high discrimination items, the optimal designs feature difficulties roughly following the distribution of the test-takers' θ's. For more moderate discrimination items, and a normal distribution of θ, both the D-optimal and maximin-optimal designs place the item difficulties all close to the centre of the distribution of θ. For a uniform distribution of θ, D-optimality leads to a bimodal distribution and maximin leads to a trimodal distribution. Berger suggests this feature arises because the uniform distribution is trimmed and relatively high information is thus needed near the boundaries. Low discrimination items have fairly flat information, so that for a normal distribution of θ setting the difficulties close to the centre gives the most information.

Berger (1994b) showed how test design for fixed-form tests, adaptive tests and testlets can all be placed into the same optimal design framework. Testlets are collections of items developed as a small test with the idea of simplifying some aspects of adaptive testing. Berger showed that optimal design could be used to select the items for the testlet, called within-testlet optimality, and to select testlets to construct a test, called between-testlet optimality. In all cases, a sequential block (with blocks possibly of size one) design is used. In an adaptive test, the θ estimate changes with each step; in a fixed-form test, the $\boldsymbol{\theta}$ vector and weights remain the same throughout.

Berger and Mathijssen (1997) have examined optimal test design for the nominal response IRT model. A new difficulty that arises in this context is that every category in an item has its own parameters that determine its relative position, within the item, on the difficulty scale. A test design will thus depend on the choice of the number of categories per item and their parameters.

An interesting result is that the maximum information for a given θ is obtained by an item for which the responses with the lowest and highest discrimination parameter a_u have a probability of selection of 0.5; that is, the item essentially reduces to an optimal dichotomous 2-PL item. In applying the D-optimality criterion, Berger and Mathijssen looked at item location vectors $\boldsymbol{b}$ for fixed discrimination vectors $\boldsymbol{a}$ for three and for four category items. As with the dichotomous model case, they found that for a normally distributed sample of examinees, the D-optimal test design required a single location vector, while for a uniform distribution of examinees a wider spread of location vectors was called for. As the category discrimination parameters increased, so did the location parameters.

Mathematical programming approaches to constrained test design appear to have been first put forth by Votaw (1952) and Feuerman and Weiss (1973). Although the textbook description of test construction as matching the test information function to the target sounds straightforward, in reality test construction includes a thicket of constraints. For example, there may be constraints on the number of items in the test, minimum and maximum number of items for specific subject areas or item types, and dependency constraints that if certain items are included others cannot be (because of overlap, or because one item gives clues to others) or must be (several items about the same reading passage, for example). A comprehensive list of constraints is given in Linden and Boekkooi-Timminga (1989).

To cast test construction as a 0–1 linear programming problem, let x_i, $i = 1 \ldots, M$ be a 0–1 variable indicating whether item i is included in the test out of the M items in the test bank. If the test construction objective is to maximize information at θ_0, then the function to be maximized is $\sum_{i=1}^{M} x_i I_i(\theta_0)$, subject to constraints on the x_i. For example, the requirement of a test of k items is simply $\sum_i x_i = k$, while a requirement of a minimum of j items of a certain type is $\sum_{i \varepsilon T} x_i \geq j$, where T is the set of items of that type. If the x_i variables were not constrained to be 0–1 variables, the resulting constrained optimization problem would be easily solved using linear programming. With the 0–1 constraint, the problem is known to be NP-hard, meaning the problem lies in the class of problems whose solution time is believed not to be bounded by a polynomial in the size of the problem. However, optimal solutions are not really required, just 'very good' ones. Approximate solutions can be found by limiting the search time of the branch-and-bound search or by using optimal rounding, and can be reasonably fast. The 0–1 programming approach was formulated by Theunissen (1985, 1986). Later work developing the concept can be found in Boekkooi-Timminga (1987, 1990), van der Linden and Boekkooi-Timminga (1989), Adema *et al.* (1991), Adema (1992), van der Linden and Adema (1998) and van der Linden (2000).

The second mathematical programming approach to test construction uses network-flow programming (Armstrong *et al.*, 1992). Here one conceptualizes supply nodes, representing items; demand nodes, representing constraints such as a minimum and maximum number of items of a certain type; directed arcs, representing the flow of items from demand nodes to supply nodes; and the cost of flow, which can be used to represent the item information. Network flow problems can be efficiently solved using the simplex algorithm. The principal advantages of a network-flow formulation of test construction are that optimal solutions can be found extremely quickly, and these solutions naturally involve only 0–1 variables. The principal disadvantage is that certain constraints, such as quantitative constraints (total word count, for example) and dependency constraints do not easily fit into the network formulation. Lagrangian relaxation is a technique to bring such constraints into the objective function by multiplying the constraint by a Lagrange multiplier and adding it to the objective function. By systematically varying the multiplier and solving the new network flow problem, near-optimal results can be very quickly found. Armstrong *et al.* (1995, 1996, 1998a, b) show how to use Lagrangian relaxation and heuristic search along with network flow to solve test construction problems with these non-network constraints.

In both mathematical programming approaches other objective functions may be more desirable than the information at a fixed value θ_0, and can be incorporated into the 0–1 linear programming model. For example, one can minimize the discrepancy between the test information function and the target over a discrete set of θ values. Van der Linden and Boekkooi-Timminga (1989) minimized the discrepancy from the shape of the target test information function, arguing that the actual values of the test information carry little meaning for test designers. Armstrong *et al.* (1998b) used classical reliability as an objective function.

In the mathematical programming paradigm it is important that a feasible solution exists; that is, that a test does exist satisfying all of the constraints. Timminga (1998) discusses diagnosing and repairing an infeasible model. It is also possible to treat the constraints as desirable conditions that can be violated if necessary. The heuristic method of test assembly now known as the weighted deviations model (Luecht and Hirsch, 1992; Swanson and Stocking, 1993; Stocking and Swanson, 1993; Luecht, 1998) incorporates constraints into the objective function by penalizing deviations from the constraints. The test is constructed using the greedy algorithm: the best item is selected, the objective function is regenerated to incorporate the prior decision, a new best item is added to the test, and the process repeats. Because constraints are incorporated into the single objective function, weights can be used to emphasize certain constraints over others. The greedy algorithm assures rapid test assembly but because the test is assembled sequentially there is no guarantee of overall optimality. Optimal design criteria could be brought into the weighted deviation model by using an optimal design criterion as the basic objective function before incorporating the constraints. Although algorithmically less sophisticated than other constrained optimization approaches, the weighted deviations model is widely used because of its speed, flexibility, ease of interpretation, and its guarantee of always finding a solution.

1.2.2 Test design for CAT

In CAT, the testing algorithm has the advantage of having an estimate of the test-taker's proficiency at each step, which leads naturally to a (constrained) sequential optimization problem. The original approach to item selection in a CAT was, at step k, to treat the current estimate $\hat{\theta}_k$ as the true value of θ and to select next the item with the highest Fisher information for $\hat{\theta}_k$. Chang and Ying (2004) showed that under the ideal conditions of an infinite item pool this design is asymptotically optimal for the 1-PL model, and with certain restrictions, principally bounds on a and c, is also asymptotically optimal for the 2-PL and 3-PL models.

Near the start of the exam, however, the estimate of θ may be quite poor. Selecting items by maximum information tends to use high discrimination items early, although these items may contain little information if $b - \theta$ is actually large (see Figure 1.2). Veerkamp and Berger (1997) thus looked at an item selection rule based on the Fisher information integrated over an interval of θ-values, using various weights. With a delta function at $\hat{\theta}_k$, one recovers the traditional maximum Fisher information rule. They also considered uniform weights over a confidence interval for θ and weights proportional to the likelihood function for θ. They found that weighting by the likelihood gave slightly superior results to either maximum information or the information integrated over a confidence interval. Chang and Ying (1996) proposed using an integrated Kullback–Leibler information criterion for item selection, again as protection against poorly estimated θ near the beginning of the test. Chen *et al.* (2000) compared all these item selection rules with simulations and, although they found differences in short tests of less than 10 items, found no substantive differences in longer tests.

Stocking and Swanson (1993) adapted the weighted deviations method to handle item constraints in CAT. Their method is based on selecting the maximum information item as modified by constraints, but any of the selection rules in the preceding paragraph could be used prior to the constraint modification.

A practical constraint in CAT goes by the name 'exposure control'. Because a test in the CAT format is generally offered frequently, there is a distinct possibility that certain items can become known in advance to certain test-takers. Unexpectedly high exposure of certain items lead to considerable embarrassment in the early administration of the CAT version of the GRE exam (Wainer and Einor, 2000). Less dramatic is the possibility of certain items being underutilized, leading to inefficiency and increased cost. The most popular approaches are based on work by Sympson and Hetter (1985) and assign selection probabilities to each item. While widespread, such methods have the disadvantage of requiring extensive simulation to assign the probabilities. A review of probabilistic methods of controlling item exposure is given in Stocking and Lewis (2000). These methods can be incorporated into the weighted deviations method and other methods of test construction.

An alternative to probabilistic methods was suggested in Chang and Ying (1999) and Chang *et al.* (2001). They propose stratifying the item pool by the a-level. Low discrimination items, with their flatter information curves, would be used early, when the proficiency is poorly estimated, and higher discrimination items, with more concentrated information, would be saved for the later stages of the test when the proficiency is better estimated. Item exposure is controlled because low discrimination items are forced into use, thus reducing the skewness of the item exposure distribution.

The speed of modern computing has enabled a promising approach to on-the-fly construction of adaptive tests satisfying the various test constraints. Van der Linden and Reese (1998) show that at each stage of an adaptive test an algorithm can select the next item by assembling a full test, fixing the items already administered. From this full test, a single next item is selected and the unselected items are returned to the pool. Following the test-taker's response, the process is repeated. Since each full test satisfies the constraints, the actual adaptive test automatically satisfies the constraints as well. Although Van der Linden and Reese used 0–1 programming in their paper, presumably the technique would apply to any method of test construction. Although computationally intensive and algorithmically sophisticated, the flexibility and potential generality of this technique make it likely to see wide use in the future.

Because item selection uses the estimate $(\hat{a}, \hat{b}, \hat{c})$ as the true parameter value, there is a potential for serious errors in CAT due to capitalization on chance. Hambleton *et al.* (1993) and van der Linden and Glas (2000a) discuss the phenomenon and outline approaches to reduce its impact.

1.3 Sampling Design

We now reverse perspective and look at designs to estimate item parameters efficiently. The problem is in some ways simpler than test design, because there are no test design constraints, yet it does involve more parameters to be estimated. The problem differs depending on whether the item calibration is to take place in a paper-and-pencil or a CAT format. In the former, at most one can specify the rough distribution of the proficiencies of the test-takers. In a CAT format, a design algorithm can take advantage of the current estimate of the test-taker's proficiency to obtain a more efficient design.

1.3.1 Paper-and-pencil calibration

Originally, items were calibrated using test-taker samples similar to the test-taker population. Lord (1962), Pandey and Carlson (1976) and Lord and Wingersky (1985) were able to show that greater precision in item parameter estimates could be obtained if the calibration sample were explicitly chosen for that purpose. In particular, they showed that giving each item to an appropriately matched subsample of test-takers was more efficient that using the entire sample for the entire set of items. Stocking (1990) considered the effect of different sampling schemes on the information for each item parameter separately in 1-PL, 2-PL and 3-PL models, showing, parameter by parameter, how to interpret the results of Wingersky and Lord (1984) that a rectangular sample of proficiencies worked well in the 3-PL model for simultaneous estimation of proficiency and item parameters.

Berger (1991) brought formal optimal design theory to the question by investigating sampling designs using the D-optimality criterion for 1-, 2-, and 3-PL models. First considering designs with only a single sample, he showed that for the 1-PL model the design efficiency drops as the standard deviation increases, because fewer examinees will have θ's near the maximally informative value of $\theta = b$. For the 2-PL model, the efficiency first increases with the standard deviation, because a can be more accurately estimated with a spread sample, and then decreases. The value of the standard deviation giving maximum efficiency is a decreasing function of a, since for high values of a even θ's moderately near to b will give values of P near 0 or 1, and so have little information for either a or b. Because of the need to estimate c, for 3-PL models the efficiency increases with the standard deviation of the sample.

Berger also explored alternative sampling designs to roughly match θ's and b's. For the 1- and 2-PL models, using a moderate number of partitions of the test-taker sample and partial test subsets roughly matching the proficiencies of the subsamples gave much higher efficiencies than giving the entire test to the full sample. The efficiency improvement depends on the discrimination: higher discrimination items show greater increases in efficiency. Except for the easiest items, in a 3-PL model the greatest efficiency comes from using the full sample.

In the 1-PL case, it is easy to show that the information for b is maximal if all test-takers are selected so that $\theta = b$. The optimal sampling of test-takers was handled by Berger (1992) for D-optimality in the 2-PL case. The optimal design for a given item has equal weights at two design points, namely θ_1 and θ_2, such that $P(\theta_1) = 0.176$ and $P(\theta_2) = 0.824$ (White, 1975; Ford, 1976). Explicitly, $\theta_i = b \pm 1.5434/a$. Again we see that as a increases the points should be closer to b. (For A-optimality, the results are not so simple, as the best design depends heavily on a – the optimal distance between the points decreases with a.) These results imply that bimodal calibration samples will have higher efficiency than a single normal sample. Samples with a standard deviation between 1 and 2, with a gap between the modes and b of about $1.5/a$, can be expected to be 60–80% efficient compared to an ideal sample. This means, for example, that an exam for 5th graders would be more efficiently calibrated (on the order of 50% more efficient) by a sample consisting of half 4th graders and half 6th graders than by a sample consisting fully of 5th graders. Berger also showed how the method of Wynn (1970) can be adapted to determine locally D-optimal sampling designs for sets of test items. The greatest gains in efficiencies, as compared to a $N(0, 1)$ sample, were seen in tests with low discriminating items and a wide range of difficulties. (This sort of test is perhaps the easiest to construct; creating items with high values of a is a surprisingly difficult task.) The typical sample design is essentially trimodal, with modes at the extremes and the centre of the range for θ. A review of optimal sampling designs for paper-and-pencil calibration can be found in Berger and van der Linden (1992).

D-optimal designs for a range of polytomous models were considered by Holman and Berger (2001), who found that D-optimal designs can reduce the calibration sample size by up to 30%.

1.3.2 CAT calibration

When an item is to be calibrated in a CAT administration, there is the possibility of a substantial efficiency gain by administering an item only to those test-takers who will add the most information to the item parameter estimates. On-line calibration easily lends itself to the sequential procedure of adapting the calibration sample as the item estimates evolve. Berger (1994a) and Jones and Jin (1994) first demonstrated this sequential D-optimal design procedure for the 2-PL model. Berger showed that there is very little loss in efficiency in making the procedure block-sequential rather than sequential. Jones and Jin demonstrated in simulations that the sequential approach yielded designs with about 70–80% of the efficiency of optimal design for known item parameters. Simple normal sampling gave efficiencies ranging from 0.75, for a proficiency equal to the sample mean, down to 0.01 for high discriminating items with extreme difficulties. As CAT administrations require a steady stream of newly calibrated items while reducing the number of items any individual test-taker will take part in calibrating, high efficiency in item calibration is quite desirable.

Buyske (1998a, b) took a different approach to a criterion function. His criterion is based on minimizing the information loss for θ that will occur due to estimation errors in the item parameters. For the 2-PL model the criterion takes the form

$$\frac{1}{a^2}\mathrm{Var}(\hat{a}) + \mathrm{const} \times a^2\mathrm{Var}(\hat{b}) \tag{1.10}$$

where the constant depends on the expected distribution of proficiencies when the item is used in production. In simulations these designs were 50% more efficient (in the sense of the calibration sample size needed for equally good proficiency estimation) than a standard normal distribution and 10% more efficient than a D-optimal design. In the 2-PL case the optimal design points are moved in slightly from D-optimal case, from θ's giving probabilities P near 0.18 and 0.82 for D-optimality to 0.25 and 0.75 for this revised criterion. A similar contraction of design points is found for the 3-PL model.

In practice, one has more than one item to calibrate at a time, and has a stream of test-takers who must be given calibration items. Thus we again have a constrained nonlinear optimization problem. This situation was first investigated by Jones *et al.*

(1997) for the 2-PL model. The problem is: given m items to be calibrated, and n test-takers, match test-takers with items to maximize information. An additional constraint is that every test-taker should receive the same number of items to calibrate and each item is calibrated by the same number of test-takers. By casting the problem as a network-flow problem and using a gradient-search algorithm, the authors were able to find exact designs quickly and with efficiency about 50% greater than random allocation. As in the two papers by Berger (1992, 1994a), the designs tend to have proficiencies cluster around the difficulty level of the items, with tighter concentrations for higher discriminating items. The same problem was investigated for the 3-PL model by Jones *et al.* (1999).

Testlets pose a slightly different calibration problem. Jones and Nediak (2004) report on D-optimality-based sequential calibration of testlets. They examined both constrained and unconstrained methods. They found that about 30% more items calibrated by optimal design than by a random sample were acceptably close to the true item characteristics, as measured by the weighted Hellinger deviance, a measure of the distance between the item response functions of the estimated and the true items.

D-optimal solutions to item calibration designs are in fact locally optimal designs, because of the dependence of the information matrix on the item parameters. Berger *et al.* (2000), working with the nominal response model, also addressed the issue of designs for calibrating items using a minimax D-optimality criterion. The minimax approach avoids the difficulty of local optimality by minimizing the worst performance of the item calibration, as measured by the D-optimal criterion, over the prespecified region of item parameters. The design points determined by this criterion are more spread out than points determined by the D-optimality criterion, yet have high efficiency relative to the local D-optimal design. Because the design does not require estimation of the item parameters, by its very nature it is more robust against misspecification of those values.

1.4 Future Directions

At this time the development of optimal test and sampling design theory for purely psychometric IRT seems mature for both dichotomous and many polytomous models. Much of the recent interest has been in the practical issues that arise in applied testing. The trend in increasing computation speed and algorithmic improvements means that constrained nonlinear optimization techniques such as 0–1 programming or network-flow algorithms will become practical competitors to the weighted deviations algorithm in test design. The on-the-fly test construction method for CAT developed in van der Linden and Reese (1998) also seems likely to see future development and wide use.

Exposure control remains an important issue in CAT, both because of the risk of cheating and the cost of developing underutilized items. A conceptual breakthrough is still needed here. Perhaps the way lies in thinking of CAT test construction not as a constrained sequential optimization problem for each test-taker, but in thinking of it as a sequential optimization problem for the overall pool of test-takers, with both the usual constraints on each test as well as constraints on the item pool limiting each item's exposure.

The very efficiency of optimal designs for online item calibration has led to some difficulties. Model-fit diagnostics for items such as multidimensionality tests and differential item functioning (different item characteristics in different groups) rely on large samples for their power. Calibration designs to date have not incorporated the need for efficient item diagnostic testing.

Many of the future challenges in optimal design theory in educational testing will follow future trends in testing itself. There is almost no design work in alternative IRT such as multidimensional traits or non-parametric models. The use of computer-generated items (item cloning) as well as innovative items, such as items requiring constructed responses with automated scoring, is gathering increased interest. As these concepts move from journal articles to actual tests there will be an increased and pressing need for effective designs incorporating them.

Acknowledgements

The author would like to thank the editors and two anonymous referees for their valuable comments.

References

Adema, J. J. (1992). Methods and models for the construction of weakly parallel tests. *Applied Psychological Measurement*, **16**, 53–63.

Adema, J. J., Boekkooi-Timminga, E. and van der Linden, W. J. (1991). Achievement test construction using 0–1 linear programming. *European Journal of Operational Research*, **55**, 103–111.

Ahuja, R. K., Magnanti, T. L. and Orlin, J. B. (1993). *Network Flows: Theory, Algorithms, and Applications*. Prentice Hall.

Armstrong, R. D., Jones, D. H. and Wu, I. L. (1992). An automated test development of parallel tests from a seed test. *Psychometrika*, **57**, 271–288.

Armstrong, R. D., Jones, D. H. and Wang, Z. (1995). Network optimization in constrained standardized test construction. In Lawrence, K. and Reeves, G. R. (eds), *Applications of Management Science: Network Applications*, vol. 8. JAI Press, pp. 189–212.

Armstrong, R. D., Jones, D. H., Li, X. and Wu, I. L. (1996). A study of a network-flow algorithm and a noncorrecting algorithm for test assembly. *Applied Psychological Measurement*, **20**, 89–98.

Armstrong, R. D., Jones, D. H. and Kunce, C. S. (1998a). IRT test assembly using network-flow programming. *Applied Psychological Measurement*, **22**, 237–247.

Armstrong, R. D., Jones, D. H. and Wang, Z. (1998b). Optimization of classical reliability in test construction. *Journal of Educational and Behavioral Statistics*, **23**, 1–17.

Berger, M. P. F. (1991). On the efficiency of IRT models when applied to different sampling designs. *Applied Psychological Measurement*, **15**, 293–306.

Berger, M. P. F. (1992). Sequential sampling designs for the two-parameter item response theory model. *Psychometrika*, **57**, 521–538.

Berger, M. P. F. (1994a). D-Optimal sequential sampling designs for item response theory models. *Journal of Educational Statistics*, **19**, 43–56.

Berger, M. P. F. (1994b). A general approach to algorithmic design of fixed-form tests, adaptive tests and testlets. *Applied Psychological Measurement*, **18**, 141–153.

Berger, M. P. F. (1998). Optimal design of test with dichotomous and polytomous items. *Applied Psychological Measurement*, **22**, 248–258.

Berger, M. P. F. and Mathijssen, E. (1997). Optimal test designs for polytomously scored items. *British Journal of Mathematical and Statistical Psychology*, **50**, 127–141.

Berger, M. P. F. and van der Linden, W. J. (1992). Optimality of sampling designs in item response theory models. In Wilson, M. (ed.), *Objective Measurement: Theory into Practice*, vol. 1. Ablex, pp. 274–288.

Berger, M. P. F., King, C. Y. J. and Wong, W. K. (2000). Minimax D-optimal designs for item response theory models. *Psychometrika*, **65**, 377–390.

Birnbaum, A. (1968). Some latent trait models and their use in inferring an examinee's ability. In Lord, F. M. and Novick, M. R. *Statistical Theories of Mental Test Scores*. Addison-Wesley.

Bock, R. D. (1972). Estimating item parameters and latent ability when responses are scored in two or more nominal categories. *Psychometrika*, **37**, 29–51.

Boekkooi-Timminga, E. (1987). Simultaneous test construction by zero–one programming. *Methodika*, **1**, 101–112.

Boekkooi-Timminga, E. (1990). The construction of parallel tests from IRT-based item banks. *Journal of Educational Statistics*, **15**, 129–145.

Buyske, S. G. (1998a). Optimal design for item calibration in computerized adaptive testing: the 2PL case. In Flournoy, N., Rosenberger, W. F. and Wong, W. K (eds), *New Developments and Applications in Experimental Design*. Berkeley: Institute of Mathematical Statistics.

Buyske, S. G. (1998b). Optimal designs for item calibration in computerized adaptive testing. Ph.D. dissertation, Rutgers University.

Chang, H. H. and Ying, Z. (1996). A global information approach to computerized adaptive testing. *Applied Psychological Measurement*, **20**, 213–229.

Chang, H. H. and Ying, Z. (1999). α-stratified multistage computerized adaptive testing. *Applied Psychological Measurement*, **23**, 211–222.

Chang, H. H. and Ying, Z. L. (2004). Nonlinear sequential designs for logistic item response theory models with applications to computerized adaptive tests. *Annals of Statistics*, to appear.

Chang, H. H., Qian, J. and Ying, Z. (2001). α-stratified multistage computerized adaptive testing with b blocking. *Applied Psychological Measurement*, **25**, 333–341.

Chen, S. Y., Ankenmann, R. D. and Chang, H. H. (2000). A comparison of item selection rules at the early stages of computerized adaptive testing. *Applied Psychological Measurement*, **24**, 241–255.

Feuerman, M. and Weiss, H. (1973). A mathematical programming model for test construction and scoring. *Management Science*, **19**, 961–966.

Ford, I. (1976). Optimal static and sequential design: a critical review. Ph.D. thesis, University of Glasgow.

Hambleton, R. K. and Swaminathan, H. (1985). *Item Response Theory: Principles and Applications*. Kluwer.

Hambleton, R. K., Jones, R. W. and Rogers, H. J. (1993). Influence of item parameter estimation errors in test development. *Journal of Educational Measurement*, **30**, 143–155.

Holman, R. and Berger, M. P. F. (2001). Optimal calibration designs for tests of polytomously scored items described by item response theory models. *Journal of Educational and Behavioral Statistics*, **26**, 361–380.

Jones, D. H. and Jin, Z. (1994). Optimal sequential designs for on-line item estimation. *Psychometrika*, **59**, 59–75.

Jones, D. H. and Nediak, M. S. (2004). Optimal online calibration of testlets, Chapter 2, this volume.

Jones, D. H., Chiang, J. and Jin, Z. (1997). Optimal designs for simultaneous item estimation. *Nonlinear Analysis. Theory, Methods and Applications*, **30**(7), 4051–4058.

Jones, D. H., Nediak, M. and Wang, X. B. (1999). *Sequential Optimal Designs for On-line Item Calibration*. Tech. rept. Rutgers University. Rutcor Research Report 2-99.

Lord, F. M. (1962). Estimating norms by item-sampling. *Educational and Psychological Measurement*, **22**, 259–267.

Lord, F. M. (1980). *Applications of Item Response Theory to Practical Testing Problems*. Lawrence Erlbaum Associates.

Lord, F. M. and Wingersky, M. S. (1985). Sampling variances and covariances of parameter estimates in item response theory. In Weiss, D. J. (ed.), *Proceedings of the 1982 IRT/CAT Conference*. University of Minnesota, Department of Psychology, CAT Laboratory.

Luecht, R. M. (1998). Computer-assisted test assembly using optimization heuristics. *Applied Psychological Measurement*, **22**, 224–236.

Luecht, R. M. and Hirsch, T. M. (1992). Item selection using an average growth approximation of target information functions. *Applied Psychological Measurement*, **16**, 41–51.

Nocedal, J. and Wright, S. (1999). *Numerical Optimization*. Springer-Verlag.

Pandey, T. N. and Carlson, D. (1976). Assessing payoffs in the estimation of the mean using multiple matrix sampling designs. In de Gruijter, D. N. M. and van der Kamp, L. J. (eds), *Advances in Psychological and Educational Measurement*. John Wiley & Sons, Ltd.

Parshall, C. G., Spray, J. A., Kalohn, J. C. and Davey, T. (2002). *Practical Considerations in Computer-based Testing*. Springer-Verlag.

Sands, W. A., Waters, B. K. and McBride, J. R. (eds). (1997). *Computerized Adaptive Testing: From Inquiry to Operation*. American Psychological Association.

Stocking, M. L. (1990). Specifying optimum examinees for item parameter estimation in item response theory. *Psychometrika*, **55**, 461–475.

Stocking, M. L. and Lewis, C. (2000). Methods of controlling the exposure of items in CAT. In van der Linden, W. J. and Glas, C. A. W. (eds), *Computerized Adaptive Testing: Theory and Practice*. Kluwer Academic Publishers.

Stocking, M. L. and Swanson, L. (1993). A method for severely constrained item selection in adaptive testing. *Applied Psychological Measurement*, **17**, 277–292.

Swanson, L. and Stocking, M. L. (1993). A model and heuristic for solving very large item selection problems. *Applied Psychological Measurement*, **17**, 151–166.

Sympson, J. B. and Hetter, R. D. (1985). Controlling item-exposure rates in computerized adaptive testing. In *Proceedings of the 27th Annual Meeting of the Military Testing Association*. San Diego: Navy Personnel Research and Development Centre.

Theunissen, T. J. (1985). Binary programming and test design. *Psychometrika*, **50**, 411–420.

Theunissen, T. J. (1986). Some applications of optimization algorithms in test design and adaptive testing. *Applied Psychological Measurement*, **10**, 381–389.

Thissen, D. and Orlando, M. (2001). Item response theory for items scored in two categories. In Thissen, D. and Wainer, H. *Test Scoring*. Lawrence Erlbaum Associates, Inc.

Thissen, D. and Steinberg, L. (1986). A taxonomy of item response models. *Psychometrika*, **51**, 567–577.

Thissen, D, Nelson, L, Rosa, K. and McLeod, L. D. (2001). Item response theory for items scored in more than two categories. In Thissen, D. and Wainer, H. *Test Scoring*. Lawrence Erlbaum Associates, Inc.

Timminga, E. (1998). Solving infeasibility problems in computerized test assembly. *Applied Psychological Measurement*, **22**, 280–291.

van der Linden, W. J. (1998). Optimal assembly of psychological and educational tests. *Applied Psychological Measurement*, **22**, 195–211.

van der Linden, W. J. (2000). Optimal assembly of tests with item sets. *Applied Psychological Measurement*, **24**, 225–240.

van der Linden, W. J. and Adema, J. J. (1998). Simultaneous assembly of multiple test forms. *Journal of Educational Measurement*, **35**, 185–198.

van der Linden, W. J. and Boekkooi-Timminga, E. (1989). A maximin model for test design with practical constraints. *Psychometrika*, **53**, 237–247.

van der Linden, W. J. and Glas, C. A. W. (2000a). Capitalization on item calibration error in adaptive testing. *Applied Measurement in Education*, **13**, 35–53.

van der Linden, W. J. and Hambleton, R. K. (eds). (1997a). *Handbook of Modern Item Response Theory*. Springer-Verlag.

van der Linden,W. J. and Hambleton, R. K. (1997b). Item response theory: brief history, common models, and extensions. In van der Linden, W. J. and Hambleton, R. K. (eds), *Handbook of Modern Item Response Theory*. Springer-Verlag.

van der Linden, W. J. and Reese, L. M. (1998). A model for optimal constrained adaptive testing. *Applied Psychological Measurement*, **22**, 259–270.

Veerkamp, W. J. J. and Berger, M. P. F. (1997). Some new item selection criteria for adaptive testing. *Journal of Educational and Behavioral Statistics*, **22**, 203–226.

Votaw, D. F. (1952). Methods of solving some personnel problems. *Psychometrika*, **17**, 255–266.

Wainer, H. (2000). *Computerized Adaptive Testing: a Primer*, 2nd edn. Lawrence Erlbaum Associates.

Wainer, H. and Einor, D. (2000). Caveats, pitfalls and unexpected consequences of implementing large-scale computerized testing. In Wainer, H., *Computerized Adaptive Testing: a Primer*, 2nd edn. Lawrence Erlbaum Associates.

White, L. V. (1975). The optimal design of experiments for estimation in nonlinear models. Ph.D. thesis, University of London.

Wingersky, M. S. and Lord, F. M. (1984). An investigation of methods for reducing sampling error in certain IRT procedures. *Applied Psychological Measurement*, **8**, 347–364.

Wynn, H. P. (1970). The sequential generation of D-optimum experimental designs. *Annals of Mathematical Statistics*, **41**, 1655–1664.

2

Optimal On-line Calibration of Testlets

Douglas H. Jones[1] **and Mikhail S. Nediak**[2]

[1]*Department of Management Science and Information Systems, Rutgers the State University, 180 University Avenue/Ackerson Hall, Newark, NJ 07102-1803, USA*

[2]*RUTCOR, McMaster University, Department of Mathematics and Statistics, 1280 Main Street West, Hamilton, Ontario, L8S 4K1, Canada*

2.1 Introduction

Several large-scale testing agencies have testing formats that include several related questions tied to a passage or common stem, also called a testlet. In general, a testlet is the stimulus that is the centre of the testlet (e.g. a reading passage, a graphic, a data table) along with a set of questions. The US Law School Admissions Test (LSAT) has two of four major sections consisting only of several reading passages along with related questions. Recently the testlet concept has been used in assessing constructs. While the concept of a testlet is old, the idea of taking special precautions in scoring testlets and extending item response theory models to testlets is relatively new. With the advent of computerized testing, researchers have been particularly interested in these issues. Wainer and Kiely (1987) started seminal work in this area, showing that the typical assumption of local independence does not hold with items (i.e. questions) in the same testlet. Bradlow *et al.* (1999) proposed a new model that extends item response theory to testlets, and have subsequently proposed estimation schemes similar to the ones used in this chapter.

Applied Optimal Designs Edited by M.P.F. Berger and W.K. Wong
 ISBN: 0-470-85697-1 (HB)

This chapter studies optimal designs for calibrating items of a testlet in a computer adaptive testing session. Since testlets consist of two or more single items, the design issues are more complicated than for singly administered items (see Jones and Jin, 1994).

Calibration of new items is an essential part of a testing system, because operational items eventually become overexposed and need replacement. Military and educational testing agencies use various protocols for gathering new data about experimental items and use this data to estimate item response theory (IRT) models (Lord, 1980). In the US military, examinees are recruited explicitly for presenting new items. In paper-and-pencil testing, items, singly or in testlets, are discreetly placed inside an operational test. If several tests contain different sets of experimental items, then each test is called a form. The forms are then spiralled together, so that upon administration, random samples of examinees are assigned to each form and the experimental items; this is called 'pure spiralling'. In computer adaptive testing, experimental items are placed near the last one-third of the testing session by randomly selecting the examinee; this is called 'random seeding'. In either pure spiralling or random seeding, the objective is the same: to assign a random sample of examinees to experimental items.

Any random method has the potential to match examinees with items inappropriately hard or easy for their abilities, and thus, there is risk that the examinees may detect experimental items, thus providing unreliable data. If the population of abilities is normal, then random matching amounts to sampling from a normal population. Recent research has shown that the precision of parameter estimates can be significantly increased from distributions other than the normal distribution (Jones and Jin, 1994; Stocking, 1990; Wingersky and Lord, 1984). Even in very large samples from a normal population, very little information may be available for calibrating some items, and that the success of a particular item calibration using item response theory depends heavily on the selection of more informative data (Stocking, 1990). Other research related to this is Berger (1991), Berger and van der Linden (1991), Vale (1986) and van der Linden (1988).

This research proposes to select abilities that are optimal for experimental testlets within a sequential sampling setting (Ford, 1976). Recent research into sequential designs for single two-parameter item response models is found in Berger (1992) and Jones and Jin (1994); for testlets of two-parameter models see Jones *et al.* (1997).

An optimal design depends on the unknown item parameters; therefore, locally optimal n-point designs must be based on parameter estimates from prior steps. It is questionable that this sequential procedure converges to the optimal design, and repeated sampling inferences are valid; however, we are encouraged by positive results shown for a nonlinear problem similar but identical to ours (Wu, 1985). In the sequential setting, the product of independent likelihoods is not the true likelihood as noted in Jones and Jin (1994, p. 66) and Silvey (1980). For small samples, sometimes the difference can be ignored (Ford *et al.* 1985). For large samples, the independent samples likelihood is valid for special nonlinear models based on data from sequentially constructed designs (Wu, 1985). Recent advances in sequential designs for nonlinear models are surveyed in Ford *et al.* (1989). Using the independent sample likelihood, we employ a fully Bayesian methodology for extracting a multivariate posterior distribution for the item parameters within a testlet using Markov chain Monte Carlo (MCMC) sampling from the posterior developed in Jones and Nediak (1999) by extending a method proposed in Patz and Junker (1999).

The design criterion is based on the celebrated notion of D-optimality using the determinant of Fisher's information (Silvey, 1980). Recent research has been successful in applying this concept to optimal designs for the two-parameter logistic model (Jones and Jin, 1994; Berger, 1992). In this chapter, we extend D-optimality to the three-parameter logistic model within the context of a testlet. Buyske (1998a, b) considers other optimality criteria for item calibration.

We begin by presenting a review of item response theory that will be used here. This is followed by the introduction of D-optimality and its extension to testlets, whereupon we formulate a mathematical programming model with transshipment constraints. We derive the projected conjugant-gradient algorithm for determining optimal designs. Finally, we present results based on the real data of 28 184 test-takers of the LSAT.

2.2 Background

2.2.1 Item response functions

Let u_i denote a response to a single item from individual i with ability level θ_i, possibly multivariate. Let $\boldsymbol{\beta}^T = (\beta_0, \beta_1 \ldots \beta_p)$ be a vector of unknown parameters associated with the item. Assume that all responses are scored either correct, $u_i = 1$, or incorrect, $u_i = 0$. An item response function (IRF), $P(\theta_i, \boldsymbol{\beta})$, is a function of θ_i, and describes the probability of correct response of an individual with ability level θ_i. The mean and variance of the parametric family are

$$\begin{aligned} E\{u_i \mid \boldsymbol{\beta}\} &= P(\theta_i; \boldsymbol{\beta}) \\ \sigma^2(\theta_i; \boldsymbol{\beta}) &= \mathrm{Var}\{u_i \mid \boldsymbol{\beta}\} = P(\theta_i; \boldsymbol{\beta})[1 - P(\theta_i; \boldsymbol{\beta})]. \end{aligned}$$

We shall focus on the family of three-parameter logistic (3PL) response functions:

$$P(\theta_i; \boldsymbol{\beta}) = \beta_2 + (1 - \beta_2) R(\theta_i; \boldsymbol{\beta}),$$

where

$$R(\theta_i; \boldsymbol{\beta}) = \frac{\mathrm{e}^{1.7\beta_1(\theta - \beta_0)}}{1 + \mathrm{e}^{1.7\beta_1(\theta - \beta_0)}}. \tag{2.1}$$

Denote three IRT characteristics of a three-parameter item to be: *difficulty*, β_0, the *discrimination power*, β_1, the *guessing*, β_2. Note: The family of two-parameter logistic (2PL)-response function is expression (2.1) with $\beta_2 = 0$. The value 1.7 is a constant, which allows R, the 2PL logistic ogive, to be a good approximation of the normal ogive. For the specific case of the 2PL item response function, see Jones and Jin (1994) and Berger (1992, 1994).

An *n-point design* for a single item is a collection of possibly replicated ability levels $\boldsymbol{\Theta} = \{\theta_i : i = 1, \ldots, n\}$. Assume the response function is differentiable with respect to $\boldsymbol{\beta}$. For each θ_I denote the $p + 1$ column vector of partial derivatives as

$$\frac{\partial P(\theta_i; \boldsymbol{\beta})}{\partial \boldsymbol{\beta}} = \left(\frac{\partial P(\theta_i; \boldsymbol{\beta})}{\partial \beta_0}, \ldots, \frac{\partial P(\theta_i; \boldsymbol{\beta})}{\partial \beta_p} \right)^T.$$

The Fisher information matrix is defined as

$$\mathbf{m}(\beta; \boldsymbol{\Theta}) = \sum_{i=1}^{n} \sigma^{-2}(\theta_i; \boldsymbol{\beta}) \frac{\partial P(\theta_i; \boldsymbol{\beta})}{\partial \boldsymbol{\beta}} \frac{\partial P(\theta_i; \boldsymbol{\beta})^T}{\partial \boldsymbol{\beta}}.$$

Under regularity conditions, the maximum likelihood estimate (MLE), $\hat{\boldsymbol{\beta}}$, is asymptotically normal with mean $\boldsymbol{\beta}$ and variance-covariance matrix $\mathbf{m}^{-1}(\beta; \boldsymbol{\Theta})$.

The expected information is defined as Fisher's information matrix. The 3PL-information matrix from an examinee with ability θ is

$$\mathbf{m}(\beta; \theta) = \frac{[1 - R(\theta; \beta)]^2}{P(\theta; \beta)[1 - P(\theta; \beta)]} \times \begin{pmatrix} (1-\beta_2)^2[R(\theta;\beta)]^2 & (1-\beta_2)^2[R(\theta;\beta)^2]\theta & (1-\beta_2)R(\theta;\beta) \\ (1-\beta_2)^2[R(\theta;\beta)]^2\theta & (1-\beta_2)^2[R(\theta;\beta)]^2\theta^2 & (1-\beta_2)R(\theta;\beta)\theta \\ (1-\beta_2)R(\theta;\beta) & (1-\beta_2)R(\theta;\beta)\theta & 1 \end{pmatrix}.$$

2.2.2 *D*-optimal design criterion

For a single item, the design problem is to choose $\boldsymbol{\Theta}$ to maximize an appropriate function of $\mathbf{m}(\beta; \boldsymbol{\Theta})$. The criterion of *D*-optimality is the determinant of the information matrix. Its square root is inversely proportional to the volume of the asymptotic probability confidence ellipsoid for $\boldsymbol{\beta}$. See Silvey (1980) and Buyske (1998a, b) for other criteria. There exist practical methods for deriving *D*-optimal designs (Donev and Atkinson, 1988; Haines, 1987; Mitchell, 1974; Welch, 1982). For the specific case of the 2PL item response function, see Jones and Jin (1994) and Berger (1992).

Suppose that we have a set T of items that is partitioned into m non-intersecting testlets $T = \{T_j : j = 1, \ldots, m\}$. We are interested in assigning test-takers to testlets so that the greatest amount of information about the item parameters can be obtained. Assume point estimates of test-takers' ability are available, denoted by $\boldsymbol{\Theta} = \{\theta_i : i = 1, \ldots, n\}$. Assume initial point estimates of item parameter vectors are available, denoted by $\mathbf{B} = \{\boldsymbol{\beta}_t : T \ni t\}$. Let $\mathbf{B}_j = \{\boldsymbol{\beta}_t : T_j \ni t\}$ be vectors of parameter items in testlet T_j.

Let $\mathbf{m}^{(i,t)} \equiv \mathbf{m}(\boldsymbol{\beta}_t; \theta_i)$ be the Fisher information matrix for the item parameters of item t given by the ability of examinee i. Denote the information matrix obtained from pairing examinee i with testlet T_j by

$$\mathbf{m}_{ij} = \mathrm{diag}(\mathbf{m}(\boldsymbol{\beta}_t; \theta_i) : t \in T_j).$$

Introduce x_{ij} equal to the number of observations taken for testlet j from examinees with ability θ_i. Then by the additive property of Fisher's information, the information matrix of testlet j is

$$\mathbf{M}_j \equiv \mathbf{M}_j(\mathbf{B}_j; \boldsymbol{\Theta}, \mathbf{x}) = \sum_{i=1}^{n} x_{ij} \mathbf{m}_{ij},$$

where $\mathbf{x} = (x_{11}, \ldots, x_{1m}; x_{21}, \ldots, x_{2m}; \ldots; x_{n1}, \ldots, x_{nm})$. If observations for different testlets are taken independently, the information matrix for all testlets taken together is

$$\mathbf{M}(\mathbf{B}; \boldsymbol{\Theta}, \mathbf{x}) = \mathrm{diag}\{\mathbf{M}_j\}. \tag{2.2}$$

Note that $\mathbf{M}$ is positive definite because each $\mathbf{M}_j$ is positive definite by the fact it is the Fisher information matrix.

The design criterion we wish to consider is

$$\log \det \mathbf{M}(\mathbf{B}; \boldsymbol{\Theta}, \mathbf{x}) = \sum_{j=1}^{m} \log \det \mathbf{M}_j(\mathbf{B}_j; \boldsymbol{\Theta}, \mathbf{x}).$$

The right-hand side follows from the block diagonal property of $\mathbf{M}$, as indicated in definition (2.2).

2.3 Solution for Optimal Designs

The reader may skip the next few technical subsections and go to the subsection on MCMC sequential estimation of item parameters.

2.3.1 Mathematical programming model

Consider the problem of selecting x_{ij}, the number of observations taken for item j from examinees with ability θ_i to maximize information. A mathematical programming model for finding an optimal design with marginal constraints on $\mathbf{x} = \{x_{ij} : i = 1 \ldots n; j = 1 \ldots m\}$ is the following:

$$\text{Maximize} \log \det \mathbf{M}(\mathbf{B}; \boldsymbol{\Theta}, \mathbf{x})$$

such that

$$\sum_{j=1}^{m} x_{ij} = s; i = 1, \ldots, n \tag{2.3}$$

$$\sum_{i=1}^{n} x_{ij} = d; j = 1, \ldots, m \tag{2.4}$$

$$x_{ij} \geq 0; i = 1, \ldots, n; j = 1, \ldots, m \tag{2.5}$$

where $md = ns$.

These constraints describe a particular type of network flow called the transportation problem (Ahuja *et al.*, 1993). The terminology of the transportation problem relies on two notions: origin and destination. In our problem, it is helpful to think of an *origin* as a warehouse of examinees with similar abilities, and a *destination* as a particular experimental testlet, and of a specific commodity, such as an examinee, to be shipped. Constraints (2.3) stipulate that at each origin i there is a supply of s test-takers at with ability θ_i for n origins. Constraints (2.4) stipulate that at each destination T_j there is a testlet that demands d test-takers. Constraints (2.5) ensure that the solution is logically consistent. The requirement $md = ns$ ensures that the total testlet demand is equal to the total supply of test-takers. These are known as trans-shipment constraints.

Typically, the transportation problem assumes different costs of shipping from origin to destination, and the objective is to minimize shipping costs. In our case, we receive a certain value of Fisher information for shipping between particular origins and destinations. Unlike the typical transportation problem, our objective is nonlinear in the Fisher information values, and cannot be solved with a compact simplex tableau.

To apply the constrained conjugate-gradient method, we need the constraints in matrix form. The constraint matrix corresponding to constraints (2.3) and (2.4) is derived as follows. Denote $\mathbf{I}_n$ as an n by n identity matrix. In addition, denote $\mathbf{e}_m$ and $\mathbf{e}_n$ as m and n column vectors of all 1's. Define the following constraint matrix:

$$\mathbf{A} = \begin{pmatrix} \mathbf{I}_n & \mathbf{I}_n & \cdots & \mathbf{I}_n \\ \mathbf{e}_n^T & 0 & \cdots & 0 \\ 0 & \mathbf{e}_n^T & \cdots & 0 \\ \vdots & \vdots & \ddots & \vdots \\ 0 & 0 & \cdots & \mathbf{e}_n^T \end{pmatrix}. \tag{2.6}$$

Define

$$\mathbf{b} = \begin{pmatrix} s\mathbf{e}_n \\ d\mathbf{e}_m \end{pmatrix}.$$

Then

$$\mathbf{A}\mathbf{x} = \mathbf{b} \tag{2.7}$$

is a restatement of the constraints (2.3) and (2.4). A point $\mathbf{x}$ is feasible if it satisfies (2.7). If $\mathbf{x}$ is feasible and α is an arbitrary scalar, then, a point $\mathbf{x} + \alpha\mathbf{r}$ is feasible if and only if $\mathbf{A}\mathbf{r} = 0$; that is, if $\mathbf{r}$ is in the null space of $\mathbf{A}$. This fact is used in the gradient search algorithm.

The objective function for this problem is known to be concave in $\mathbf{x}$ (Fedorov, 1972). Therefore, we use the conjugate-gradient method for linearly constrained problems and logarithmic penalty functions for the non-negativity constraints (2.5) to get approximate designs (for an overview of optimality procedures, see Gill *et al.*, 1981). For the penalty function method, our objective function becomes:

$$F(\mathbf{x}) \equiv F_\mu(\mathbf{x}) = \log\det \mathbf{M}(\mathbf{B}; \boldsymbol{\Theta}, \mathbf{x}) + \mu \sum_{i,j} \log\ x_{ij}$$

where μ is a penalty parameter; this parameter is sequentially reduced to a sufficiently small value.

2.3.2 Unconstrained conjugate-gradient method

The unconstrained conjugate-gradient method is described as follows. Assume that $\mathbf{x}_k$ approximates the maximum of F. Given a direction of search $\mathbf{p}_k$, let

$$\alpha^* = \arg\ \max_a\ F(\mathbf{x}_k + \alpha\mathbf{p}_k).$$

The next approximation to the maximum of F is

$$\mathbf{x}_{k+1} = \mathbf{x}_k + \alpha^*\mathbf{p}_k.$$

Denote the gradient of F at $\mathbf{x}_k$ as $\mathbf{g}(\mathbf{x}_k)$ (see Appendix A and the note that follows here). The direction of search, obtained from the method of conjugate gradient, is defined as:

$$\mathbf{p}_k = \mathbf{g}(\mathbf{x}_k) + \mathbf{p}_{k-1}\frac{\mathbf{g}(\mathbf{x}_k)^T[\mathbf{g}(\mathbf{x}_k) - \mathbf{g}(\mathbf{x}_{k-1})]}{\|\ \mathbf{g}(\mathbf{x}_{k-1})\ \|^2} \tag{2.8}$$

where $\mathbf{p}_{k-1}$ and $\mathbf{x}_{k-1}$ denote the previous search direction and approximation, respectively. The conjugate gradient must be restarted in the direction of steepest ascent every nm steps.

2.3.3 Constrained conjugate-gradient method

The constrained conjugate-gradient method is an extension of the foregoing method. The linearly constrained problem uses the projection of the direction (2.8) on the null space of the constraint matrix (2.6). Denote $\mathbf{p}$ as an nm dimensional direction vector. In Appendix B we derive the following simple expression for the ij^{th} component of the projection $\mathbf{r}$ of $\mathbf{p}$ on $\{\mathbf{x} : \mathbf{A}\mathbf{x} = 0\}$:

$$r_{ij} = p_{ij} - \frac{1}{m}\sum_{l=1}^{m} p_{il} - \frac{1}{n}\sum_{k=1}^{n} p_{kj} + \frac{1}{mn}\sum_{k,l} p_{kl}.$$

This result is not very surprising in light of the theory of analysis of variance in statistics using p_{ij} for data, as these elements are known as *interactions* in the two-way model with main effects for item and ability. They are the least-squares residuals from the fitted values in the main effects model, and thus must lie in the null space of the constraints (2.3) and (2.4). Note that searching from a feasible point in the direction of $\mathbf{r}$ ensures feasibility of the solution.

2.3.4 Gradient of $\log\det \mathbf{M}(\mathbf{B}; \boldsymbol{\Theta}, \mathbf{x})$

Since the information matrix is block-diagonal with respect to individual items, the gradient of $\mathbf{M}(\mathbf{B}; \boldsymbol{\Theta}, \mathbf{x})$ with respect to $\mathbf{x}$ is easy to obtain. We show the result here for three-parameter item response models. Introduce the information matrix $\mathbf{M}^{(t)}$ of item t belonging to testlet T_j:

$$\mathbf{M}^{(t)} \equiv \mathbf{M}^{(t)}(\boldsymbol{\beta}_t; \boldsymbol{\Theta}, \mathbf{x}) = \sum_{i=1}^{n} x_{ij}\mathbf{m}^{(i,t)}.$$

(Note that x_{ij} is the same for all items t belonging to testlet T_j.) The criterion can also be written as

$$\log\det \mathbf{M}(\mathbf{B}; \boldsymbol{\Theta}, \mathbf{x}) = \sum_{j=1}^{n}\sum_{t\in T_j} \log\det \mathbf{M}^{(t)}(\boldsymbol{\beta}_t; \boldsymbol{\Theta}, \mathbf{x}).$$

Then

$$\frac{\partial}{\partial \mathbf{x}_{ij}} \log\det \mathbf{M}(\mathbf{B}; \boldsymbol{\Theta}, \mathbf{x}) = \sum_{t\in T_j} \frac{\partial}{\partial \mathbf{x}_{ij}} \log\det \mathbf{M}^{(t)}(\boldsymbol{\beta}_t; \boldsymbol{\Theta}, \mathbf{x}),$$

where

$$\frac{\partial \log \det \mathbf{M}^{(t)}(\boldsymbol{\beta}_t;\boldsymbol{\Theta},\mathbf{X})}{\partial x_{ij}} = \frac{1}{\det \mathbf{M}^{(t)}}$$

$$\times \left\{ \begin{array}{c} \det \begin{pmatrix} \mathbf{m}_{11}^{(i,t)} & \mathbf{m}_{12}^{(i,t)} & \mathbf{m}_{13}^{(i,t)} \\ \mathbf{M}_{21}^{(t)} & \mathbf{M}_{22}^{(t)} & \mathbf{M}_{23}^{(t)} \\ \mathbf{M}_{31}^{(t)} & \mathbf{M}_{32}^{(t)} & \mathbf{M}_{33}^{(t)} \end{pmatrix} + \det \begin{pmatrix} \mathbf{M}_{11}^{(t)} & \mathbf{M}_{12}^{(t)} & \mathbf{M}_{13}^{(t)} \\ \mathbf{m}_{21}^{(i,t)} & \mathbf{m}_{22}^{(i,t)} & \mathbf{m}_{23}^{(i,t)} \\ \mathbf{M}_{31}^{(t)} & \mathbf{M}_{32}^{(t)} & \mathbf{M}_{33}^{(t)} \end{pmatrix} \\ + \det \begin{pmatrix} \mathbf{M}_{11}^{(t)} & \mathbf{M}_{12}^{(t)} & \mathbf{M}_{13}^{(t)} \\ \mathbf{M}_{21}^{(t)} & \mathbf{M}_{22}^{(t)} & \mathbf{M}_{23}^{(t)} \\ \mathbf{m}_{31}^{(i,t)} & \mathbf{m}_{32}^{(i,t)} & \mathbf{m}_{33}^{(i,t)} \end{pmatrix} \end{array} \right\}. \tag{2.9}$$

2.3.5 MCMC sequential estimation of item parameters

In item response theory, optimal designs depend on knowing the item parameters. For obtaining designs for the next stage of observations, a sequential estimation scheme supplies item parameters that should converge to the true parameters. By continually improving the item parameters, the solutions of mathematical program yield better designs for obtaining data that are more efficient.

This chapter explores a method for item calibration using stochastic approximation of the posterior density. Stochastic approximation consists of Monte Carlo sampling directly from the posterior distribution. The sampled process is obtained by the Hastings–Metropolis algorithm that generates a Markov chain with a stationary distribution the same as the desired posterior distribution. In this case, stochastic approximation is called Markov chain Monte Carlo (MCMC) simulation. Estimation of the posterior density and its moments is based on these sampled data. The reader is referred to Patz and Junker (1999), who outlined general MCMC sampling methods for obtaining estimates of parameters in the three-parameter logistic IRT model.

In this chapter, we study the performance of optimal design methods using MCMC Bayesian estimators incorporating uninformative priors. In particular, we employ MCMC estimation incorporating the Metropolis–Hastings algorithm and a sample reuse scheme (Jones and Nediak, 1999). The use of an uninformative prior allows the data to have the greatest influence on the posterior distribution of the parameter estimates. Unless we knew we had some specific knowledge about how difficult and discriminant a particular item may be, we have no choice but to use an uninformative prior.

We also reuse the MCMC samples to average the criterion function over the joint posterior of all the ability parameters. Define the joint posterior density of the test-taker abilities, given N observations $\mathbf{U}_N$, as $\pi(\boldsymbol{\Theta} \mid \mathbf{U}_N)$. We use MCMC sample from this density to average the criterion (2.2) and all its derivatives.

2.3.6 Note on performance measures

Although the intent of this methodology is to create optimal designs, the final objective is to efficiently estimate the item response functions (IRFs) of experimental items. It is by measuring the quality of the estimated IRFs that one can evaluate the practical worth of optimal design employment. Therefore we compare the IRFs of the sequentially estimated parameters to the BILOG estimated parameters based on all the data (see Mislevy and Bock, 1990). BILOG is widely used by testing agencies to calibrate item parameters, and thus it could be considered a standard. MULTILOG (Thissen, 1991) and PARSCALE (Muraki and Bock, 2003) are similar packages in wide use as well. Therefore, we will benchmark our MCMC estimators using optimal or spiralling samples against the BILOG estimators using all the data. The BILOG estimators are obtained using the marginal maximum likelihood criterion with the expectation-maximization (EM) algorithm. The marginal likelihood is the likelihood of the observed data integrated over the assumed distribution of the abilities.

We use the Hellinger distance (Prakasa Rao, 1987) defined as follows. The Hellinger distance between two IRFs, P_1 and P_2, for instance, is

$$H(P_1, P_2) = \int \left\{ \left[|P_1(\theta) - P_2(\theta)|^{1/2} \right]^2 + \left[|\bar{P}_1(\theta) - \bar{P}_2(\theta)|^{1/2} \right]^2 \right\} \mathrm{d}\Phi(\theta)$$

where $\bar{P}_j = 1 - P_j$ and Φ is a weight function. Hellinger distance is non-negative and bounded by 2.

The estimated ability scale is standardized. We assume the distribution is normal and we wish to investigate differences between our estimators and the BILOG estimators of IRF in the region of $\theta = 0$. For this reason, in the following we use a normal weight function with mean 0 and standard deviation 2. For investigations about a cut-point on ability that is different from 0, other weight functions may be desirable.

We prefer the Hellinger distance for determining the closeness of two IRFs over the usual Euclidean metric based on item parameters. We have found that the Hellinger distance is more predictive of the empirical behaviour of tests based on two sets of item parameters, than the Euclidean distance would indicate. It can be shown that the two metrics are equivalent in convergence. That is, if a sequence of item parameter estimates converges in the Euclidean sense, the sequence converges in the Hellinger sense as well, and vice versa.

Frequently two or more valid estimation methodologies can result in different item parameters for the same item but appear to be greatly different in the Euclidean sense, but are close in the Hellinger sense. Therefore, the Hellinger distance is useful for comparing the performance of different methodologies and sample design schemes. The other performance measure we use is based on the posterior confidence bands provided by the Bayes methodology. This comparison is primarily graphical. We use the two measures, Hellinger and posterior confidence band, in tandem; that is, if the Hellinger distances are extreme, then we inspect the graph of the posterior confidence band.

2.4 Simulation Results

We present some results using actual data from paper-and-pencil administration of the LSAT. The data consist of the responses to 23 items from 28 184 test-takers. We use these data to simulate an on-line item calibration. In the simulation, the optimal and spiralling design chose 32 000 from the 28 184 test-takers to receive testlets. Note that this is not a usual simulation situation where responses are generated under a known model.

Batches of 40 test-takers are randomly chosen. Each test-taker receives exactly one testlet. Initial estimates of **B** are fed to the mathematical program and a design is derived. Two mathematical programming models were implemented. One model, denoted as 'constrained', using the original model with constraints (2.3)–(2.5), and the other model, denoted as 'unconstrained', using the original model but dropping constraint (2.4). The latter model results in unequal numbers of observations for testlets. Therefore, if a testlet is 'demanding' more observations than other testlets, it may have them if this increases the Fisher information matrix for that testlet.

The following is a summary of the simulation components: we have separated 23 items into 4 testlets, of size 6, 5, 6 and 6 items, that are calibrated simultaneously. Table 2.1 displays the parameters of the testlets under the three columns with heading 'BILOG parameters'. The parameters β_0, β_1 and β_2 correspond to the estimated difficulty, discrimination and guessing parameters of the three-parameter logistic response function.

The spiralling design consists of assigning randomly selected test-takers to testlets in equal numbers. This was done with batches of 40 randomly selected test-takers, so that each batch of 40 test-takers yields 10 new observations for each of the 4 testlets. The total sample size is 80 batches of 40, yielding 800 observations for each testlet. We call this design 'pure spiralling', since, as shall see, we use some spiralling for the optimal designs.

The optimal designs use the spiralling design for the first 40 batches of test-takers, but each of the last 40 batches obtains the D-optimal information matrix for item parameters. After each batch, estimates of item parameters are updated to feed to the optimal design for the next batch. For the constrained case, each batch of 40 test-takers yields 10 new observations for each of the 4 testlets. For the unconstrained case, each batch of 40 test-takers yields a variable number of new observations for each testlet. Eighty batches were generated, resulting in 800 observations per testlet for the constrained designs and approximately

Table 2.1 Item parameter estimates

					Sequential MCMC estimates											
		BILOG parameters			Constrained optimal				Unconstrained optimal				Pure spiralling			
Testlet	ID	β_0	β_1	β_2	b_0	b_1	b_2	H^a	b_0	b_1	b_2	H^b	b_0	b_1	b_2	H^c
1	1	0.17	0.80	0.13	–0.02	0.90	0.18	0.0034	0.09	0.82	0.17	0.0012	-0.33	0.98	0.32	0.0311
	2	0.63	0.96	0.30	1.05	1.15	0.18	0.0221	0.87	1.25	0.23	0.0093	1.01	1.20	0.18	0.0218
	3	−0.24	1.17	0.05	−0.20	1.32	0.03	0.0016	−0.15	1.25	0.02	0.0022	−0.16	1.31	0.04	0.0012
	4	−1.05	1.18	0.14	−1.11	1.31	0.11	0.0010	−0.89	1.16	0.09	0.0029	−0.85	1.19	0.07	0.0051
	5	0.44	0.70	0.14	0.54	0.78	0.16	0.0002	0.24	0.93	0.24	0.0079	0.42	0.77	0.23	0.0033
	6	1.16	0.57	0.17	0.90	0.68	0.23	0.0034	0.80	0.72	0.27	0.0083	0.87	0.71	0.29	0.0093
2	7	1.28	0.68	0.00	0.79	0.71	0.23	0.0542	0.70	0.77	0.29	0.0768	0.78	0.78	0.21	0.0461
	8	−1.62	1.27	0.11	−2.06	1.54	0.12	0.0125	−1.89	1.54	0.12	0.0069	−1.42	0.97	0.09	0.0077
	9	0.82	0.84	0.02	0.45	0.82	0.19	0.0332	0.19	0.88	0.24	0.0583	0.39	0.81	0.19	0.0359
	10	−1.29	1.16	0.24	−0.84	0.91	0.12	0.0208	−1.23	1.17	0.18	0.0013	−1.36	1.06	0.24	0.0002
	11	0.57	0.73	0.22	0.24	0.72	0.36	0.0145	0.14	0.78	0.34	0.0138	0.19	0.98	0.34	0.0141
3	12	−2.27	1.31	0.17	−2.88	1.69	0.19	0.0185	−2.91	1.68	0.19	0.0194	−3.32	1.89	0.21	0.0448
	13	−1.02	1.01	0.24	−1.28	1.22	0.24	0.0046	−1.48	1.33	0.25	0.0128	−1.45	1.33	0.27	0.0129
	14	−1.96	1.05	0.27	−2.98	1.53	0.34	0.0415	−3.67	1.81	0.35	**0.0872**	−3.98	2.04	0.35	**0.1104**
	15	−2.24	1.10	0.14	−2.13	1.10	0.12	0.0005	−2.07	1.01	0.12	0.0021	−1.93	0.88	0.09	0.0099
	16	−3.45	1.43	0.10	−4.04	1.66	0.09	0.0084	−3.98	1.78	0.10	0.0092	−4.73	1.89	0.11	0.0292
	17	−1.54	0.91	0.20	−2.09	1.15	0.23	0.0152	−2.34	1.26	0.24	0.0307	−2.19	1.15	0.22	0.0180
4	18	0.27	0.56	0.01	−0.27	0.83	0.17	0.0283	−0.30	0.77	0.20	0.0328	−0.16	0.82	0.18	0.0249
	19	−0.82	0.89	0.12	−0.93	0.97	0.17	0.0022	−0.75	0.83	0.11	0.0006	−0.71	0.86	0.07	0.0019
	20	−1.96	0.99	0.10	−1.77	0.97	0.07	0.0018	−1.69	0.82	0.06	0.0073	−1.92	1.02	0.08	0.0002
	21	−3.00	1.18	0.09	−2.67	1.04	0.08	0.0052	−2.99	1.09	0.10	0.0003	−3.01	1.23	0.09	0.0001
	22	−2.45	0.99	0.14	−1.95	0.80	0.10	0.0138	−2.00	0.78	0.08	0.0148	−2.14	0.80	0.13	0.0074
	23	−2.07	0.90	0.24	−3.81	1.61	0.30	**0.0902**	−3.37	1.44	0.30	0.0598	−2.99	1.16	0.28	0.0270

[a]Hellinger distance between constrained optimal and BILOG estimates.
[b]Hellinger distance between unconstrained optimal and BILOG estimates.
[c]Hellinger distance between spiralling and BILOG estimates.

800 observations per testlet for the unconstrained designs. Each optimal design is a mixture of pure spiralling and optimal designs. Therefore, the optimal designs get to choose only 400 of the observations optimally. A more extensive simulation could have studied the best mixture of spiralling and optimal designs.

Table 2.1 displays the item parameter estimates for each of the designs. In addition, Table 2.1 displays the Hellinger distances between the estimate and BILOG IRFs. A Hellinger distance of 0.001 or smaller indicates that the two IRFs are essentially on top of one another and that the items parameters are close as well. In the other extreme, we have learned that if the Hellinger distance is 0.02 or larger, the IRFs are different over one or more intervals of θ, and the item parameters are very different as well.

Table 2.2 displays the quartiles of the Hellinger fits over the items of all testlets combined for each design. Roughly 75% of the Hellinger fits coming from constrained optimal designs are less than 0.0221, while those of unconstrained optimal and spiralling designs are 0.0307 and 0.0292, respectively. In other words, each estimation scheme provides poor fits for about 25% of the items, with constrained optimal designs producing a few less poor fitting IRFs. Perhaps the sample size is not large enough for the affected items.

Table 2.2 Summary of Hellinger distances

		H^a	H^b	H^c
N		23	23	23
Percentiles	25	0.0022	0.0022	0.0033
	50	0.0125	0.0092	0.0129
	75	0.0221	0.0307	0.0292

Figure 2.1 displays the box plot of the Hellinger distances for all the items for each design. The box plot displays the quartiles in Table 2.2 as well as extreme values. The vertical line at 0.02 has been superimposed: values to the left of the line are good, while values to the right are poor. The reference line reveals the same information as Table 2.2: that constrained optimal design is a little better than either unconstrained optimal or spiralling. In addition, the unconstrained optimal design is a little better than spiralling. Although the unconstrained optimal design is a little worse than the constrained optimal design in terms of fits over 0.02, the median and first quartile fits of the unconstrained optimal design are better than the fits of the constrained optimal design, and are substantially better than fits from the spiralling design.

Figure 2.2 displays the box plot of the Hellinger distances for all the items for each design and each testlet. No one design appears to dominate over all testlets. However, there may be features of the testlets that enable the optimal designs to dominate spiralling as seen here. Note that in case of testlets 1 and 3, Figure 2.2

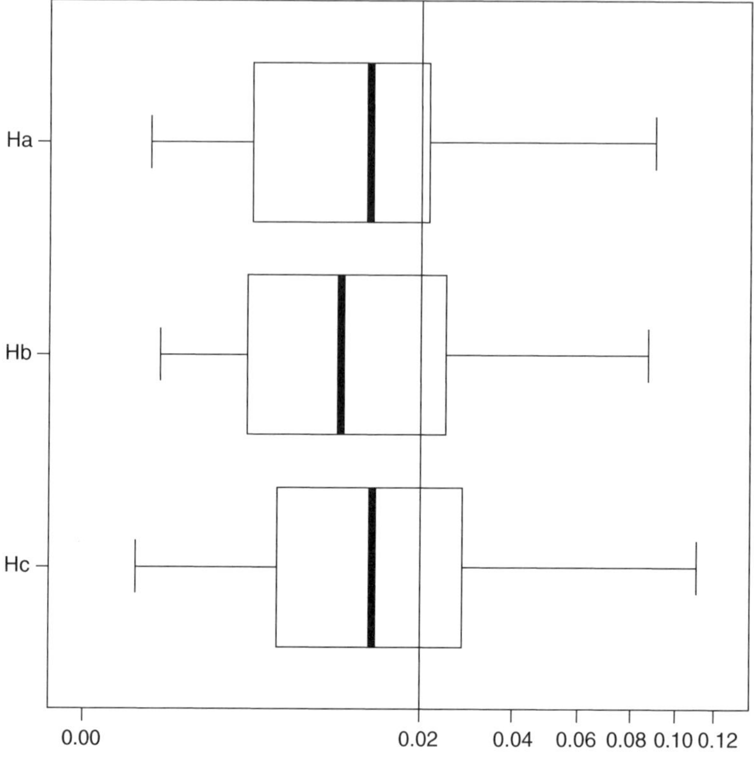

Figure 2.1 Box plot of Hellinger distances

reveals that optimal designs did a substantially better than pure spiralling. Moreover, in case of testlet 2, where the constrained optimal design is poor, pure spiralling does not do a lot better.

For testlet 3, Figures 2.3(a) and (b) display the total data BILOG IRF (denoted by the plus line) and the sequentially estimated IRF with the posterior 95% confidence bands superimposed. Denoted in bold in Table 2.1 are the three largest Hellinger distances: 0.0902 for constrained optimal for item 23 in testlet 4, 0.0872 for the unconstrained optimal design, and 0.1104 for pure spiralling, both for item 14 in testlet 3. We note that the graphs in Figures 2.3(a) and (b) agree with the Hellinger distances about the poor quality of the fit for item 14 with the unconstrained optimal and spiralling design. The graphs of the other testlets do not reveal much more information, therefore they have been suppressed.

We do not display any of the optimal designs. In the final analysis, the quality of the estimated IRF is important and the actual sample of test-takers is secondary.

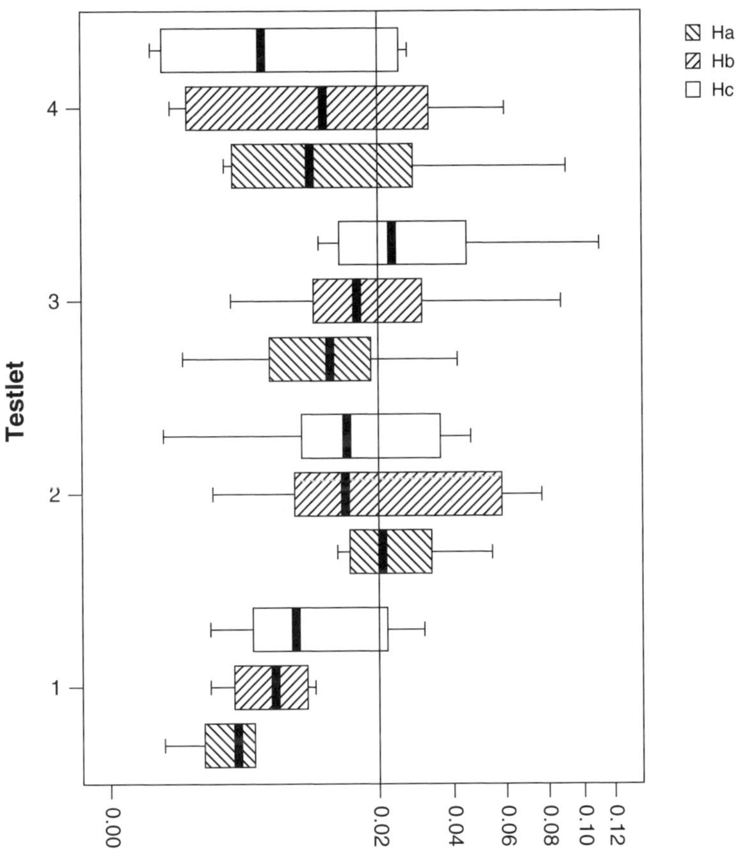

Figure 2.2 Box plot of Hellinger distances by testlet

2.5 Discussion

The goal of this chapter was to study the feasibility and superiority of sequential estimation with *D*-optimal designs using testlets. Our results found that sequential estimation with *D*-optimal designs is as feasible as the standard spiralling design. Computational times for deriving optimal designs were well within bounds for implementation in an actual setting. The issue of superiority of optimal design over spiralling is less clear; overall however, the optimal designs were marginally better.

In practice, test constructors would design testlets with homogeneous items; in contrast, our testlets were merely random sets of items, and could therefore be heterogeneous. Optimal *D*-design schemes would tend to select sets of test-takers reflecting a broad range of ability to satisfy the heterogeneous items within testlets.

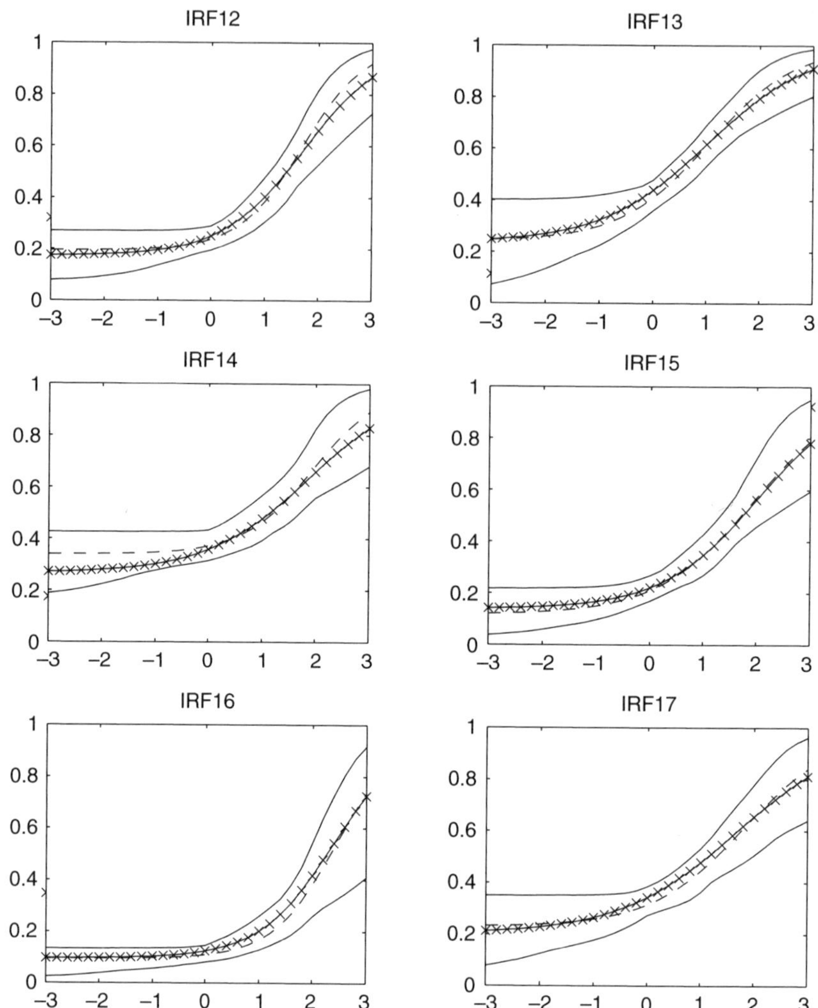

Figure 2.3(a) Testlet 3 – constrained optimal designs

The spiralling design scheme would also tend to select similar sets of test-takers. This characteristic is seen with the results for testlets 2 and 4, which are quite heterogeneous, and the optimal design was mediocre as compared to spiralling. However, for testlets 1 and 3, which are homogeneous, the optimal designs were substantially better than spiralling.

Extensions of this research would be promising and theoretically interesting. Estimation should be better with informative priors. Linear hierarchical priors could

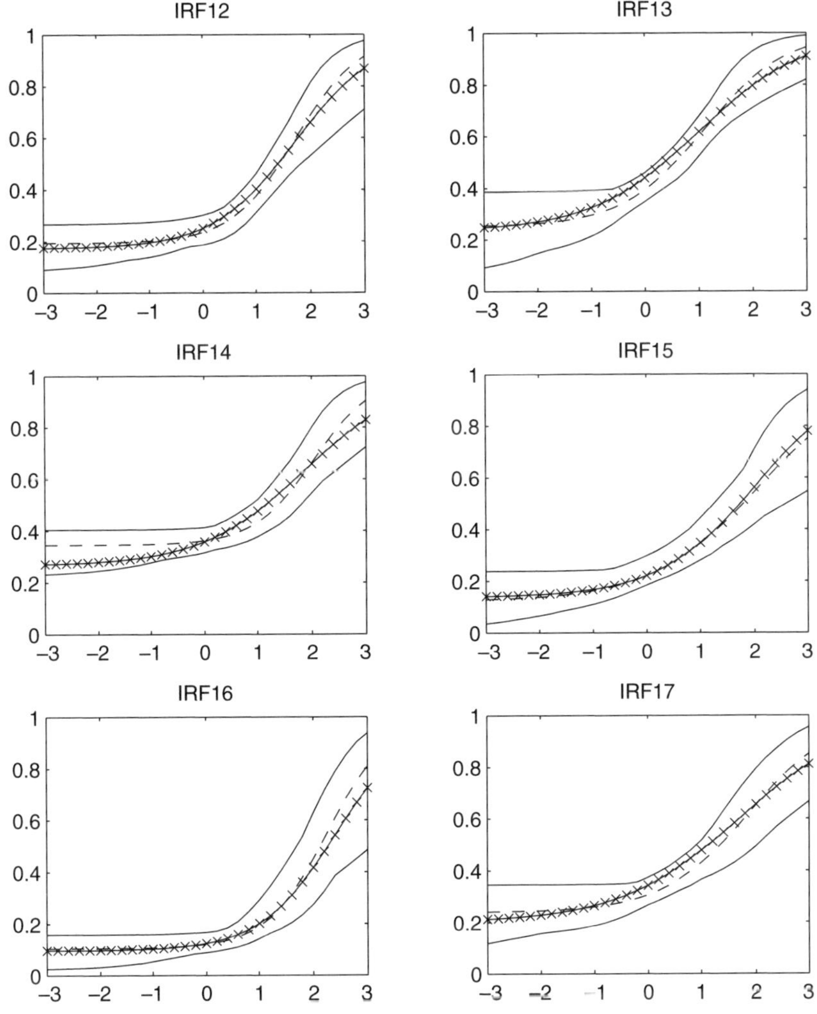

Figure 2.3(b) Testlet 3 – unconstrained optimal designs

be employed to obtain priors that are more informative. These may be obtained from past calibrations and, possibly, information about the item. Other design criteria could be employed, e.g. Buyske (1998a, b). Stopping rules based on the variation of measures in the posterior could be employed as well. Clearly, some items were calibrated more effectively than others. In actual implementation of on-line item calibration, the more easily calibrated testlets could be removed from sampling earlier.

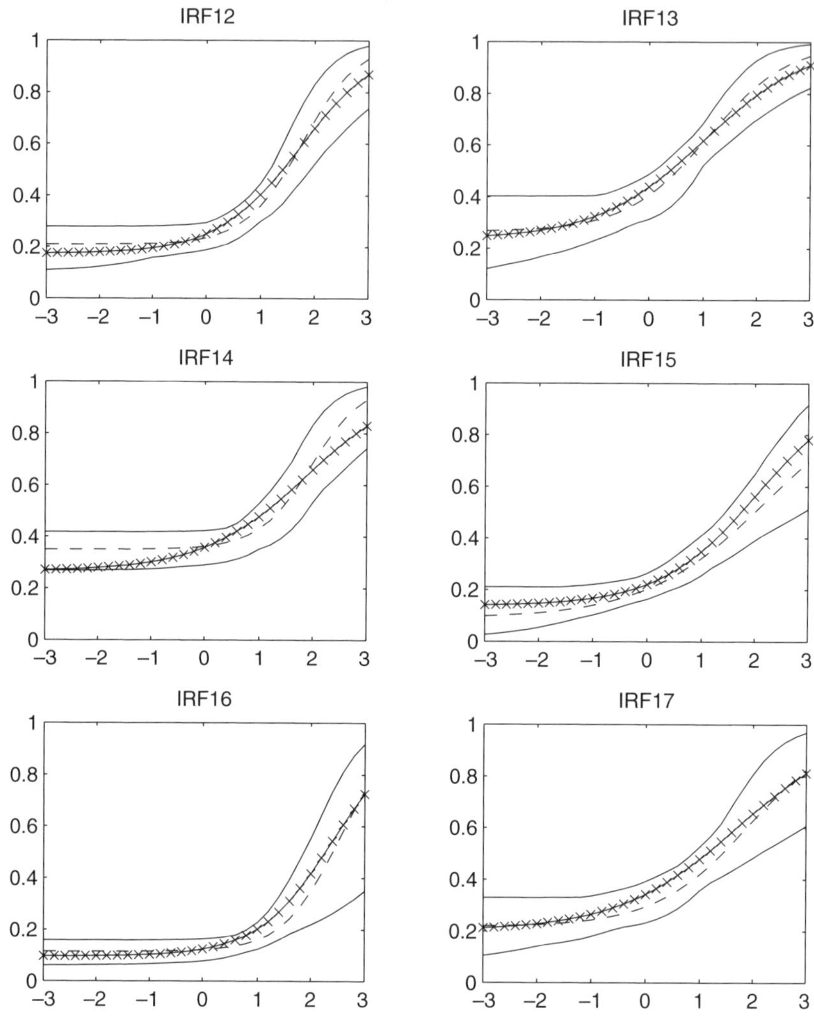

Figure 2.3(c) Testlet 3 – spiraling designs

Appendix A Derivation of the Gradient of $\log \det \mathbf{M}(\mathbf{B}; \boldsymbol{\Theta}, \mathbf{x})$

In the following, we sketch the derivation for the gradient. In general, consider the matrix

$$\mathbf{H}(\mathbf{y}) = \sum_{j=1}^{N} y_j \mathbf{h}^{(j)}, \tag{2.10}$$

where $h^{(j)}$ is a k by k matrix. For the 3PL model, $N = 3$. Let $(i_1, \ldots, i_k)$ denote the permutation of the numbers from 1 to k and $\sigma_(i_1, \ldots, i_k)$ its sign. By definition,

$$\det \mathbf{H}(\mathbf{y}) = \sum_{(i_1,\ldots,i_k)} (-1)^{\sigma(i_1,\ldots,i_k)} \left[\left(\sum_{j_1=1}^{N} y_{j_1} \mathbf{h}_{1i_1}^{(j_1)} \right) \cdots \left(\sum_{j_k=1}^{N} y_{j_k} \mathbf{h}_{1i_k}^{(j_k)} \right) \right]. \tag{2.11}$$

Applying the chain rule for differentation,

$$\begin{aligned}
\frac{\partial \det \mathbf{H}(\mathbf{y})}{\partial y_j} &= \sum_{(i_1,\ldots,i_k)} (-1)^{\sigma(i_1,\ldots,i_k)} \left\{ \sum_{l-1}^{k} \left[\left(\sum_{j_1=1}^{N} y_{j_1} \mathbf{h}_{li_1}^{(j_1)} \right) \cdots g_{li_l}^{(j)} \cdots \left(\sum_{j_k=1}^{N} y_{j_k} \mathbf{h}_{1i_k}^{(j_k)} \right) \right] \right\} \\
&= \sum_{l=1}^{k} \sum_{(i_1,\ldots,i_k)} (-1)^{\sigma(i_1,\ldots,i_k)} \left[\left(\sum_{j_1=1}^{N} y_{j_1} \mathbf{h}_{1i_1}^{(j_1)} \right) \cdots g_{li_l}^{(j)} \cdots \left(\sum_{j_k=1}^{N} y_{j_k} \mathbf{h}_{1i_k}^{(j_k)} \right) \right] \\
&= \sum_{l=1}^{k} \det \mathbf{C}^{(l,j)},
\end{aligned} \tag{2.12}$$

where is $\mathbf{C}^{(l,j)}$ a k by k matrix of the form

$$\mathbf{C}^{(l,j)} = \begin{pmatrix} \mathbf{H}_{11}(\mathbf{y}) & \cdots & \mathbf{H}_{1k}(\mathbf{y}) \\ \vdots & & \vdots \\ \mathbf{h}_{l1}^{j} & \cdots & \mathbf{h}_{lk}^{j} \\ \vdots & & \vdots \\ \mathbf{H}_{k1}(\mathbf{y}) & \cdots & \mathbf{H}_{kk}(\mathbf{y}) \end{pmatrix}. \tag{2.13}$$

Applying this result to our objective function, we get expression (2.9).

Appendix B Projection on the Null Space of the Constraint Matrix

Denote $\mathbf{p}$ as an nm dimensional direction vector. Denote $\mathbf{I}_m$ and $\mathbf{I}_n$ as m by m and n by n identity matrices, respectively. In addition, denote $\mathbf{e}_m$ and $\mathbf{e}_n$ as m and n column vectors of all 1's. Define the following constraint matrix:

$$\mathbf{A} = \begin{pmatrix} \mathbf{I}_n & \mathbf{I}_n & \cdots & \mathbf{I}_n \\ \mathbf{e}_n^T & 0 & \cdots & 0 \\ 0 & \mathbf{e}_n^T & \cdots & 0 \\ \vdots & \vdots & \ddots & \vdots \\ 0 & 0 & \cdots & \mathbf{e}_n^T \end{pmatrix}. \tag{2.14}$$

The projection $\mathbf{r}$ of $\mathbf{p}$ on $\{\mathbf{y} : \mathbf{A}\mathbf{y} = 0\}$, the null space of $\mathbf{A}$, is given by

$$\mathbf{r} = \mathbf{p} - \mathbf{A}^T\mathbf{y}, \tag{2.15}$$

where $\mathbf{y}$ is a solution to

$$(\mathbf{A}\mathbf{A}^T)\mathbf{y} = \mathbf{A}\mathbf{p}. \tag{2.16}$$

Introduce the following notation:

$$\mathbf{y} = \begin{pmatrix} \mathbf{y}_1 \\ \mathbf{y}_2 \end{pmatrix}; \mathbf{A}\mathbf{p} = \begin{pmatrix} \tilde{\mathbf{p}}_1 \\ \tilde{\mathbf{p}}_2 \end{pmatrix}; \mathbf{y}_1, \tilde{\mathbf{p}}_1 \in \mathbf{R}^n; \mathbf{y}_2, \tilde{\mathbf{p}}_2 \in \mathbf{R}^m. \tag{2.17}$$

We have the following simplifications:

$$\tilde{\mathbf{p}}_1 = (p_{1\bullet}, \ldots, p_{n\bullet})^T; \tilde{\mathbf{p}}_2 = (p_{\bullet 1}, \ldots, p_{\bullet m})^T, \tag{2.18}$$

where

$$p_{i\bullet} = \sum_{j=1}^{m} p_{ij}; p_{\bullet j} = \sum_{i=1}^{n} p_{ij}, \tag{2.19}$$

and

$$\mathbf{A}\mathbf{A}^T = \begin{pmatrix} m\mathbf{I}_n & \mathbf{E}^T \\ \mathbf{E} & n\mathbf{I}_m \end{pmatrix}, \tag{2.20}$$

where $\mathbf{E}$ is an m by n matrix of all ones. Consequently, Equation (2.16) can be written as

$$\begin{aligned} m\mathbf{y}_1 + \mathbf{e}_n(\mathbf{e}_m^T\mathbf{y}_2) &= \tilde{\mathbf{p}}_1 \\ \mathbf{e}_m(\mathbf{e}_n^T\mathbf{y}_1) + n\mathbf{y}_2 &= \tilde{\mathbf{p}}_2, \end{aligned} \tag{2.21}$$

implying

$$\begin{aligned} \frac{\mathbf{e}_n^T\mathbf{y}_1}{n} + \frac{\mathbf{e}_m^T\mathbf{y}_2}{m} &= \frac{\sum_{ij} p_{ij}}{nm} \\ \mathbf{y}_1 &= \frac{1}{m}[\tilde{\mathbf{p}}_1 - e_n(\mathbf{e}_m^T\mathbf{y}_2)] \\ \mathbf{y}_2 &= \frac{1}{n}[\tilde{\mathbf{p}}_2 - \mathbf{e}_m(\mathbf{e}_n^T\mathbf{y}_1)]. \end{aligned} \tag{2.22}$$

Thus

$$\mathbf{A}^T\mathbf{y} = \mathbf{A}^T\begin{pmatrix}\frac{1}{m}\tilde{\mathbf{p}}_1\\ \frac{1}{n}\tilde{\mathbf{p}}_2\end{pmatrix} - \mathbf{A}^T\begin{pmatrix}\mathbf{e}_n\left(\frac{\mathbf{e}_m^T\mathbf{y}_2}{m}\right)\\ \mathbf{e}_m\left(\frac{e_n^T\mathbf{y}_1}{n}\right)\end{pmatrix}$$
$$= \mathbf{A}^T\begin{pmatrix}\frac{1}{m}\tilde{\mathbf{p}}_1\\ \frac{1}{n}\tilde{\mathbf{p}}_2\end{pmatrix} - \mathbf{e}_{n+m}\left(\frac{\mathbf{e}_m^T\mathbf{y}_2}{m} + \frac{\mathbf{e}_n^T\mathbf{y}_1}{n}\right). \tag{2.23}$$

Denote $\mathbf{z} = \mathbf{A}^T\mathbf{y}$, then by (2.23),

$$z_{ij} = \frac{1}{m}p_{i\bullet} + \frac{1}{n}p_{\bullet j} - \frac{1}{nm}\sum_{i,j} p_{ij}. \tag{2.24}$$

Since $\mathbf{r} = \mathbf{p} - \mathbf{z}$, we have

$$\mathbf{r}_{ij} = p_{ij} - \frac{1}{m}p_{i\bullet} - \frac{1}{n}p_{\bullet j} + \frac{1}{nm}\sum_{i,j} p_{ij}. \tag{2.25}$$

Acknowledgements

This study received funding from the Law School Admission Council (LSAC). The opinions and conclusions contained in this report are those of the authors and do not necessarily reflect the position or policy of LSAC. The work of D.H. Jones was partially supported by the Cognitive Science Program of the Office of the Chief of Naval Research under Contact N00012-87-0696.

References

Ahuja, R. K., Magnanti, T. L. and Orlin, J. B. (1993). *Network Flows: Theory, Algorithms, and Applications*. Englewood Cliffs, NJ: Prentice-Hall.

Berger, M. P. F. (1991). On the efficiency of IRT models when applied to different sampling designs. *Applied Psychological Measurement*, **15**, 283–306.

Berger, M. P. F. (1992). Sequential sampling designs for the two-parameter item response theory model. *Psychometrika*, **57**, 521–538.

Berger, M. P. F. (1994). D-Optimal sequential designs for item response throe models. *Journal of Educational Statistics*, **19**, 43–56.

Berger, M. P. F. and van der Linden, W. J. (1991). Optimality of sampling designs in item response theory models. In M. Wilson (ed.), *Objective Measurement: Theory into Practice*. Norwood, NJ: Ablex Publishing.

Bradlow, E. T., Wainer, H. and Wang, X. (1999). A Bayesian random effects model for testlets. *Psychometrika*, **64**, 153–168.

Buyske, S. G. (1998a). Item calibration in computerized adaptive testing using minimal information loss. Preprint.

Buyske, S. G. (1998b). Optimal designs for item calibration in computerized adaptive testing. In N. Flournoy, W. F. Rosenberger and W. K. Wong (eds), *New Developments and Applications in Experimental Design*. Hayward, Calif.: Inst. of Math. Statistics.

Donev, A. N. and Atkinson, A. C. (1988). An adjustment algorithm for the construction of exact D-optimum experimental designs. *Technometrics*, **30**, 429–433.

Fedorov, V. V. (1972). *Theory of Optimal Experiments*. New York: Academic Press. (Federov, 1972).

Ford, I. (1976). Optimal static and sequential design: a critical review. Unpublished doctoral dissertation, University of Glasgow.

Ford, I., Kitsos, D. M. and Titterington, C. P. (1989). Recent advances in nonlinear experimental design. *Technometrics*, **31**, 49–60.

Ford, I., Titterington, C. P. and Wu, C. F. J. (1985). Inference and sequential design. *Biometrika*, **72**, 545–551.

Gill, P. E., Murray, W. and Wright, M. H. (1981). *Practical Optimization*. London: Academic Press.

Haines, L. M. (1987). The application of the annealing algorithm to the construction of exact optimal designs for linear-regression models. *Technometrics*, **29**, 439–447.

Jones, D. H., Chiang, J. and Jin, Z. (1997). Optimal designs for simultaneous item estimation. *Nonlinear Analysis, Methods and Applications*, **30**(7), 4051–4058.

Jones, D. H. and Jin, Z. (1994). Optimal sequential designs for on-line item estimation. *Psychometrika*, **59**, 57–75.

Jones, D. H. and Nediak, M. S. (1999). *Item Parameter Calibration of LSAT Items Using MCMC Approximation of Bayes Posterior Distributions*. RUTCOR Research Report, Rutgers the State University of New Jersey: Faculty of Management and RUTCOR.

Lord, F. M. (1980). *Applications of Item Response Theory to Practical Testing Problems*. Hillside, NJ: Erlbaum.

Mislevy, R. and Bock, D. (1990). *BILOG 3.06*. Lincolnwood, Ill.: Scientific Software International.

Mitchell, T. J. (1974). An algorithm for the construction of D-optimal experimental designs. *Technometrics*, **16**, 203–210.

Muraki, E. and Bock, D. (2003) *PARSCALE*. Lincolnwood, Ill.: Scientific Software International.

Patz, R. J., and Junker, B. W. (1999). Applications and extensions of MCMC for IRT: multiple item types, missing data, and rated responses. *Journal of Educational and Behavioral Statistics*, **24**, 342–366.

Prakasa Rao, B. L. S. (1987). *Asymptotic Theory of Statistical Inference*. New York: John Wiley & Sons, Ltd.

Silvey, S. D. (1980). *Optimal Design*. New York: Chapman and Hall.

Stocking, M. L. (1990). Specifying optimum examinees for item parameter estimation in item response theory. *Psychometrika*, **55**, 461–475.

Thissen, D. (1991). *MULTILOG User's Guide – Version 6*. Chicago, Ill.: Scientific Software, Inc.

Vale, C. D. (1986). Linking item parameters onto a common scale. *Applied Psychological Measurement*, **10**, 333–344.

van der Linden, W. J. (1988). *Optimizing Incomplete Sampling Designs for Item Response Model Parameters*. Research Report No. 88–5. Enschede, The Netherlands: University of Twente.

Wainer, H. and Kiely, G. (1987). Item clusters and computer adaptive testing: A case for testlets. *Journal of Educational Measurement*, **24**, 185–201.

Welch, W. J. (1982). Branch-and-bound search for experimental designs based on D optimality and other criteria. *Technometrics*, **24**, 41–48.

Wingersky, M. S. and Lord, F. M. (1984). An investigation of methods for reducing sampling error in certain IRT procedures. *Applied Psychological Measurement*, **8**, 347–364.

Wu, C. F. J. (1985). Asymptotic inference from sequential design in a nonlinear situation. *Biometrika*, **72**, 553–558.

3

On the Empirical Relevance of Optimal Designs for the Measurement of Preferences

Heiko Großmann, Heinz Holling, Michaela Brocke, Ulrike Graßhoff and Rainer Schwabe

Westfälische Wilhelms-Universität Munster, Psychologisches Institut IV, Fliednerstr. 21, D-48149, Münster, Germany

3.1 Introduction

Over the past century psychologists have undertaken a multitude of efforts to put the measurement of psychological variables on a firm grounding. As a result, the importance of formal models has continuously increased. This is illustrated, for example, by the widely accepted application of item response theory models to the construction of today's achievement tests. The utilization of a clearly stated statistical model offers the opportunity to improve the estimation and testing procedures related to the particular model with respect to well-defined and meaningful statistical criteria. Design considerations constitute one starting point in this direction. Applications of statistical design theory to psychological test theory can be found in the contributions by Buyske as well as Jones and Nediak in this volume.

In this chapter we focus on a different field of psychological application, namely the measurement of preferences. Preferences are assumed to be indicative of the utilities or values people associate with objects or features of the objects. According to many decision theories these utilities in turn are important when a decision has to

Applied Optimal Designs Edited by M.P.F. Berger and W.K. Wong
 ISBN: 0-470-85697-1 (HB)

be reached, for example when it has to be decided which course of action should be taken.

The practical significance of preference elicitation techniques derives from their widespread use for new product or concept development in areas such as marketing, transportation, health care, and many others. In marketing research, for example, preference elicitation techniques may provide answers to questions such as which product will be successful or which product attributes drive a consumer's purchasing decision and may thus serve as a valuable aid for managerial decisions.

In this context, methods for quantifying the joint effect of a set of attributes on the overall evaluation of objects, collectively known under the name of conjoint analysis (Green and Srinivasan, 1978), have become particularly popular. In conjoint analysis, respondents evaluate a set of objects that are represented by factorial combinations of the levels of certain attributes. The combinations of attribute levels are usually referred to as profiles. Typically, the profiles are presented in the form of verbal descriptions, but sometimes pictorial representations are used as well. For example, suppose the object was a car, and the attributes were colour, size, price and fuel economy. The attribute levels of size were 4, 5, or 6-passenger and the attribute levels for colour were red, blue, etc. Every individual has to state preferences for different profiles. These responses are then used to determine utility values for each attribute level.

The profiles have to be rank ordered or rated on some graded scale. Alternatively, graded paired comparisons are also commonly used. That is, two profiles are presented simultaneously and the respondent's task is to indicate how much s/he prefers one profile to the other. For instance, the respondent is asked a series of questions such as how much s/he prefers a red 4-passenger car to a blue 6-passenger. Methods that employ choices among two or more profiles are sometimes also considered as a type of conjoint analysis but are more frequently treated within the framework of choice models (see e.g. Louviere *et al.*, 2000). Within each method it is assumed that the preference judgements are based on the overall utility of each profile, which is taken to be a linear combination of the unknown utility values of the considered profile's attribute levels. These unknown parameters are then estimated from the data. Because each respondent usually evaluates several profiles, this analysis can be performed at the individual level, that is, a vector of parameter estimates can be computed for every single person. If the data consists of rankings, techniques from linear programming (Srinivasan and Shocker, 1973), non-metric versions of ANOVA (Kruskal, 1965) or, more generally, techniques from optimal scaling such as MORALS (Gifi, 1990) can be used. Variants of conjoint analysis that use rating scales have been subsumed under the heading of metric conjoint analysis in the literature (Louviere, 1988a). Here, the utility values are usually estimated by least squares procedures.

Using all combinations of attribute levels, that is, a full factorial design, the number of evaluations required from each respondent soon becomes prohibitively large as the number of attributes and/or levels increases. To deal with this problem, the application of formal experimental designs was suggested. Green and co-workers

(Green, 1974; Green *et al.*, 1978) proposed the use of orthogonal arrays, incomplete block designs and fractional factorial designs to reduce the number of evaluations to be performed. In addition to these classical reduction principles, techniques from optimal design theory can be applied. In particular, due to the metric response format and the linear relationship between preference judgements and attributes, these techniques are especially suited for models of metric conjoint analysis.

The optimality of a design can only be established within the framework of a particular model and with respect to a particular criterion. Finding a design that is best with regard to the optimality criterion under consideration often represents a challenging task. From the viewpoint of a statistician, the job is finished when the optimization problem has been solved. The user, however, is mainly interested in whether the theoretical superiority of a design will hold up when it is actually employed. As the user of a design may not necessarily be an expert in optimal design theory it is crucial that the criterion chosen by the statistician reflects those aspects of a design that are considered important by the user. Moreover, since any one model cannot capture all characteristics of a situation, it might well be the case that a design that is optimal in theory does not turn out to be best in practice. This is particularly true for conjoint analysis where factors like respondent fatigue, boredom or information overload might impact on the results. For example, suppose a respondent has already evaluated a portion of a series of profiles, but has not yet finished the entire series. Because comparing and trading off attribute levels represents a cognitively demanding task attention might decrease, and when evaluating the remaining profiles the person might only glimpse at the descriptions. In addition, the evaluation task might be perceived as monotonous which could also lead to a cursory consideration of the profiles. The effect of information overload appears to be particularly relevant with large attribute sets. Here, the respondent might not read the complete description of an object, skipping some of the attributes and basing the preference judgement on only a subset of the features.

These peculiarities of subjective assessments are not easily accommodated by a model. In particular, the aforementioned effects might invalidate some of the assumptions of the model for which an optimal design is generated. Therefore, the conclusion that the profiles in a conjoint analysis should be assembled according to an optimal design might not be as justified as the adoption of optimal designs in other fields of application like engineering or the sciences where the validity of a model can be assumed with higher confidence. As a result, it is not clear whether optimal designs are best suited for practical conjoint applications. For instance, suppose a suboptimal design was able to produce a high level of involvement in the evaluation task on the side of the respondents by presenting profiles that are perceived as more interesting or challenging compared to an optimal design. It is conceivable that this psychological advantage of the former design can offset the statistical benefits of the latter. It is therefore necessary to put optimal designs to an empirical test.

In what follows, we concentrate on design problems for paired comparisons in metric conjoint analysis. After a brief sketch of the history of conjoint analysis, a linear model for paired comparisons that is similar to the model considered by Scheffé (1952) is presented. This will be followed by a discussion of design criteria. Besides the well-known criteria of D-, A- and E-optimality, we consider the criterion of stochastic distance optimality originally proposed by Sinha (1970) which has attracted attention recently (e.g. Liski *et al.*, 1999; SahaRay and Bhandari, 2003). The idea underlying this criterion is to minimize the distance between the vector of parameters and its corresponding estimate in a stochastic sense. From an applied point of view, this criterion appears to be particularly appropriate, since it matches the conception of a good design as intuitively conceived by many practitioners of conjoint analysis: a good design should yield estimates which are close to the parameters. Recently developed D-optimal paired comparison designs (Graßhoff *et al.*, 2004) with a manageable number of evaluations also happen to be optimal with regard to the distance criterion. Empirical tests of these designs are performed in two experiments. In the first experiment, optimal paired comparison designs are compared with a design derived from procedures presented by Green (1974), whereas in the second the standard of comparison is defined by a design heuristic that is used to generate paired comparison designs by ACA (1994), the market leader for conjoint analysis software over recent years. The name ACA is an acronym for adaptive conjoint analysis and signifies that in the course of a conjoint analysis interview previous evaluations are taken into account when the objects for future pairs have to be chosen. A special feature of the second experiment consists in the utilization of incomplete descriptions of the objects. That is, the objects presented for evaluation are characterized by only a subset of the attributes. These so-called partial profiles are often employed in conjoint analysis as a means to prevent information overload.

3.2 Conjoint Analysis

Historically, the development of conjoint methods was inspired by what was considered a breakthrough in the branch of mathematical psychology called measurement theory. Luce and Tukey (1964) named this invention conjoint measurement. The core of conjoint measurement is a representation theorem concerning the preference relation among objects that are characterized by the levels of certain attributes. The theorem provides a set of sufficient conditions that guarantee the existence of an additive representation of the preference relation (for a general formulation of the theorem, see Krantz *et al.*, 1971, p. 302). That is, when all conditions are met, there is a utility value for every attribute level so that given any two objects one is preferred over the other if and only if the sum of the utilities corresponding to the levels of the preferred object exceeds the respective sum for the other object. Moreover, the utility values are unique up to certain linear transformations, that is, they are measured on an interval scale. More precisely, the utility values corresponding to the levels of an attribute are unique up to a linear transformation with an attribute specific intercept and a slope parameter which is the same for all attributes. The important consequence of this uniqueness property is that contrasts between utility values within attributes can be compared across attributes.

The potential to yield interval scaled utilities from ordinal preference judgements contributed much to the method's initial appeal when conjoint measurement was discovered by marketing researchers. In this vein, Green and Rao (1971) introduced conjoint measurement to a larger audience in marketing, stressing the fact that 'its procedures require only rank-ordered input, yet yield interval-scaled output' (p. 355). With regard to the original formulation of conjoint measurement's representation theorem, this statement is only valid, however, when all attributes possess an infinite number of levels and the complete preference relation is known. In order to determine this relation it would be necessary to rank order the infinite number of objects that can be generated by combining the attribute levels or to compare all pairs of objects. Clearly, this is not possible in practice where only a finite number of levels can be used. For attributes with a finite number of levels the uniqueness problem of the measurement scale was solved by Fishburn and Roberts (1988).

Despite the aforementioned limitation, conjoint measurement and its numerous modifications attracted much attention. Since many of these modifications were only loosely connected with the original conjoint measurement approach, Green and Srinivasan (1978) introduced the name of conjoint analysis as a collective term 'to cover models and techniques that emphasize the transformation of subjective responses into estimated parameters' (p. 103). A detailed account of the theory and methods of conjoint analysis is given in Louviere (1988b) where, among other things, the relationship between conjoint measurement and conjoint analysis is clarified. Green and Srinivasan (1990) summarize the development of conjoint methods during the 1980s. In addition to reviewing technical developments, they address the issues of reliability and validity of measurement. Carroll and Green (1995) provide a taxonomy of the host of conjoint methods. The commercial use of conjoint analysis has been documented by Cattin and Wittink (1982), Wittink and Cattin (1989) as well as Wittink *et al.* (1994). These papers indicate a continuous increase of applications in marketing and related areas. Moreover, it can be seen that the general availability of software for conducting conjoint analysis by the mid-1980s has fostered the propagation of some methods, whereas the usage of other methods has steadily declined. In particular, graded paired comparisons play a prominent role since they represent a major portion of the interview conducted by the popular ACA program. An edited book by Gustaffson *et al.* (2000) attests to the continuing interest in conjoint analysis. Green *et al.* (2001) integrate findings from over 30 years of research.

3.3 Paired Comparison Models in Conjoint Analysis

Direct quantitative evaluation of a set of objects is often deemed to yield imprecise results when the objects are judged with respect to a subjective criterion as is the case with preferences. For example, to assess the sweetness of a number of different sweeteners represents a considerably difficult task when specimens are tasted one at a time. Here, the difficulty arises from the lack of a reference standard. A technique that circumvents this problem is the method of paired comparisons.

In a paired comparison task, objects are presented in pairs and the respondent has to trade off one alternative against the other. The method was originally invented in the field of psychophysics in the middle of the nineteenth century, where it was used, for example, to measure the perceived heaviness of vessels (see e.g. David, 1988). However, as was already pointed out by Thurstone (1928) who employed paired comparisons in connection with his famous law of comparative judgement for the measurement of attitudes, the method's domain of application is much wider. Today's fields of application include, among others, psychology, economics and sensory analysis.

Models for data arising from paired comparisons can be broadly classified with regard to the kind of response. Most of the more common models like the one by Bradley and Terry (1952) assume a qualitative response. That is, it is only observed which of the objects in a pair dominates the other with respect to the criterion. As has already been explained, in conjoint analysis, a different type of paired comparisons is most common. In addition to requiring the respondent to indicate the preferred object, these graded paired comparisons demand a quantitative response. That is, the strength of preference has to be indicated. For such paired comparisons with a quantitative response, a linear model approach is appropriate (Scheffé, 1952).

The objects that have to be evaluated in a conjoint analysis differ in a variety of attributes. These attributes can be formally identified with influential factors occurring in an engineering set-up. Thus it seems attractive to apply methods of multi-factor approaches from standard linear design theory (Schwabe, 1996). For that purpose a general linear model has to be developed to fit the data structure of paired comparisons arising from conjoint analysis. In particular, such a model should deal simultaneously with both quantitative (continuous) and qualitative (discrete) types of attributes covering the work by van Berkum (1987) and Großmann *et al.* (2001) as well as the more traditional approach based on incomplete block designs (for recent treatments see Mukerjee *et al.*, 2002, and Street *et al.*, 2001).

As a starting point we assume a general linear model

$$\tilde{Y}(\mathbf{t}) = \mathbf{f}(\mathbf{t})'\boldsymbol{\beta} + \tilde{Z}(\mathbf{t})$$

for the overall utility $\tilde{Y}(\mathbf{t})$ of a single object $\mathbf{t}$ which is chosen from a set $\mathcal{T}$ of possible realizations for the object. In general, each object $\mathbf{t} = (t_1, \ldots, t_K)$ consists of a variety of K components that represent the attributes. In this setting, $\mathbf{f}$ is a vector of known regression functions which describe the form of the functional relationship between the possible $\mathbf{t}$ and the corresponding mean response $E(\tilde{Y}(\mathbf{t}))$, and $\boldsymbol{\beta}$ is a vector of parameters. Throughout, transposition of vectors and matrices is indicated by a prime.

In contrast to standard design problems, the utilities of the objects are usually not directly measurable in paired comparisons. Only preferences can be observed for one or the other of two objects presented. We assume that the preference is numerically given as the difference between utilities $Y(\mathbf{s}, \mathbf{t}) = \tilde{Y}(\mathbf{s}) - \tilde{Y}(\mathbf{t})$ when a pair of objects $\mathbf{s}$ and $\mathbf{t}$ are compared. In the given case, the observations are properly described by the linear model

$$Y(\mathbf{x}) = \mathbf{F}(\mathbf{x})'\boldsymbol{\beta} + Z(\mathbf{x}) = (\mathbf{f}(\mathbf{s}) - \mathbf{f}(\mathbf{t}))'\boldsymbol{\beta} + Z(\mathbf{s}, \mathbf{t}) \tag{3.1}$$

with settings $\mathbf{x} = (\mathbf{s}, \mathbf{t})$ for the paired comparisons chosen from the design region $\mathcal{X} = \mathcal{T} \times \mathcal{T}$ of possible pairs. Here, $\mathbf{F}$ is the derived regression function defined by $\mathbf{F}(\mathbf{x}) = \mathbf{F}(\mathbf{s}, \mathbf{t}) = \mathbf{f}(\mathbf{s}) - \mathbf{f}(\mathbf{t})$. The comparisons are assumed to result in uncorrelated and homoscedastic errors $Z = Z(\mathbf{s}, \mathbf{t})$ with a mean of zero. Note that the above difference model can be interpreted as a linearization of the Bradley–Terry model in the particular case of equal utilities (Großmann *et al.*, 2002).

The model can easily accommodate variants of conjoint analysis models like the so-called vector model, the ideal point model and the part-worth function model (Green and Srinivasan, 1978). For example, the part-worth function model posits that the utility of a single object $\mathbf{t}$ equals the sum of the utilities corresponding to its K attribute levels $t_1, \ldots, t_K$. Formally, this situation can be identified with the K-way layout without interactions in an ANOVA setting. Here, the attributes are represented by K factors with v_k levels $i_k = 1, \ldots, v_k$, $k = 1, \ldots, K$, each. Let $1_{\{i\}}(j)$ denote the indicator function which is 1 when j equals i and 0 otherwise. Then, for

$$\mathbf{f}(\mathbf{t}) = (1_{\{1\}}(t_1), \ldots, 1_{\{v_1\}}(t_1), \ldots, 1_{\{1\}}(t_K), \ldots, 1_{\{v_K\}}(t_K))'$$

and

$$\boldsymbol{\beta} = (\beta_{1,1}, \ldots, \beta_{1,v_1}, \ldots, \beta_{K,1}, \ldots, \beta_{K,v_K})'$$

the overall utility of object $\mathbf{t} = (t_1, \ldots, t_K)$ with the level t_k chosen from the set $\mathcal{T}_k = \{1, \ldots, v_k\}$ becomes

$$\tilde{Y}(\mathbf{t}) = \beta_{1,t_1} + \beta_{2,t_2} + \cdots + \beta_{K,t_K} + \tilde{Z}(\mathbf{t}),$$

where β_{k,t_k} represents the utility of the level t_k of attribute k. Accordingly, in the part-worth function model of conjoint analysis the paired comparison of objects $\mathbf{s}$ and $\mathbf{t}$ in $\mathcal{T} = \mathcal{T}_1 \times \ldots \times \mathcal{T}_K$ is represented by the equation

$$Y(\mathbf{s}, \mathbf{t}) = (\beta_{1,s_1} - \beta_{1,t_1}) + \cdots + (\beta_{K,s_K} - \beta_{K,t_K}) + Z. \tag{3.2}$$

In this setting the representation of the complete vector $\boldsymbol{\beta}$ and, hence, its least squares estimator $\hat{\boldsymbol{\beta}}$ is not unique. This is not a drawback, however, since in conjoint analysis one is usually interested in comparing certain within-attribute contrasts across attributes. Examples of such contrasts are the set of effect sizes

$$\theta_{k,i_k} = \beta_{k,i_k} - \bar{\beta}_k, \quad k = 1, \ldots, K, i_k = 1, \ldots, v_k,$$

where $\bar{\beta}_k = v_k^{-1} \sum_{i_k=1}^{v_k} \beta_{k,i_k}$, or the set of effect differences

$$\theta_{k,i_k,j_k} = \beta_{k,i_k} - \beta_{k,j_k}, \quad k = 1, \ldots, K, 1 \leq i_k < j_k \leq v_k.$$

Uniqueness of the parameters $\boldsymbol{\beta}$ could be accomplished by imposing the identifiability conditions

$$\sum_{i_1=1}^{v_1} \beta_{1,i_1} = \ldots = \sum_{i_K=1}^{v_K} \beta_{K,i_K} = 0$$

on the parameters. The vector of contrasts

$$\boldsymbol{\theta} = (\theta_{1,1}, \ldots, \theta_{1,v_1}, \ldots, \theta_{K,1}, \ldots, \theta_{K,v_K})'$$

is a linear transformation $\boldsymbol{\theta} = \mathbf{L}\boldsymbol{\beta}$ of the parameter vector $\boldsymbol{\beta}$ with a suitable matrix $\mathbf{L}$. Then, the unique least squares estimator $\hat{\boldsymbol{\theta}}$ of $\boldsymbol{\theta}$ is obtained by applying the transformation $\mathbf{L}$ to any least squares estimator $\hat{\boldsymbol{\beta}}$ of $\boldsymbol{\beta}$, that is, $\hat{\boldsymbol{\theta}} = \mathbf{L}\hat{\boldsymbol{\beta}}$.

Thus far, only the situation has been considered where all attributes are used to describe the objects. A peculiarity of the paired comparison model lies in the fact that it can also deal with so-called partial profiles, that is, incomplete descriptions of the objects. Partial profiles are primarily used to make the evaluation task less demanding. In this setting, the number of attributes which are presented within a single paired comparison is restricted by the so-called profile strength. For example, when there are five attributes and a profile strength of three is employed, only three out of the five attributes are used to describe the objects in every pair while the other attributes remain indeterminate. The selection of the attributes that are presented varies between the single comparisons. However, the same set of attributes is used to describe both objects in a given pair. Of course, for these partial profiles no direct valuations are possible while paired comparisons still lead to meaningful results. To cope with partial profiles the regression functions $\mathbf{F}$ have to be modified to keep track of the selection and, more importantly, restrictions have to be imposed onto the set $\mathcal{X}$ of possible pairs to be presented. The necessary ramifications are detailed in Graßhoff *et al.* (2004).

3.4 Design Issues

The quality of the statistical analysis based on a paired comparison experiment heavily depends on the pairs of alternatives presented. The choice of such pairs

$$\xi = (\mathbf{x}_1, \ldots, \mathbf{x}_N) = ((\mathbf{s}_1, \mathbf{t}_1), \ldots, (\mathbf{s}_N, \mathbf{t}_N))$$

is called a design. Under the general model (3.1) the design performance is reflected by the corresponding standardized information matrix

$$\mathbf{M}(\xi) = \frac{1}{N}\sum_{n=1}^{N} \mathbf{F}(\mathbf{x}_n)\mathbf{F}(\mathbf{x}_n)' = \frac{1}{N}\sum_{n=1}^{N} (\mathbf{f}(\mathbf{s}_n) - \mathbf{f}(\mathbf{t}_n))(\mathbf{f}(\mathbf{s}_n) - \mathbf{f}(\mathbf{t}_n))'.$$

Note that for every identifiable linear aspect $\boldsymbol{\theta} = \mathbf{L}\boldsymbol{\beta}$ the covariance matrix of the least squares estimator $\hat{\boldsymbol{\theta}} = \mathbf{L}\hat{\boldsymbol{\beta}}$ is proportional to $\mathbf{L}\mathbf{M}(\xi)^{-}\mathbf{L}'$ where $\mathbf{M}(\xi)^{-}$ is an arbitrary generalized inverse of $\mathbf{M}(\xi)$.

Many common performance measures are based on the non-zero eigenvalues of the information matrix $\mathbf{M}(\xi)$. For example, the most popular criterion of D-optimality aims at maximizing the determinant of the information matrix in the case of a minimal parameterization when $\mathbf{M}(\xi)$ is of full rank. In general, the D-criterion is given by the product of the p (non-zero) eigenvalues of $\mathbf{M}(\xi)$, where p is the dimension of the linear space spanned by the design locus $\{\mathbf{F}(\mathbf{x}); \mathbf{x} \in \mathcal{X}\}$.

In the following discussion we will assume normally distributed observations with equal variances σ^2 for simplicity. Then a D-optimal design minimizes the product of the p positive eigenvalues of the covariance matrix and, hence, the volume of a confidence ellipsoid for $\boldsymbol{\beta}$ in the corresponding p-dimensional subspace if identifiability conditions are imposed. Apart from other classical criteria like A-, MV- or E-optimality, which aim at minimizing the trace, the maximal diagonal element or the largest eigenvalue of the covariance matrix for $\hat{\boldsymbol{\theta}}$, we are particularly interested in the so-called stochastic distance criterion, or DS-criterion in short, introduced by Sinha (1970). For any design $\xi = (\mathbf{x}_1, \ldots, \mathbf{x}_N)$ let $\hat{\boldsymbol{\theta}}_\xi$ denote the least squares estimator of a linear transformation $\boldsymbol{\theta} = \mathbf{L}\boldsymbol{\beta}$ of the original parameters $\boldsymbol{\beta}$. Then, a design $\xi^* = (\mathbf{x}_1^*, \ldots, \mathbf{x}_N^*)$ is said to be DS-optimal for $\boldsymbol{\theta}$, if

$$P\left(\| \hat{\boldsymbol{\theta}}_{\xi^*} - \boldsymbol{\theta} \| \leq \varepsilon\right) \geq P\left(\| \hat{\boldsymbol{\theta}}_{\xi} - \boldsymbol{\theta} \| \leq \varepsilon\right)$$

for all $\varepsilon > 0$ and all competing designs ξ where the symbol $\| \cdot \|$ denotes Euclidean distance. Hence, a design ξ^* is DS-optimal if the corresponding estimator $\hat{\boldsymbol{\theta}}_{\xi^*}$ yields estimates that tend to be closer to the vector of interest $\boldsymbol{\theta}$ than those of any other design ξ . As was already noted in the introduction, this criterion very well reflects what according to our experience users of conjoint analysis consider an important characteristic of an optimal design, namely that it should allow for more precise measurement of the utilities than any other design of the same size. Clearly, in a certain sense all other design criteria represent a notion of precise estimation as well. It is our conviction, however, that the definition of precision in terms of distance is more acceptable to practitioners than a definition that refers to more abstract concepts like, for example, the volume of a confidence ellipsoid. For more details on *DS*-optimality and its connection to other criteria we refer the reader to Liski *et al.* (2002, Chapter 5).

In the following we consider the part-worth function paired comparison model (3.2) of the previous section. In particular, we are interested in finding *DS*-optimal designs for the vector $\boldsymbol{\theta} = \mathbf{L}\boldsymbol{\beta}$ introduced with that model where $\boldsymbol{\theta}$ is either the vector of effect sizes

$$\theta_{k,i_k} = \beta_{k,i_k} - \bar{\beta}_k, \quad k = 1, \ldots, K, i_k = 1, \ldots, v_k,$$

or the vector of effect differences

$$\theta_{k,i_k,j_k} = \beta_{k,i_k} - \beta_{k,j_k}, \quad k = 1, \ldots, K, 1 \leq i_k < j_k \leq v_k.$$

For both, effect sizes and effect differences, a permutation of the levels of an attribute leads to a rearrangement of the entries in the vector $\boldsymbol{\theta}$. As the *DS*-criterion is not affected by the particular order of the entries in $\boldsymbol{\theta}$ it is invariant with respect to the group of permutations of the levels for each attribute. Thus, it can be seen that the design which is uniform on all pairs of objects differing in every single attribute has the property of being *DS*-optimal. Moreover, every design for which the (standardized) information matrix is diagonal and coincides with the information matrix

$$\mathbf{M}(\xi) = \operatorname{diag}\left(\tfrac{1}{v_1}\mathbf{I}_{v_1}, \ldots, \tfrac{1}{v_K}\mathbf{I}_{v_K}\right)$$

of that uniform design, where $\mathbf{I}_{v_k}$ denotes the identity matrix of order v_k, is also *DS*-optimal.

Obviously, the uniform design requires $\prod_{k=1}^{K} v_k(v_k - 1)$ comparisons and is not applicable in practice even for moderate numbers of levels and attributes. Therefore, it is essential to construct fractions of the uniform design which share the information matrix and which can be realized with a reasonable number of comparisons. Combinatorial methods for reducing the number of comparisons are exhibited by Graßhoff *et al.* (2004), Mukerjee *et al.* (2002) and Street and Burgess (2004) for various situations. The results can be generalized to partial profiles. In this situation, optimal designs require the objects in each pair to have distinct levels for all attributes selected (Graßhoff *et al.*, 2004).

3.5 Experiments

To assess the empirical benefits that can be achieved by the application of optimal designs we present the results of two experiments in which the part-worth function model for paired comparisons was employed. Because of its clear interpretation the *DS*-criterion is considered in the following. Put simply, this criterion states that the vector of utilities estimated from a *DS*-optimal design tends to be closer to the true utilities than estimates obtained from any other design. This can be rephrased more precisely as follows: given any ball that has the vector of true utilities at its centre, it is more likely for a *DS*-optimal design to produce estimates that will fall within that ball than for any other design.

As has been noted in the previous section, designs that are optimal with regard to the *DS*-criterion can be constructed according to the methods described by Graßhoff *et al.* (2004). In the first experiment, a *DS*-optimal design and a paired comparison design generated according to a design approach proposed by Green (1974) are compared. In the second experiment, a design heuristic which was introduced by the conjoint analysis software ACA serves as the standard of comparison.

In both experiments, we face the problem that in order to compare estimated and true utilities as required by the *DS* criterion the latter set of values has to be known. Moreover, the true values are usually heterogeneous, that is, they vary between respondents. To resolve these problems two different strategies are adopted. In the first study, true utilities are 'induced' through a learning process. This approach tries to ensure that the utility structure is known and the same across all respondents.

The strategy adopted for the second experiment is different. Here, we try to establish true utilities for every individual respondent on the basis of a number of measurements over several weeks. That is, every individual has to perform a conjoint analysis task for the same scenario in several sessions. From each of these tasks utilities can be estimated. Because the least squares estimator is unbiased, it is argued that the average of these estimates over sessions approximates the true utilities of each person. These averages are then treated as if they were equal to the true utilities. Also, the second experiment differs from the first in that only partial profiles are presented for evaluation.

3.5.1 Experiment 1

In this experiment, true utility scores were induced according to the principal–agent paradigm (see e.g. West, 1996). Here, participants (the agents) had to learn preferences of a fictitious person (the principal) and make decisions for this person thereafter. That is, the agents' task consisted of predicting the principal's actions. This procedure has been applied several times before (Reiners, 1996; Teichert, 2000; Huber *et al.*, 2002).

In our experiment, participants had to book flights as employees of a company for their superior. The flights were characterized by nine attributes with two levels each. The attributes and levels are shown in Table 3.1. The table also contains the part-worth values that represent the true utilities of the levels. To keep things simple, the utility of one level of each attribute was set to zero.

Table 3.1 Attributes used in Experiment 1

Attribute	Levels	True utilities
Delay	4 hours/1 hour	0/600
Intermediate landing	with/without	0/500
Reservation	90%/50%	0/400
Time until departure	4 hours/1 hour	0/300
Service	average/very good	0/300
Airline	KLM/Swiss Air	0/200
Aircraft	Airbus A320/Boeing 747	0/200
Miles account	without/with	0/100
Alternatives of meals	one/two	0/100

The attributes of this scenario were partially taken from an example due to Green (1974). He used this example as an illustrative device for demonstrating how experimental designs can be used to reduce the number of evaluations required from respondents by different variants of conjoint analysis.

For paired comparisons, Green (1974) proposed a two-step procedure based on orthogonal arrays and partially balanced incomplete block designs. For the present example, this procedure is as follows: in the first step a 16-run orthogonal array for the 9 attributes is found. Each row of this array represents the description of a different flight. The descriptions are then arranged into pairs according to a partially balanced incomplete design in the second step. The final design results in 48 paired comparisons. Großmann *et al.* (2002) have shown that this approach may yield designs that are far from efficient with respect to the D-criterion. Yet, they did not perform an empirical test. The present experiment compared the design constructed according to Green (1974) to an optimal design with 24 pairs derived from the procedures in Graßhoff *et al.* (2004) that was administered twice. Both designs can be found in Großmann *et al.* (2002, pp. 106–107).

The experiment took place on two consecutive days. Seventeen undergraduate students were first introduced to the general scenario. True part-worth values were not presented in numerical form, rather participants learned the principal's ranking of the attributes and his preferred level of each attribute. In the next experimental phase, the utility values for the non-zero levels were presented graphically by circles. The area of each circle was proportional to the corresponding true utility in Table 3.1. During this training phase participants had to work on several exercises to internalize the true utilities. The difficulty of the exercises increased during the training. The first exercises only required ordinal judgements of flight descriptions while later preferences had to be expressed on metric scales. In order to induce a high motivation a financial award of 50 euros was announced for the three participants who achieved the best results.

After the training was completed, participants worked on both sets of paired comparisons in balanced order. All exercises and paired comparisons were carried out with the software ALASCA (Holling *et al.*, 2000). Preference for one of the two flights in each pair was indicated with a slider on a metric scale.

The learning success was tested statistically by Hotelling's one-sample T^2-test. If the hypothesis that the mean vector of the least squares estimator for the part-worth utilities is equal to the vector of true utilities is not rejected, we assume that the participants sufficiently internalized the true part-worths. For each of the 17 participants utilities were estimated from all 96 paired comparisons by multiple regression with differences of effect coded attributes as predictors excluding an intercept. The result of $T^2 = 37.70 < T^2_{.05,9,16} = 60.99$ indicates a satisfactory learning success at the 0.05 significance level.

To assess the performance of the designs with regard to the *DS*-criterion, an estimate of the utilities based on the 48 evaluations from the optimal design and the respective estimate based on the 48 comparisons from the benchmark design was calculated for each respondent. Figure 3.1 depicts the cumulative frequency curves of the distances of these estimates from the true utilities for both designs.

Except for large deviations of the estimated utilities from the true scores the optimal design proves to be superior. For large deviations both designs perform almost equally well.

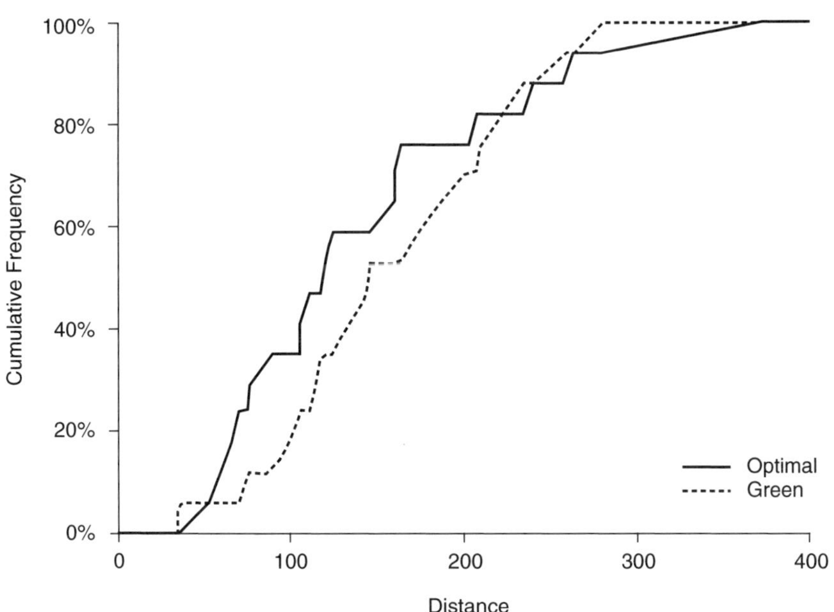

Figure 3.1 Cumulative frequency curves for deviations from true utilities

3.5.2 Experiment 2

The purpose of the second experiment was to compare optimal designs with a class of widely used designs that are generated according to a heuristic approach. The popularity of this approach derives from its implementation in the conjoint analysis software ACA.

In practice, the alternatives that must be judged in a paired comparison are usually only described on a subset of the attributes, that is, they represent partial profiles. The profile strength may vary between comparisons, but is typically kept fixed throughout the whole interview. To decide what attributes and levels are chosen, the design algorithm proceeds in an adaptive manner. That is, in the course of an interview the composition of every consecutive pair depends on the previous responses. The design heuristic strives to balance the levels and attributes. To this end, the algorithm keeps track of which attributes and corresponding levels have occurred together least frequently in previous comparisons. For example, when there are five attributes and a profile strength of two is employed, the two attributes to be presented in the next comparison are chosen at random from among those attributes that have been used together the fewest times before. The corresponding levels are selected in a similar way. These levels do not uniquely determine the next comparison, since there are several ways to combine the levels into pairs of objects. For instance, with a profile strength of two there are essentially two different pairs of objects that can be generated by combining the chosen levels of the selected attributes. To decide which of these pairs is actually chosen for presentation the algorithm proceeds in three steps. First, using the evaluations from all previous paired comparisons an estimate of the vector of attribute level utilities is computed. Second, the overall utility of every object in a candidate pair is computed by summing the estimated utilities corresponding to its levels. Third, the pair is selected whose objects possess overall utilities that are most nearly equal. As a result of this procedure, the design of the conjoint analysis task is not known until the interview has been completed. In the following, designs that were constructed according to the above outlined heuristic strategy will be referred to as adaptive.

Experiment 2 was conducted in seven sessions. Initially, 72 students volunteered to participate in the study. Thirty of these students only showed up at a few sessions and were discarded from the analysis. The remaining 42 participants missed only two sessions at most. Twenty attended all sessions and another 19 missed only one.

At each session, the participants had to perform a conjoint analysis task concerning apartments. This scenario was chosen because presumably all participants had prior house-hunting experience. Also, apartments have been the subject of many published conjoint analysis studies (e.g. Green *et al.*, 1988; Kohli and Mahajan, 1991). The apartments were described to the participants by using six attributes with five levels each, as shown in Table 3.2.

All conjoint analysis interviews were conducted with the software ALASCA (Holling *et al.*, 2000) which allows the administration of optimal as well as adaptive paired comparison designs. A schedule of the sessions is shown in Table 3.3.

Table 3.2 Attributes used in Experiment 2

Attribute	Levels
Size of apartment	15, 20, 25, 30, 35 m^2
Time to city by bike	5, 10, 15, 20, 25 minutes
Light level of apartment	dim, quite dim, average, quite bright, bright
Noise level of apartment	noisy, quite noisy, average, quite quiet, quiet
Condition of apartment	extensive renovation required, renovation required, partially renovated, renovated, completely renovated
Special features	washing machine, parking place, bicycle cellar, sauna, garden

Table 3.3 Schedule of Experiment 2

	Session						
	1	2	3	4	5	6	7
N	39	39	40	39	40	40	32
Comparisons	120	120	60	120	120	120	120
Monetary response scale	No	No	Yes	No	Yes	Yes	No

Participants were introduced to the general scenario at the first meeting. Also, to become acquainted with the conjoint analysis task and the handling of the computer program, every student completed an interview of 120 paired comparisons. The strength of preference for one of the apartments had to be indicated on a metric scale. Here and in all subsequent sessions, paired comparisons with a profile strength of two and a slider for recording the responses were used.

The comparison of optimal and adaptive designs was performed at two points in time. During the second and seventh sessions, the participants completed 60 paired comparisons of an adaptive as well as an optimal design. Both sets of paired comparisons were separated by an intermediate task. The order of designs was counterbalanced between participants. Optimal designs were constructed according to the methods described in Graßhoff *et al.* (2004).

The four intermediate sessions served to determine part-worth utilities for the levels in Table 3.2 that could be treated as true utilities in the comparison of designs with respect to the *DS*-criterion as in Experiment 1. To obtain these reference values for every single respondent, additional interviews were conducted during sessions three to six. Except for the third session at which only 60 evaluations had to be performed, each additional interview required 120 paired comparisons. At three of the intermediate meetings preferences had to be expressed in monetary units in order to make the task more realistic. Under the assumption of stability of preferences across time, which for apartments might not be too critical to retain, the vector of utilities estimated from every intermediate session provides information with regard to a participant's latent utility structure. The average of these vectors across sessions may then serve as a reasonable approximation to the vector of true part-worth utilities.

Nine of the 42 participants did not show up on one of the intermediate sessions. In order to ensure that the computation of reference vectors was based on the same number of sessions for all participants, judgements from session one were substituted for these respondents. From the data provided separately at each of the total of four sessions, part-worth utilities were estimated by least squares with effect coded predictors. The estimate for the fifth level of each attribute was obtained as the negative sum of the estimates for the remaining levels. Prior to averaging, the estimated parameters were divided by the standard deviation of the dependent variable to account for heterogeneous variances that might have been caused by the use of monetary units.

For sessions two and seven, parameter estimates under the optimal and adaptive designs were computed and standardized in the same way. Cumulative frequency curves for the differences between the estimates and the averaged vectors for the two sessions are shown in Figures 3.2 and 3.3, respectively. Note that the figures are based on different numbers of participants due to respondents missing either the second or seventh session. However, for participants who attended both meetings distances refer to the same averaged vector.

The optimal designs exhibit a superior performance at both points in time. The difference between optimal and adaptive designs is particularly pronounced at the

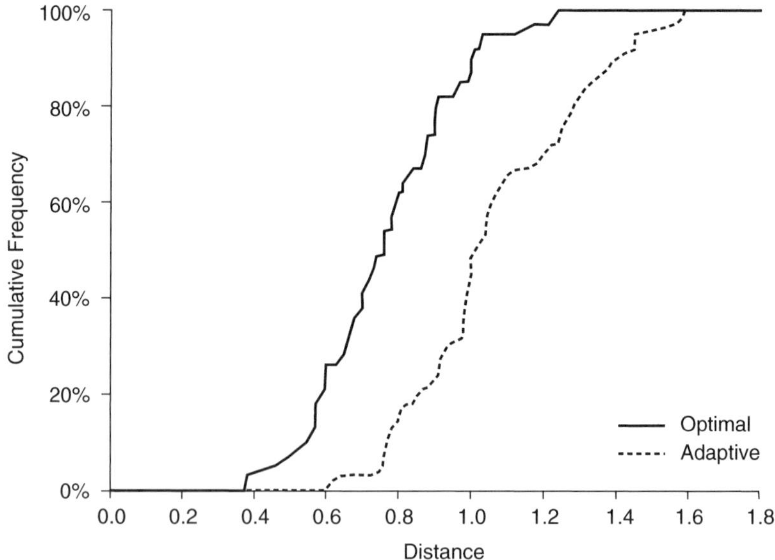

Figure 3.2 Cumulative frequency curves for distances at session 2

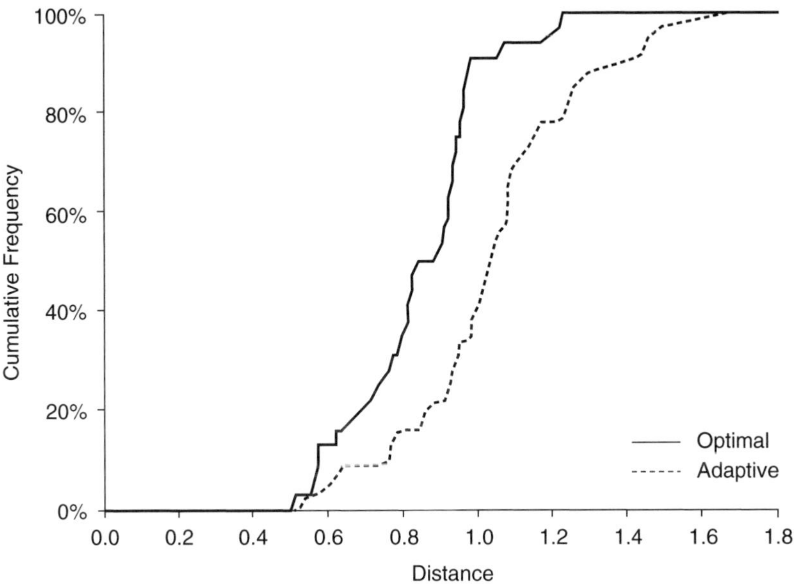

Figure 3.3 Cumulative frequency curves for distances at session 7

earlier session. In contrast to the findings from the first experiment, the frequency curves do not intersect. This might be due to the larger sample size.

3.6 Discussion

The idea of applying techniques from optimal design theory to the measurement of preferences has been previously advocated by several authors. Formal efficiency gains through improved designs have been demonstrated for conjoint analysis and related methods (Großmann *et al.*, 2002; Kuhfeld *et al.*, 1994). However, little is known about the positive empirical effects that can be achieved. In our experiments, we addressed this issue for paired comparisons. The perspective to compare designs in terms of the distance optimality criterion is new. This criterion conveys the differences between designs in a more easily interpretable way than more traditional and more abstract comparisons in terms of D-efficiency.

In the first experiment, a DS-optimal design was compared to a design constructed according to a method proposed by Green (1974). Even with the small sample size of this experiment, the optimal design yielded more accurate estimates than the competing design. This result was corroborated in the second experiment where optimal designs were compared to adaptive designs. Moreover, this effect was stable over time. The adaptive design heuristic has often been claimed to make the conjoint analysis task more interesting and challenging. However, as our second experiment demonstrates these alleged psychological advantages do not manifest themselves empirically.

The critical point for the application of the *DS*-criterion in an empirical comparison of designs is that the 'true' vector of utilities that represents the latent utility structure of a participant has to be known or at least must be approximated in some way. In the first experiment, we relied on the principal–agent paradigm to induce a certain utility structure. This approach appears to be promising. However, further research is needed to investigate whether the learning task is mastered successfully by respondents. To infer the learning success from the non-rejection of a null hypothesis clearly represents a necessary yet insufficient condition. Also, at this point it is not clear how the induction of a preference structure can best be accomplished.

As an alternative to the method used in the first experiment we propose a training procedure that presents the participants only with, and asks only for, qualitative information concerning the principal's preferences. For example, at the beginning of the training ordinal information concerning the preferences for the attributes and corresponding levels could be given in a narrative description of the principal. This would be followed by a sequence of learning tasks where among other things the respondents would have to make vicarious choices. Feedback following each response would indicate the principal's preference. The reasons for proposing this approach are twofold. First, the process by which the external preferences are learned resembles the way people develop their own preferences. In contrast, when numerical utility values for the attribute levels are provided it is not clear whether evaluations are based on preference or computation. Second, a criterion for the learning success can be derived from a theorem of Fishburn and Roberts (1988) concerning the uniqueness of utilities in so-called finite conjoint measurement. This uniqueness theorem can be used to construct a rank ordering of objects that determines a vector of numerical utilities which are unique up to certain linear transformations. In contrast to the classical conjoint measurement theorem, the Fishburn and Roberts theorem applies even in situations where every attribute possesses a finite number of levels. Hence, if a respondent can produce a respective rank ordering after the training, it can be concluded that the numerical utility values have been successfully learned.

The approach used in the second experiment to approximate the utility structures of participants based on repeated performances of conjoint analysis interviews for the same scenario has not previously been investigated. Here, the idea is that measurement errors cancel out when several estimates are averaged. Also, this new approach deserves further study.

In summary, we were able to demonstrate the superiority of optimal designs with regard to different standards of comparison, scenarios, operationalizations of 'true' utilities, and over time. Even in the presence of potentially disturbing psychological influences the *DS*-optimal designs yield the most precise estimates. These findings demonstrate that optimal designs are not only a matter of academic interest but of practical relevance. However, to further clarify the empirical status of optimal designs in the context of preference measurement more experiments are needed.

Acknowledgements

This work was partly supported by the research grants Ho 1286/2-1 and -2 of the Deutsche Forschungsgemeinschaft. We thank two anonymous reviewers for helpful comments.

References

ACA system (Version 4.0) [Computer software] (1994). Sawtooth Software, Evanston, Ill.

Bradley, R. A. and Terry, M. E. (1952). Rank analysis of incomplete block designs. I. The method of paired comparisons. *Biometrika*, **39**, 324–345.

Carroll, J. D. and Green, P. E. (1995). Psychometric methods in marketing research: Part I, conjoint analysis. *Journal of Marketing Research*, **32**, 385–391.

Cattin, P. and Wittink, D. R. (1982). Commercial use of conjoint analysis. A survey. *Journal of Marketing*, **46**, 44–53.

David, H. A. (1988). *The Method of Paired Comparisons*, 2nd edn. Charles Griffin, London.

Fishburn, P. C. and Roberts, F. S. (1988). Unique finite conjoint measurement. *Mathematical Social Sciences*, **16**, 107–143.

Gifi, A. (1990). *Nonlinear Multivariate Analysis*. John Wiley & Sons, Ltd, New York.

Graßhoff, U., Großmann, H., Holling, H. and Schwabe, R. (2004). Optimal designs for main effects in linear paired comparison models. *Journal of Statistical Planning and Inference*, **126**, 361–376.

Green, P. E. (1974). On the design of choice experiments involving multifactor alternatives. *Journal of Consumer Research*, **1**, 61–68.

Green, P. E., Carroll, J. D. and Carmone, F. J. (1978). Some new types of fractional factorial designs for marketing experiments. In: J. N. Sheth (ed.), *Research in Marketing* Vol. 1, pp. 99–122. JAI Press, Greenwich, Conn.

Green, P. E., Helsen, K. and Shandler, B. (1988). Conjoint internal validity under alternative profile presentations. *Journal of Consumer Research*, **15**, 392–397.

Green, P. E., Krieger, A. M. and Wind, Y. (2001). Thirty years of conjoint analysis: reflections and prospects. *Interfaces*, **31**(3), 56–73.

Green, P. E. and Rao, V. R. (1971). Conjoint measurement for quantifying judgmental data. *Journal of Marketing Research*, **8**, 355–363.

Green, P. E. and Srinivasan, V. (1978). Conjoint analysis in consumer research: issues and outlook. *Journal of Consumer Research*, **5**, 103–123.

Green, P. E. and Srinivasan, V. (1990). Conjoint analysis in marketing: new developments with implications for research and practice. *Journal of Marketing*, **54**, 3–19.

Großmann, H., Holling, H., Graßhoff, U. and Schwabe, R. (2001). Efficient designs for paired comparisons with a polynomial factor. In: A. Atkinson, B. Bogacka and A. Zhigljavsky (eds), *Optimum Design 2000*, pp. 45–56. Kluwer, Dordrecht.

Großmann, H., Holling, H. and Schwabe, R. (2002). Advances in optimum experimental design for conjoint analysis and discrete choice models. In: P. H. Franses and A. L. Montgomery (eds), *Econometric Models in Marketing, Advances in Econometrics*, Vol. 16, pp. 93–117. JAI Press, Amsterdam.

Gustafsson, A., Herrmann, A. and Huber, F. (eds). (2000). *Conjoint Measurement: Methods and Applications*. Springer, Berlin.

Holling, H., Jütting, A. and Großmann, H. (2000). ALASCA: Computergestützte Entscheidungs- und Nutzenanalyse [ALASCA: Computerized decision and utility analysis]. Hogrefe, Göttingen.

Huber, J., Ariely, D. and Fischer, G. (2002). Expressing preferences in a principal–agent task: a comparison of choice, rating, and matching. *Organizational Behavior and Human Decision Processes*, **87**, 66–90.

Kohli, R. and Mahajan, V. (1991). A reservation-price model for optimal pricing of multiattribute products in conjoint analysis. *Journal of Marketing Research*, **28**, 347–354.

Krantz, D. H., Luce, R. D., Suppes, P. and Tversky, A. (1971). *Foundations of Measurement*. Vol. I: *Additive and Polynomial Representations*. Academic Press, New York.

Kruskal, J. B. (1965). Analysis of factorial experiments by estimating monotone transformations of the data. *Journal of the Royal Statistical Society, Ser. B*, **27**, 251–263.

Kuhfeld, W. F., Tobias, R. D. and Garratt, M. (1994). Efficient experimental design with marketing research applications. *Journal of Marketing Research*, **31**, 545–557.

Liski, E. P., Luoma, A. and Zaigraev, A. (1999). Distance optimality design criterion in linear models. *Metrika*, **49**, 193–211.

Liski, E. P., Mandal, N. K., Shah, K. R. and Sinha, B. K. (2002). *Topics in Optimal Design*. Lecture Notes in Statistics, **163**. Springer, New York.

Louviere, J. J. (1988a). *Analyzing Decision Making. Metric Conjoint Analysis*. Sage, Beverly Hills.

Louviere, J. J. (1988b). Conjoint analysis modelling of stated preferences: a review of theory, methods, recent developments and external validity. *Journal of Transport Economics and Policy*, **10**, 93–119.

Louviere, J. J., Hensher, D. A. and Swait, J. D. (2000). *Stated Choice Methods. Analysis and Application*. Cambridge University Press, Cambridge, UK.

Luce, R. D. and Tukey, J. W. (1964). Simultaneous conjoint measurement: a new type of fundamental measurement. *Journal of Mathematical Psychology*, **1**, 1–27.

Mukerjee, R., Dey, A. and Chatterjee, K. (2002). Optimal main effect plans with nonorthogonal blocking. *Biometrika*, **89**, 225–229.

Reiners, W. (1996). *Multiattributive Präferenzstrukturmodellierung durch die Conjoint Analyse* [Multiattribute preference structure modelling by conjoint analysis]. LIT, Münster.

SahaRay, R. and Bhandari, S. K. (2003). DS-optimal designs in one way ANOVA. *Metrika*, **57**, 115–125.

Scheffé, H. (1952). An analysis of variance for paired comparisons. *Journal of the American Statistical Association*, **47**, 381–400.

Schwabe, R. (1996). *Optimum Designs for Multi-factor Models*. Lecture Notes in Statistics **113**. Springer, New York.

Sinha, B. K. (1970). On the optimality of some designs. *Calcutta Statistical Association Bulletin*, **20**, 1–20.

Srinivasan, V. and Shocker, A. D. (1973). Linear programming techniques for multidimensional analysis of preferences. *Psychometrika*, **38**, 337–369.

Street, D. J., Bunch, D. S. and Moore, B. J. (2001). Optimal designs for 2^k paired comparison experiments. *Communications in Statistics – Theory and Methods*, **30**, 2149–2171.

Street, D. J. and Burgess, L. (2004). Optimal and near-optimal pairs for the estimation of effects in 2-level choice experiments. *Journal of Statistical Planning and Inference*, **118**, 185–199.

Teichert, T. (2000). Auswirkungen von Verfahrensalternativen bei der Erhebung von Präferenzurteilen [Effects of alternative methods on the elicitation of preference judgments]. *Marketing ZFP*, **22**(2), 145–159.

Thurstone, L. L. (1928). Attitudes can be measured. *American Journal of Sociology*, **33**, 529–554.

van Berkum, E. E. M. (1987). *Optimal Paired Comparison Designs for Factorial Experiments*. CWI Tract **31**. Amsterdam, Centrum voor Wiskunde en Informatica.

West, P. M. (1996). Predicting preferences: an examination of agent learning. *Journal of Consumer Research*, **23**, 68–80.

Wittink, D. R. and Cattin, P. (1989). Commercial use of conjoint analysis: an update. *Journal of Marketing*, **53**, 91–96.

Wittink, D. R., Vriens, M and Burhenne, W. (1994). Commercial use of conjoint analysis in Europe: results and critical reflections. *International Journal of Research in Marketing*, **11**, 41–52.

4

Designing Optimal Two-stage Epidemiological Studies

Marie Reilly[1] **and Agus Salim**[2]

[1]*Department of Medical Epidemiology and Biostatistics, Karolinska Institutet, Stockholm, SE 17177, Sweden*

[2]*National Centre for Epidemiology and Population Health, The Australian National University, Canberra, Australia*

4.1 Introduction

Epidemiologists have been aware of efficient study designs for some time. The classical case-control study design has been widely adopted for its efficiency in studying rare diseases. When a disease is rare, a cross-sectional or cohort study of the population will yield few diseased 'cases', whereas a simple case-control design selects a random sample of cases and a random sample of healthy 'controls'. To obtain the estimates of risk (expressed as odds ratios) associated with the various factors under study, one can analyse the case-control data using the same logistic regression model as is appropriate for the cross-sectional or cohort designs (Prentice and Pyke, 1979). The case-control design does not allow the prevalence of disease to be estimated, but has the advantage of greater efficiency due to the better representation of cases. A variation of the simple case-control design, which samples equal numbers of cases and controls within predefined strata in the population, is referred to as a 'balanced design'. While such balance seems intuitively efficient, it is not necessarily optimum when sampling costs are different for cases and controls and perhaps also across strata.

Applied Optimal Designs Edited by M.P.F. Berger and W.K. Wong
 ISBN: 0-470-85697-1 (HB)

Since the 1960s there have been a number of developments in designing more efficient epidemiological studies, with many researchers also addressing the important consideration of cost. The early work by Cochran (1963) and Nam (1973) shows that when the cost per control is C_0 and cost per case is C_1, the optimum ratio of controls to cases is $n_0/n_1 = \sqrt{C_1/C_0}$, i.e. the numbers are inversely proportional to the square root of the costs. Considerations of efficient design have also been addressed in linear regression models with measurement error, where, for example, the cost of measuring the true covariate of interest is different from the cost of obtaining a 'surrogate'. Buonaccorsi (1990) derived optimal 'double sampling' schemes where the surrogate data is obtained for all subjects and the true covariate measured only on a random validation sample. These optimal designs minimise the variance of the coefficient of interest in the linear regression model subject to a fixed total cost. Similar developments for epidemiological applications were provided by Tosteson and Ware (1990) who investigated the logistic regression model. Assuming that prior surrogates U and Z are available for the dichotomous outcome (Y) and the exposure (X) respectively, they derived a method to choose which of three study designs would give minimum variance for the coefficient of interest in the logistic model, subject to the total sample size being fixed: depending on the prevalence of the outcome (e.g. disease), one chooses between a case-control design, a prospective design, or a hybrid of these two. This work did not consider the relative costs of different sampling mechanisms nor of sampling the different individuals. Likewise, the balanced design proposed by Cain and Breslow (1988) as most efficient for a logistic regression framework did not consider costs.

In this chapter, we will present work (Reilly, 1996) that provides further refinement of these designs. Traditional epidemiological studies gather all exposure variables on each of the study subjects and typically, a small number of these exposures will be of key interest in the final analysis. However, we will concentrate on two-stage designs, where we assume that the outcome variable and a number of easy-to-get or cheap exposure variables are measured for all subjects at the first stage, while the more expensive or hard-to-get exposures are ascertained only for a subsample of subjects at the second stage. The first-stage data may include cheap 'surrogates' for the expensive 'true' exposure to be measured at the second stage. Such designs are intuitively appealing to researchers with limited resources who need to gather expensive specialised data, and indeed are often implemented, although appropriate analysis of such data requires non-standard methods. What we present here is a method of guiding the sampling in such circumstances, by identifying the most 'informative' subjects to include in the second-stage sample, so that the experimenter implements an efficient study. We also allow for differing costs for first- and second-stage data, so that one may design cost-efficient studies subject to budgetary constraints.

These designs are primarily directed at epidemiological and clinical applications, and so they focus on studies of the relationship between covariates (predictor variables) and a dichotomous outcome, modelled by the logistic regression model.

Hence, although the methodology has been derived for the general case and applies to any likelihood function (Reilly and Pepe, 1995), the software we have designed is only for the logistic regression setting and the illustrations presented here will concentrate on applications where the first-stage data is obtained from a cross-sectional, cohort or case-control study. This first-stage sample may also be a 'balanced' sample, with equal numbers of cases and controls within predefined strata, provided estimates are available of the prevalence of the various strata in the population. A special case of balanced design is a 'matched case-control' design, where one or more controls are chosen to match each case on variables that we wish to adjust for in the analysis. Even with very refined matching (e.g. twin studies, or sib-pairs), conditional logistic regression provides estimates of odds ratios. However, we do not consider such first-stage data here, as the strata are too sparse and further second-stage sampling within strata is rarely practical or feasible. However, for studies that aim to estimate the effects of genetic markers, adjusted for known phenotypic factors (such as age, sex, etc.), our methodology has clear relevance, as the expense of laboratory investigations makes it critically important to identify 'informative' subsamples of subjects.

4.2 Illustrative Examples

The optimal sampling methods presented here are based on the Meanscore method of analysis of incomplete data (Reilly and Pepe, 1995). A real example serves to illustrate that although the incomplete data may not have been gathered in a planned study, it can nonetheless often be regarded as a two-stage sample. Furthermore, the context of an example enables one to readily see that an optimal design can be thought of as one that has 'intentionally missing' data.

4.2.1 Example 1

A study was undertaken at Karolinska Institute in 1993 to examine the impact of lifestyle factors and genetic variants on breast cancer risk. An initial sample of 3200 breast cancer cases and 3200 age-matched controls were identified, and exposure to menopausal hormone treatment (HRT) recorded for all these women. Additional variables that were measured included reproductive history and body size. Then 900 cases and 900 controls were randomly sampled for genotyping. Subsequently, all women with long duration of exposure to menopausal hormone treatment (four or more years) were sampled, resulting in more cases than controls. Finally, from among the remaining subjects, a further random sample of 600 cases and 600 controls were selected for genotyping. This apparently complex sampling scheme can be represented by the sketch in Figure 4.1, where for simplicity of presentation we will disregard the age and other covariates that were recorded in the study.

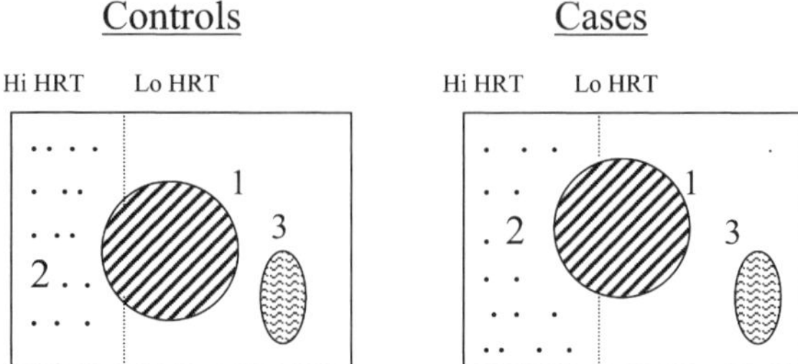

Figure 4.1 Schematic representation of sampling of subjects at several stages. (1) random sample of 900 cases and 900 controls; (2) sampling of all subjects with high exposure to HRT; (3) random sample of 600 cases and 600 controls from the remainder

Since case/control status and HRT exposure are known for all subjects, the same design would have been obtained by regarding this information as the first-stage data, which defines four strata (two levels of disease status, two levels of HRT use).The second-stage data is then obtained by sampling 100% of subjects in the two strata with high exposure to HRT and an appropriate fraction of the subjects in the two strata with low exposure.

The sequential sampling done in this study suggests that the investigators had surplus funds after the first round of sampling, and that they considered the subjects with high exposure to HRT to be potentially more informative. These considerations raise a number of issues that are relevant to the methods that we will present in this chapter. Firstly, could sufficient information have been obtained from fewer subjects in the first round of sampling? Secondly, given the information available from the first round, what additional subjects are most informative to genotype? Thirdly, if there was a fixed budget available from the outset for genotyping, how might one design the total sample size of this study (perhaps more than 6400, perhaps less!) and the numbers of cases and controls in each HRT level to be genotyped, so that the investment in this study 'buys' the maximum possible precision for the estimate of genetic effect. Clearly, the genotype data in this example is missing by design (albeit a staggered one), but we are posing the question as to how else one could have designed which subjects to sample so that the study is optimal.

4.2.2 Example 2

A common design in epidemiological investigations is the 'nested case-control study', where cases and controls are identified within a large cohort or population who are being monitored for the incidence of the disease of interest. Further particulars will then usually be gathered for this case-control sample in order to address the research question. This design is particularly useful where the population under study is routinely recorded in electronic 'registers'. One example of such a resource is the register of the Coronary Artery Surgery Study (CASS) described by Vliestra *et al.* (1980), which recorded details of treatment and outcome on approximately 25 000 patients who underwent coronary arteriography. As an illustrative example we will limit our attention to the operative mortality of the 8096 subjects who had coronary bypass surgery. From Table 4.1, we see that there are 144 male cases, and 58 female cases in this cohort.

Table 4.1 Numbers (and proportions) of Coronary Artery Surgery Study (CASS) coronary bypass subjects, stratified by sex and operative mortality

	Males	Females
Alive ($Y = 0$)	6666 (0.823)	1228 (0.152)
Deceased ($Y = 1$)	144 (0.018)	58 (0.007)

Let us suppose that our task is now to select a case-control sample from this data, and to ascertain their age, so that we can estimate the age-adjusted association between sex and mortality from an appropriate logistic regression model. We will show in the following sections, that if we have age available for a pilot sample in each of the four strata of Table 4.1, then we can compute the second-stage sampling fractions for each of these strata that will maximise the precision of the estimate of the age-adjusted coefficient.

4.2.3 Example 3

This example is also based on the CASS data described above, but we consider the more complex and realistic logistic regression model of Cain and Breslow (1988), which includes sex, age, body weight, angina, congestive heart failure (CHF) score, blood pressure, and urgency of surgery. Let us assume that we wish to estimate the association between urgency of surgery and mortality, adjusted for the other factors, and that the only first-stage variables available on the register are mortality, sex, and a three-category weight variable (as in Cain and Breslow). The breakdown of the 8096 first-stage subjects by these variables is presented in Table 4.2.

Note that there were 8 subjects with missing information, so that the total sample size is now 8088. Let us assume in addition that we have all the variables of interest recorded for a pilot sample of observations in each of the 12 strata of Table 4.2. We will show below how a second-stage sample of a predetermined size can be selected from these various strata in such a way as to maximise the precision of the coefficient of interest (urgency of surgery) in the logistic regression model. We will also show how one can derive the total study size and the second-stage sampling fractions that will maximise the precision of this estimate subject to a fixed available budget.

Table 4.2 Numbers (and proportions) of Coronary Artery Surgery Study (CASS) coronary bypass subjects, stratified by sex, operative mortality and categorical weight

Sex and weight (kg)	Alive	Deceased
Males		
<60	160 (0.02)	8 (0.001)
60–70	1083 (0.134)	33 (0.004)
≥70	5418 (0.670)	103 (0.013)
Females		
<60	440 (0.054)	18 (0.002)
60–70	407 (0.050)	26 (0.003)
≥70	378 (0.047)	14 (0.002)
Total	7886	202

In truth, all variables are available on the CASS register for *all* subjects in Examples 2 and 3. However, for the purpose of illustration, we 'manufactured' the pilot data sets by selecting 25 observations at random from each of the 4 strata in Table 4.1, and 10 observations at random in each of the 12 strata in Table 4.2, yielding pilot data sets of 100 and 118 observations respectively (note that one of the strata in Table 4.2 has only 8 observations, so all of these were included in the pilot sample). These are the same pilot data sets used in Reilly (1996). Having all the data available means that the optimal designs that we derive using our pilot data sets can be 'implemented' by sampling the required number of subjects. Thus, by repeated sampling, we can investigate the average performance of any design relative to an analysis of the whole cohort. In addition, we can investigate the sensitivity of the optimal designs to sampling variation in the pilot sample.

In the following sections, we will introduce and briefly describe the Meanscore method for the analysis of incomplete or two-stage data. We will then outline how the variance expression of the Meanscore estimate leads to the derivation of study designs that are optimal with respect to considerations of precision and cost. We have developed software tools in R, S-PLUS and STATA for the computation of these 'optimal' study designs for the logistic regression setting, and these are freely available on the Internet (Reilly and Salim, 2001). We will introduce these tools and illustrate their use. Readers interested only in the application of the methods to practical examples may wish to skip directly to the Appendix and work through the examples presented there.

4.3 Meanscore

In this section, we introduce the problem of missing covariates in regression models. Perhaps the most common approach when confronted with missing data is to exclude the incomplete cases from analysis and proceed to analyse the complete cases using standard methods. This approach is valid when the incomplete cases are a simple random sample of all cases, referred to as missing completely at random (MCAR, see Rubin, 1987). If the incomplete cases are a random sample within the strata defined by the completely known variables (i.e. missing at random, MAR), an analysis of complete cases is still valid *provided missingness does not depend on the outcome*. However, if the missingness is MAR and can depend on outcome, or if missingness depends on the unobserved values of the incomplete covariates (known as 'informative' missingness), analysis of only the complete cases is not valid. It would seem reasonable to assume MAR and missingness independent of outcome in Example 1 above, since both case and control patients were chosen for genotyping either completely randomly, or based on their HRT exposure. Assumptions regarding MAR or non-informative missingness are more difficult to justify in situations where incomplete data has arisen but where one cannot easily discern/describe the underlying process.

Even when valid, an analysis of only the complete cases results in a loss of precision due to the ignored observations. In recent years, a number of methods that accommodate all available cases have been developed. These methods include pseudo-likelihood (Breslow and Cain, 1988), weighted likelihood (Flanders and Greenland, 1991), Meanscore (Reilly and Pepe, 1995) and non-parametric maximum likelihood (Breslow and Holubkov, 1997). Here we will introduce the Meanscore method and show how it can be used to derive optimal two-stage designs. The Meanscore method, which incorporates information from all available cases into the regression model, is likelihood-based. For completely random missingness, it results in an improvement in efficiency over the analysis of complete cases only. More importantly, the method is applicable to MAR (whether or not missingness depends on outcome) which covers a wide range of patterns of missingness.

For simplicity of notation, let Y denote the response variable, Z the complete covariates (which must be categorical) and X the covariates of interest in the regression model, where some components of X are missing. The complete covariates Z may contain some auxiliary or surrogate variables that are informative about the missing components of X. Such was the case in Example 3 above, where categorical weight was available at the first stage, but actual weight, in kilograms, was recorded at the second stage. In that example, Y is the indicator for operative mortality, Z is a six-level variable defined by sex and categorical weight, and X includes sex, age, weight, angina, CHF score, blood pressure and urgency of surgery. In Example 1, Y is an indicator for case/control status (i.e. breast cancer), Z includes HRT exposure and age, and X contains the results of the genotyping.

Interest is focused on estimating the parameters in the regression model P_β $(Y \mid X)$, with some components of β often of particular importance. For our Example 3 above, P_β $(Y \mid X)$ will be a logistic regression model, so that β is the log-odds of the risk associated with the covariates X. Estimates will usually be reported as odds ratios (e^β), and in this example, attention is focused on the odds ratio for urgency of surgery.

By modelling the relationship between Z and X as $P_\theta\ (X \mid Z)$, the likelihood can be written:

$$L(\beta, \theta) = \prod_{i \in V} P_\beta(Y_i \mid X_i) P_\theta(X_i \mid Z_i) \prod_{j \in V} P_{\beta,\theta}(Y_j \mid Z_j), \tag{4.1}$$

where $P_{\beta,\theta}(Y \mid Z) = \int P_\theta(X \mid Z) P_\beta(Y \mid X) \mathrm{d}X$, or equivalently, the log-likelihood is

$$\log L(\beta, \theta) = \sum_{i \in V} \log P_\beta(Y_i \mid X_i) + \sum_{j \in \overline{V}} \log P_{\beta,\theta}(Y_j \mid Z_j), \tag{4.2}$$

where V denotes the set of complete (validation) cases available at the second stage, and $\bar{V}$ denotes the set of incomplete (non-validation) cases with only first-stage data recorded. Maximisation of this log-likelihood to obtain the maximum likelihood estimate (MLE) of β is not common in practice, due to the difficulty of specifying the nuisance function $P_\theta(X \mid Z)$, and the possibility of inconsistent estimates due to misspecification.

If the likelihood components of the complete observations are factored as $P_\theta(X \mid Z) P_\beta(Y \mid X)$, then the EM algorithm (Dempster *et al.*, 1977) provides the maximum likelihood estimator of β by the iterative maximisation of

$$Q(\beta) = \sum_{i \in V} \log P_\beta(Y_i \mid X_i) + \sum_{j \in \overline{V}} E\{\log P_\beta(Y_j \mid X) \mid \beta^c, Y_j, Z_j\}, \tag{4.3}$$

where β^c denotes the current estimate, provided θ and $P_\theta(X \mid Z)$ are known so that the expectation on the right-hand side can be evaluated. Because the exact relationship between X and Z is unknown, the Meanscore method uses a non-parametric estimate of the conditional expectation in (4.3), so that an estimate of β is obtained by maximising

$$\hat{Q}(\beta) = \sum_{i \in V} \log P_\beta(Y_i \mid X_i) + \sum_{j \in \overline{V}} \left\{ \sum_{i \in V(Z_j Y_j)} \frac{\log P_\beta(Y_i \mid X_i)}{n^{V(Z_j Y_j)}} \right\}, \tag{4.4}$$

where $V(Z_j Y_j)$ denotes the complete (validation) cases with $Z = Z_j$ and $Y = Y_j$ and $n^{V(Z_j Y_j)}$ is the number of such cases. Thus the Meanscore estimate of β is the solution to

$$\sum_{i \in V} S_\beta(Y_i \mid X_i) + \sum_{j \in \overline{V}} \left\{ \sum_{i \in V(Z_j Y_j)} \frac{S_\beta(Y_i \mid X_i)}{n^{V(Z_j V_j)}} \right\} = 0, \tag{4.5}$$

where

$$S_\beta(Y_i \mid X_i) = \frac{\partial P_\beta(Y \mid X)}{\partial \beta},$$

the usual score statistic. We see that observations that are incomplete (and thus have unknown score) are assigned the average score of complete observations with matching Y and Z, hence the name 'Meanscore'. A little algebra shows that the Meanscore estimator is thus the solution to the score equation

$$\sum_{i \in V} \frac{n^{Z_i Y_i}}{n^{V(Z_i Y_i)}} S_\beta(Y_i \mid X_i) = 0. \tag{4.6}$$

In agreement with intuition, each validation sample member is being 'upweighted' to represent the non-validation sample members that match on the observed variables. The 'weighted likelihood' equation we solve here is in fact the same as that of Flanders and Greenland (1991), but we have found a different variance expression, which is the key to our ability to derive optimal designs.

The Meanscore estimator has been shown (Reilly and Pepe, 1995) to be unbiased with asymptotic variance given by

$$\frac{1}{n}(I^{-1} + I^{-1}\Omega I^{-1}), \tag{4.7}$$

where

$$I = E\frac{-\partial^2}{\partial \beta^2} \log P_\beta(Y \mid X),$$

the observed Fisher information matrix, and

$$\Omega = \sum_{(Z,Y)} \rho^{Z,Y} \frac{\rho^{\bar{V}(Z,Y)}}{\rho^{V(Z,Y)}} \mathrm{Var}[S_\beta(Y \mid X, Z) \mid Y, Z], \tag{4.8}$$

where n is the total number of observations (study size), $\rho^{Z,Y}$ is the probability that a subject is in the stratum (Z, Y), while $\rho^{V(Z,Y)}$ and $\rho^{\bar{V}(Z,Y)}$ are the probabilities that a subject in stratum (Z, Y) is in the validation and non-validation samples respectively. The second term in the variance expression is the loss of precision over full maximum likelihood due to the incomplete observations. This 'sandwich' penalty term depends on the extent of the missingness in the various (Z, Y) strata, as might be expected, and on the variability of the score within those strata. We can estimate I and Ω as

$$\hat{I} = \frac{1}{n}\sum_{i\in V}\frac{n^{Z_iY_i}}{n^{V(Z_iY_i)}}\left[\frac{-\partial^2}{\partial\beta^2}\log P_\beta(Y_i \mid X_i)\right]_{\beta=\hat{\beta}}, \tag{4.9}$$

$$\hat{\Omega} = \sum_{(Z,Y)}\frac{n^{ZY}}{n}\frac{n^{\overline{V}(ZY)}}{n^{V(ZY)}}\mathrm{Var}[S_\beta(Y \mid X, Z) \mid Y, Z], \tag{4.10}$$

where the term $n^{V(ZY)}/n^{ZY}$ is referred as the 'validation sampling fraction' or 'second-stage sampling fraction' for the (Z, Y) stratum.

4.3.1 Example of meanscore

As an illustration of the Meanscore method, we return to Example 3, where the outcome Y is the operative mortality of patients undergoing coronary bypass surgery. The first-stage covariate Z consists of sex and a three-level categorical weight variable. The numbers (and proportions) of subjects in the 12 strata defined by Y and Z were already presented in Table 4.2. Let us suppose that we wish to fit the same logistic regression model as Cain and Breslow (1988) and that we can select a random sample of 1000 second-stage observations from which to ascertain the X covariates age, weight, angina, CHF score, blood pressure, and urgency of surgery. In practice, we simply have to sample these observations from the database, as all observations are in fact complete, but this gives us the opportunity to investigate how Meanscore compares with an analysis of only the 1000 complete cases, or an analysis of all cases in the full data set. The logistic regression model is:

$$\begin{aligned}\mathrm{logit}\,P(Y_i = 1) = \beta_0 + \beta_1X_{1i} + \beta_2X_{2i} + \beta_3X_{3i} + \beta_4X_{4i}\\ + \beta_5X_{5i} + \beta_6X_{6i} + \beta_7X_{7i},\end{aligned} \tag{4.11}$$

Table 4.3 Coefficients and standard errors from logistic regression analyses of random second-stage sample only and of full CASS data, compared to estimates obtained by Meanscore analysis of first-stage and second-stage data (M. Reilly, Optimal sampling strategies for two-stage studies, *American Journal of Epidemiology*, **143**(1): 92–100, by permission of Oxford University Press)

	Random subsample $n = 1000$	Meanscore $n = 8088$ $n_V = 1000$	Full data $n = 8088$
	β (SE)	β (SE)	β (SE)
Sex	0.745 (0.569)	0.200 (0.265)	0.244 (0.194)
Weight	0.0002 (0.021)	−0.014 (0.014)	−0.015 (0.007)
Age	0.064 (0.030)	0.055 (0.026)	0.049 (0.009)
Angina	1.391 (0.527)	1.260 (0.416)	0.218 (0.239)
CHF	0.275 (0.259)	0.401 (0.198)	0.336 (0.072)
LVEDP	0.002 (0.033)	−0.013 (0.024)	0.031 (0.010)
Urgency	0.947 (0.481)	0.791 (0.496)	0.797 (0.154)

where $X_{1i}, X_{2i}, \ldots X_{7i}$ denote the sex, weight, age, CHF score, angina, LVEDP (left ventricular end diastolic blood pressure) and urgency of surgery respectively of the ith study subject, and Y_i denotes the outcome (operative mortality). The coefficients and standard errors from our analyses are presented in Table 4.3, where we see the improvement in precision from including all available data in a Meanscore analysis.

4.4 Optimal Design and Meanscore

In the previous section, we noted that the variance of the Meanscore estimate is a function of the total number of observations and the validation sampling fractions in all the (Z, Y) strata. However, as explained earlier, the missing covariates scenario can also be viewed as a two-stage design, where the response variable Y and the complete variables Z are regarded as the first-stage information and the incomplete components of X are regarded as the second-stage information. Hence the dependence of the variance on the study size and the validation sampling fractions means that *it is possible to minimise the variance using an appropriate study design*. That is, it is possible to minimise the variance by an astute choice of sample size and sampling fractions. How this can be implemented in practice is the subject of the remainder of this chapter. We will outline how one can derive optimal sampling designs for two-stage studies for the following three scenarios:

1. Where we already have the first-stage data and would like to sample a specified number of observations at the second stage. For example, if we already have a database or register, and we wish to gather additional information on some subjects in order to address a research question.
2. Where a fixed budget is available and we wish to design a study that will minimise the variance of an estimate subject to this budget.
3. Where a coefficient of interest is to be estimated with a specified precision and we wish to design a study that will achieve this for minimum cost.

4.4.1 Optimal design derivation for fixed second-stage sample size

Suppose we wish to select n_2 subjects at the second stage in such a way as to minimise the variance of the kth component of the regression coefficient vector $\boldsymbol{\beta}$. This is equivalent to minimising the $[k, k]$ element in the variance–covariance matrix $\mathbf{V}$:

$$V_{kk} = \frac{1}{n}(I_{kk}^{-1} + [I^{-1}\Omega I^{-1}]_{kk}), \tag{4.12}$$

with the constraint that the second-stage sample size n_2 is fixed.

Note that we can write the constraint as

$$\sum_{(Z,Y)} \rho^{V(Z,Y)} \rho^{Z,Y} = \frac{n_2}{n}, \tag{4.13}$$

where $\rho^{Z,Y}$ is the prevalence (probability) of the (Z, Y) stratum, $\rho^{V(Z,Y)}$ is the second-stage sampling fraction for the (Z, Y) stratum, and n is the total number of observations. A Lagrange multiplier (λ) can be used to accommodate the constraint so that in this case we would minimise

$$V_{kk} - \lambda \left(\sum_{(Z,Y)} \rho^{V(Z,Y)} \rho^{Z,Y} - n_2/n \right). \tag{4.14}$$

Taking the first derivative of this function with respect to $\rho^{V(Z,Y)}$ and setting it to zero yields the optimal second-stage sampling fractions. After some algebraic manipulation it can be shown that the optimal second-stage sampling fraction in the (Z, Y) stratum is given by

$$\rho^{V(Z,Y)} = \frac{\frac{n_2}{n}\sqrt{[I^{-1}\mathrm{Var}(S_\beta \mid Z, Y) I^{-1}]_{kk}}}{\sum_{(Z,Y)} \rho^{Z,Y} \sqrt{[I^{-1}\mathrm{Var}(S_\beta \mid Z, Y) I^{-1}]_{kk}}}. \tag{4.15}$$

4.4.2 Optimal design derivation for fixed budget

Assume now that we wish to minimise the variance of the kth component of the regression coefficient vector $\boldsymbol{\beta}$, given that we have a fixed budget $\mathbf{B}$ available and the first- and second-stage costs per observation are known to be $\mathbf{c}_1$ and $\mathbf{c}_2$ respectively.

Using the fact that $n\sum_{(Z,Y)} \rho^{V(Z,Y)} \rho^{Z,Y} = n_2$, note that we can rewrite the constraint as

$$B = n \left(c_1 + c_2 \sum_{(Z,Y)} \rho^{V(Z,Y)} \rho^{Z,Y} \right), \tag{4.16}$$

where n is the study size and $\rho^{Z,Y}$ and $\rho^{V(Z,Y)}$ are the prevalence (probability) and the second-stage sampling fraction for the (Z, Y) stratum. Again using a Lagrange multiplier, our task is to minimise

$$V_{kk} - \lambda \left(nc_1 + nc_2 \sum_{(Z,Y)} \rho^{V(Z,Y)} \rho^{Z,Y} - B \right), \tag{4.17}$$

where V_{kk} is the $[k, k]$ element of the variance–covariance matrix $\mathbf{V}$.

The optimal study size and second-stage sampling fractions can be obtained by taking the derivatives of the function in 4.17 with respect to n and $\rho^{V(Z,Y)}$, setting these to zero and solving them simultaneously. It can be shown (see Reilly and Pepe, 1995 for more details of the derivation) that the optimal study size is given by

$$n = B\left[\frac{c_1 + \sqrt{c_1c_2}\sum_{(Z,Y)} \rho^{Z,Y}\sqrt{[I^{-1}\mathrm{Var}(S_\beta \mid Z,Y)I^{-1}]_{kk}}}{\sqrt{[I^{-1}]_{kk} - \sum_{(Z,Y)} \rho^{Z,Y}[I^{-1}\mathrm{Var}(S_\beta \mid Z,Y)I^{-1}]_{kk}}}\right]^{-1}, \tag{4.18}$$

and the optimal second-stage sampling fraction for the *(Z,Y)* stratum is given by

$$\rho^{V(Z,Y)} = \frac{B - nc_1}{nc_2}\frac{\sqrt{[I^{-1}\mathrm{Var}(S_\beta \mid Z,Y)I^{-1}]_{kk}}}{\sum_{(Z,Y)} \rho^{Z,Y}\sqrt{[I^{-1}\mathrm{Var}(S_\beta \mid Z,Y)I^{-1}]_{kk}}}. \tag{4.19}$$

4.4.3 Optimal design derivation for fixed precision

In this case we would like to achieve a fixed variance estimate, say δ for the kth component of the regression coefficient vector $\boldsymbol{\beta}$, while minimising the study cost. This is equivalent to designing a study with a specified statistical power, or to yield a confidence interval of a specified width for the parameter of primary interest.

Assume again that the first-stage cost is $\mathbf{c}_1$ per observation and second-stage cost is $\mathbf{c}_2$ per observation. Note that as in the fixed budget case above we can write the total study cost as $n(c_1 + c_2\sum_{(Z,Y)} \rho^{V(Z,Y)}\rho^{Z,Y})$. Using a Lagrange multiplier, we now wish to minimise the following function:

$$nc_1 + nc_2\sum_{(Z,Y)} \rho^{V(Z,Y)}\rho^{Z,Y} - \lambda(V_{kk} - \delta). \tag{4.20}$$

The optimal solution can be obtained by taking the first derivatives of this function with respect to n and $\rho^{V(Z,Y)}$, setting these to zero and solving the resultant equations simultaneously.

It can be shown (Reilly and Pepe, 1995) that the optimal study size is given by

$$n = \frac{\left[I^{-1} - \sum_{(Z,Y)} \rho^{Z,Y}W_{ZY}\right]_{kk}}{\delta} + \frac{1}{\delta}\sqrt{\frac{c_2}{c_1}}\sum_{(Z,Y)} [\rho^{Z,Y}\sqrt{W_{ZY}}]_{kk} \times \sqrt{\left[I^{-1} - \sum_{(Z,Y)} \rho^{Z,Y}W_{ZY}\right]_{kk}}, \tag{4.21}$$

and the optimal-second stage sampling fraction for the (Z, Y) stratum is

$$\rho^{V(Z,Y)} = \sqrt{\frac{c_1}{c_2}} \sqrt{\frac{[W_{ZY}]_{kk}}{[I^{-1} - \sum_{(Z,Y)} \rho^{Z,Y} W_{ZY}]_{kk}}}, \tag{4.22}$$

where

$$W_{ZY} = I^{-1} \mathrm{Var}(S_\beta \mid Z, Y) I^{-1}.$$

It is interesting to note the similarity between the contribution of costs to the sampling fractions here, and the simple case-control set-up of Cochran (1963) and Nam (1973) where the cost per control is C_0 and cost per case is C_1, and the optimum ratio of controls to cases is $n_0/n_1 = \sqrt{C_1/C_0}$.

4.4.4 Computational issues

The derivation of the optimal designs above was carried out without constraining the second-stage sampling fractions to be less than or equal to 1 ($\rho^{V(Z,Y)} \leq 1$). As a result the 'optimal' second-stage sampling fractions computed with these formulae can be greater than 1. Reilly and Pepe (1995) proposed the following 'ad hoc' method to overcome this problem: (i) sample 100% from the stratum with the largest $\rho^{V(Z,Y)} > 1$, and optimally sample from the remaining strata, (ii) repeat this step iteratively until all $\rho^{V(Z,Y)} \leq 1$. We have since performed numerical simulation studies that show that the constrained objective function is minimised at our proposed solution.

In the fixed budget and fixed precision scenarios the problem is more complicated since in addition to the second-stage sampling fractions we have to estimate the optimal study size (total number of observations). Note that the formula for this quantity involves the square root of

$$\left[I^{-1} - \sum_{Z,Y} \rho^{Z,Y} (I^{-1} \mathrm{Var}(S_\beta \mid Z, Y) I^{-1}) \right]_{kk}.$$

In practical examples this term can be negative so that the square root is undefined and an optimal design cannot be found. In such situations, we investigated a procedure which is analogous to that used when the second-stage sampling fractions are greater than 1: (i) sample 100% from the stratum with maximum $\rho^{Z,Y} [I^{-1} \mathrm{Var}(S_\beta \mid Z, Y) I^{-1}]_{kk}$ and optimally sample from the remaining strata, (ii) repeat this step iteratively until

$$\left[I^{-1} - \sum_{Z,Y} \rho^{Z,Y} (I^{-1} \mathrm{Var}(S_\beta \mid Z, Y) I^{-1}) \right]_{kk} > 0.$$

Numerical investigation of this method has also shown it to yield the constrained optimum solution.

4.5 Deriving Optimal Designs in Practice

4.5.1 Data needed to compute optimal designs

It can be seen from the above expressions that, in addition to information on costs and/or precision, it is also necessary to have estimates of the likelihood-derived quantities I and $\mathrm{Var}(S_\beta \mid Z, Y)$ before one can compute an optimal design. This is akin to the 'usual' methods of sample-size calculation, where one must have estimates of the variance of the variable under study to input into the sample-size formula. In that simpler case, investigators will often be able to provide estimates from their knowledge of the subject matter. However, in the two-stage design case that we are dealing with here, knowledge of underlying parameter values (even with some idea about the extent of confounding) cannot easily yield estimates of I and $\mathrm{Var}(S_\beta \mid Z, Y)$, which are matrix quantities. Hence, estimation of these quantities requires that we have available a random sample of subjects in each of the strata defined by the outcome variable and the first-stage covariates of interest. This 'pilot' data set must have a minimum of two (and preferably three) observations in each of the strata. We will assume that users of our methods will have such pilot data available, and our software algorithms expect this data as input, together with the design parameters such as sample size, cost or precision.

In some cases, pilot data will be available as the experimenters may already have unwittingly embarked on a two-stage design due to limited resources, but the second-stage data collected so far will not be 'designed'. We envisage that some experimenters considering a two-stage design will have access to first-stage data, and so can produce a pilot data set simply by sampling a few second-stage observations in each of the strata defined by the outcome and the first-stage covariate(s) of main interest. With pilot data of either of these types, running the optimal sampling functions illustrated below will then guide the remainder of the data collection for the experiment.

If the experiment is being planned before any data has been collected, then careful thought should be given to identifying the important stratification variables before collecting a few pilot observations in each stratum. In the absence of first-stage data, one must now provide estimates of the prevalence of the strata in the population being studied, which should not be too difficult a task for a researcher familiar with the subject area.

Finally, if collecting real pilot data is not possible, one could of course create fictitious pilot data by computer simulation, and examine the optimal designs that are proposed under a range of settings.

4.5.2 Examples of optimal design

Consider the simple CASS example (Example 2), where the numbers and proportion of subjects in each of the four first-stage strata are given in Table 4.1. Let us suppose that only this information is available and that resources are available to ascertain age for a second-stage sample of 1000 subjects, in order to estimate the age-adjusted effect of sex on the risk of operative mortality. Since the calculation of optimal designs requires estimates of the score and information components in the various strata, pilot data must be available. In practice, one could use a little of the available resources to obtain this pilot sample, and then ascertain how the remainder of the 1000 should be sampled to ensure an efficient study. However, for this illustration, we have selected a random sample of 25 observations in each of the four strata, and application of Equation (4.15) to this pilot data suggests that a second-stage sample of 1000 should be chosen as indicated in Table 4.4. This design indicates that at the second stage one should sample all cases, 11% of male controls and 4% of female controls.

Table 4.4 Numbers (n) of CASS subjects, stratified by sex and operative mortality, and the optimal sampling fractions (ρ^{opt}) and sample sizes (n^{opt}) for a second-stage sample of 1000. Optimality is with respect to the variance of the age-adjusted effect of sex on the risk of operative mortality, estimated from a logistic regression model

	Males	Females
Alive ($Y = 0$)		
n	6666	1228
ρ^{opt}	0.113	0.037
n^{opt}	753	45
Deceased ($Y = 1$)		
n	144	58
ρ^{opt}	1.0	1.0
n^{opt}	144	58

We return again to Example 3 using the CASS data, where the first-stage data define 12 strata (see Table 4.2) and our objective is to fit the logistic regression model of Cain and Breslow (1988), i.e.

$$\begin{aligned}\operatorname{logit} P(Y = 1) = \beta_0 + \beta_1 X_1 + \beta_2 X_2 + \beta_3 X_3 + \beta_4 X_4 \\ + \beta_5 X_5 + \beta_6 X_6 + \beta_7 X_7,\end{aligned} \tag{4.23}$$

where $X_1, X_2, \ldots X_7$ denote sex, weight, age, CHF score, angina, LVEDP (left ventricular end diastolic blood pressure) and urgency of surgery, and again the response variable Y is operative mortality. In contrast to the Meanscore analysis presented in Table 4.3, where we randomly sampled 1000 second-stage subjects, we now wish to *optimally* sample these second-stage observations. We will assume that the covariate of primary interest is urgency of surgery, so that the optimality we seek is with respect to the standard error of the estimate of this coefficient. For a pilot sample, we have taken a random sample of 10 subjects in each of the 12 first-stage strata (taking all 8 in the stratum that has only 8 available). Using this pilot data, Equation (4.15) yields the sampling fractions and sample sizes displayed in Table 4.5. As in the previous simple example, the design indicates that all cases

Table 4.5 Numbers of CASS subjects stratified by sex, operative mortality and categorical weight, with the optimal sampling fractions (ρ^{opt}) and sample sizes (n^{opt}) for a second-stage sample of 1000 in parentheses. Optimality is with respect to the variance of the independent effect of urgency of surgery on the risk of operative mortality, estimated from a logistic regression model which adjusts for sex, age, weight, angina, CHF score and LVEDP

Sex and weight (kg)	Alive	Deceased
Males		
<60	160 [0.225, 36]	8 [1, 8]
60–70	1083 [0.186, 201]	33 [1, 33]
≥70	5418 [0.039, 209]	103 [1, 103]
Females		
<60	440 [0.391, 172]	18 [1, 18]
60–70	407 [0.053, 22]	26 [1, 26]
≥70	378 [0.419, 159]	14 [1, 14]
Total	7886	202

should be sampled at the second stage, with varying proportions of controls in the different sex and weight strata. The coefficient of urgency of surgery will have an estimated standard error of 0.2 with this design, a significant improvement over the standard error of 0.481 obtained from a simple random sample, and quite close to the maximum precision (0.154) one can achieve with the full data set (see Table 4.3).

Optimality with respect to cost

The examples above both assume that the first-stage data is already available and that a fixed number of subjects (1000) are to be sampled at the second stage. Thus the only consideration is how to best choose these second-stage observations. However, another situation that arises frequently in practice is where a fixed budget is available for a study, and *prior to data collection* one wishes to determine the total study size and the second-stage sampling fractions that are optimal subject to this budget. Let us assume that we have a budget of £10 000, and that first- and second-stage observations cost £1.0 and £0.5 respectively, and we wish to find the optimal design corresponding to Table 4.5. Using the same pilot data set and Equations (4.18) and (4.19) we obtain the optimal design displayed in columns 2 and 3 of Table 4.6. Thus the optimal study would recruit 8756 subjects, among whom we expect 175 ($= 8756 \times 0.02$) male controls in the lowest weight category, 1173 ($= 8756 \times 0.134$) in the middle weight category, etc., using the proportions from Table 4.2. The second-stage information would then be obtained for 106 ($= 0.604 \times 175$), 591 ($= 0.504 \times 1173$) etc. subjects as indicated in Table 4.6, resulting in a total second-stage sample size of 2490, and a total study cost of $8756 + 0.5(2490) = 10\,001$ due to some rounding error. Repeating the calculations for a second-stage cost of £5 gives the design in columns 4 and 5, which has a total study size of 6341 subjects, with 734 of these sampled at the second stage.

Table 4.6 Total study size and optimal second-stage sampling fractions (with sample sizes in brackets), for a budget of £10 000, where C_1 and C_2 are the costs per first-stage and second-stage observation respectively. Optimality is with respect to the variance of the independent effect of urgency in the same logistic regression model as in Table 4.5

Sex and weight (kg)	$C_1 = £1, C_2 = £0.5$ $N = 8756$		$C_1 = £1, C_2 = £5$ $N = 6341$	
	Alive	Deceased	Alive	Deceased
Males				
<60	0.604 [106]	1 [9]	0.202 [26]	1 [6]
60–70	0.504 [591]	1 [35]	0.169 [144]	1 [25]
≥70	0.106 [624]	1 [114]	0.036 [153]	1 [82]
Females				
<60	1 [473]	1 [18]	0.351 [120]	1 [13]
60–70	0.146 [64]	1 [26]	0.049 [16]	1 [19]
≥70	1 [412]	1 [18]	0.393 [117]	1 [13]

Both of the optimal designs in Table 4.6 recommend sampling all cases at the second stage, and the study with less expensive second-stage data also samples all women in the highest and lowest weight categories. However, when the second-stage covariate information is more expensive, it is more efficient to conduct a smaller study and to sample less intensively at the second stage in all the control strata.

The design on the left of Table 4.6 yields a variance estimate of 0.031 for the urgency coefficient. If our objective had been to achieve a variance of 0.031 for this coefficient at minimum cost, then Equations (4.21) and (4.22) would have arrived at the same design above, at the same total cost of £10 000! Thus the optimal designs with respect to cost and precision are simply two sides of the same coin, as one would expect intuitively.

4.5.3 The optimal sampling package

We have written the *optimal sampling* algorithms for logistic regression models in three programming languages: *R*, *S-PLUS* and *STATA* (see bibliography for website addresses for these packages). A brief introduction to each of these statistical packages can be found in the Appendix, together with illustrations of how the optimal sampling functions are implemented in each.

R was chosen because it is an open, programmable package with most statistical functions already available, and it is *freely available*. For both R and S-PLUS, there is a large body of material contributed by users, which is easily obtained/shared via the Internet.

We chose STATA because it is widely used by biostatisticians and epidemiologists. This package also has a large library of functions contributed by users, and a journal (both hard-copy and electronic) in which these contributions are discussed. Another important attribute of STATA is its web compatibility, which greatly simplifies the process of finding and installing new functions such as ours.

The optimal sampling package consists of three functions that calculate the optimal two-stage designs under the following three scenarios:

1. *Fixed sample size*: here it is assumed that first-stage data is already available and it has been decided to sample a fixed number of subjects at the second stage.
2. *Fixed budget*: this function calculates the total study size and the second-stage sampling fractions that will maximise precision of a specified coefficient in the model.
3. *Fixed precision*: this function calculates the total study size and the second-stage sampling fractions that will achieve a specified precision for minimum cost.

The fixed sample size and fixed budget functions were used to produce the results shown in Tables 4.4–4.6 above.

4.5.4 Sensitivity of design to sampling variation in pilot data

The optimal sampling fractions derived above use estimates of score and information components obtained from pilot data. Thus, the computed optimal design will be subject to variation due to the sampling variation in the pilot sample. In order to investigate the sensitivity of the design to this sampling variation, we conducted a study using the CASS registry data. We focused our efforts on the scenario where we already have the first-stage data and wish to sample a fixed number of observations at the second stage.

For our initial investigation, we considered the simple situation of Example 2, where we wish to estimate the effect of sex and age on operative mortality using a logistic regression model. As before, the first stage consists only of operative mortality and sex, and we wish to ascertain age for a subset of 1000 patients at the second stage.

Two sampling schemes were used to take a pilot sample of 1000 subjects from the population. The first was a balanced sampling scheme, where ideally, we wished to sample 250 patients from each stratum defined by operative mortality and sex (see Table 4.1). However, since the total number of cases is less than 250, we decided to sample all the cases and sample the remaining patients from controls in such a way that we would have 500 males and 500 females in the pilot sample. In the second sampling scheme, we randomly sampled 1000 subjects from the total population of 8096. The sampling was repeated to yield 100 different pilot samples under each sampling scheme. For each pilot sample, we calculated the optimal second-stage sampling fractions for a fixed second-stage sample size (1000), where optimality was defined as maximising the precision of the sex-adjusted effect of *age* on operative mortality. This simulation study was implemented in R using the optimal sampling package, and the estimate and standard error of the age coefficient were recorded for each realisation.

The average second-stage optimal sampling fraction and its standard error for each stratum are given in Table 4.7. These results were compared to the 'true' optimal sampling fraction obtained by using all 8096 subjects as pilot information. It can be seen that both sampling schemes produce unbiased estimates of the optimal sampling fractions, but that the estimates from the balanced scheme have lower variance.

Table 4.7 The average second stage optimal sampling fractions (with standard deviations in parenthes), from 100 repetitions of selecting a pilot sample, under both a balanced and a random sampling scheme. The 'true' optimal sampling fractions were obtained using all 8096 subjects as pilot sample

	'True'	Balanced	Random	'True'	Balanced	Random
Alive	0.0887	0.0885	0.0895	0.1685	0.1693	0.1642
		(0.0023)	(0.0029)		(0.0123)	(0.0159)
Dead	1.0	1.0000	1.0000	1.0	1.0000	0.9983
		(0.000)	(0.0000)		(0.0000)	(0.0172)

The variation of the estimated optimal sampling fractions is displayed graphically in Figure 4.2. For each stratum, it is clear that the balanced scheme has less sampling variation. It is interesting to note that under the random sampling scheme, one pilot sample gave an estimate of less than 1 for the optimal sampling fraction for the female cases.

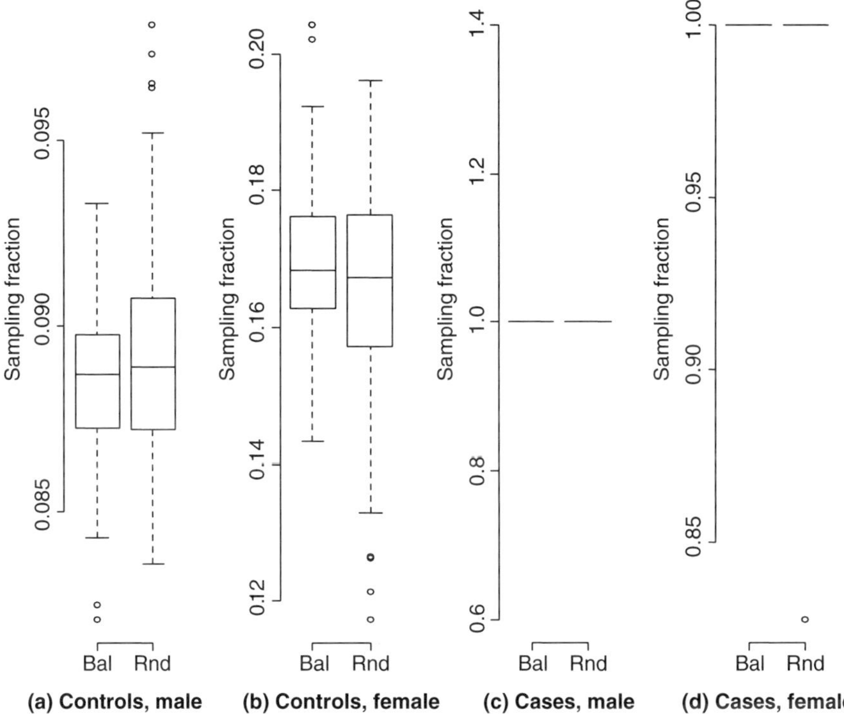

Figure 4.2 Sensitivity of optimal sampling fractions to sampling variation in the pilot data, using the CASS data from Example 1. Bal = balanced sampling, Rnd = random sampling

The variation in optimal sampling fractions is only of concern if it impacts on the estimates of interest. Hence, we proceeded in Figure 4.3 to display the variation of the estimates of the coefficients of age and sex under both sampling schemes. Again, although both sampling schemes produced unbiased estimates of the coefficients, as would be expected, the balanced scheme clearly has lower variation. More importantly, if we look at the Z-value (where Z-value is defined as the estimate of an effect divided by its standard error), under random sampling there are two pilot samples that produce Z-values less than 1.96 (i.e. not significant at the 5% level), which is contrary to the estimates produced by using all 8096 patients, which clearly show that age and sex are significant. Thus the stability of the sampling fractions can have important consequences for the ultimate estimates obtained from the two-stage analysis. These preliminary investigations suggest that one should take a balanced pilot sample whenever possible in order to help minimise the effects of sampling variation on the optimal design computed, and ultimately on the stability of the estimates from the regression model.

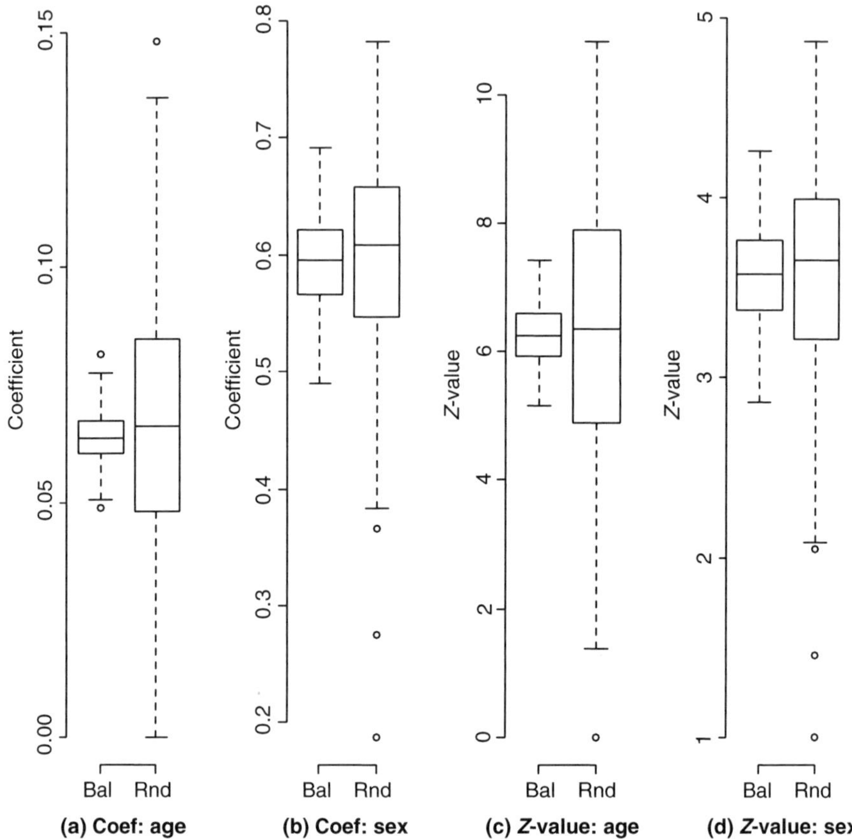

Figure 4.3 Variation in the regression coefficients achieved by optimal designs computed from repeated sampling of the pilot data, under both random and balanced sampling scheme

4.6 Summary

It is quite common for epidemiologists to gather some covariate data on a non-random subsample of their total study sample. This is particularly true for covariate information that is expensive or difficult to measure, or where the investigators have a perception that certain study subjects are more 'informative' regarding the research hypothesis. We have outlined here how the incomplete data generated by many such studies can be viewed as a 'two-stage' sample. Provided the incomplete observations are MAR, there are a number of methods (including Meanscore) that can perform a valid analysis of such data. A close look at the variance expression of the Meanscore estimate reveals that if one chooses the subsample of complete (i.e. second-stage) subjects in a judicious way, then the precision of an estimate of interest can be maximised subject to constraints of study size or cost. We have presented here the formulae for the calculation of such 'optimal' two-stage designs, and provided software tools for computing them for the logistic regression setting that is common in epidemiological studies. These software tools are available as functions in several widely used statistical packages and can be freely downloaded from the Internet.

We have provided here some illustrative examples of the calculation of optimal two-stage designs. Although optimality is with respect to a single coefficient of interest in the logistic regression model, we have found in numerous applications that the coefficients of the other variables in the model also improve when the optimal design is implemented. In epidemiological applications, there is often one covariate of primary interest, and investigators would readily choose a study design to maximise the precision of this coefficient. However, for other applications, it may be of interest to derive designs that simultaneously minimise the variance of several coefficients.

Our programs require some pilot data in order to derive optimal designs, in the same way that power and sample size calculations require estimates of effect size and standard deviation. These pilot data are used to estimate the score and information components in the first-stage strata. This raises the question of whether the computed optimal design is sensitive to misspecification of these components. Our preliminary investigations suggest that the sensitivity to sampling variation in the pilot data is reduced by sampling a balanced pilot sample. This agrees with intuition, as a more balanced sample protects against sparse data cells and consequent high variability of adjusted estimates. A closer study of the sampling variation of the precision achieved by optimal designs is an interesting area for further study, particularly for the more complex models encountered in practice.

4.7 Appendix 1: Brief Description of Software Used

4.7.1 R language

R is a language and environment for statistical computing and graphics. It was chosen because it is an open, programmable, extendable package with most statistical functions already available, is *freely available* and has a large user base of academic statisticians committed to constant improvement and update of the package. R is similar to the S language and environment (Chambers, 1998), and in fact R can be viewed as an implementation of S. There are some important differences, but much code written for S runs unaltered under R.

R runs on many operating systems including Windows 9x/NTXP, Windows 2000, Windows Me, UNIX and Mac. R can be downloaded at the R project website at http://www.r-project.org or other mirror sites around the world. Once installed the software includes a reference manual and help files from which one can learn. The version used for the latest updates of our optimal sampling modules is R 1.7.0.

4.7.2 S-PLUS

The S-PLUS software is based on the S language, originally designed by John Chambers and colleagues at Bell Laboratories (Chambers, 1998). S-PLUS is sold by Mathsoft, Inc. and more details about the package can be found on the S-PLUS website http://www.mathsoft.com/splus. A good introduction to the application of S-PLUS in many statistical areas can be found in Venables and Ripley (1999). There is a large amount of user-contributed code for S, available at the http://lib.stat.cmu.edu/DOS/S/ at Carnegie Mellon University.

We used S-PLUS version 4.0 to develop our software.

4.7.3 STATA

STATA is a statistical software package sold by STATA corporation (www.stata.com). It has attracted a lot of interest from biostatisticians, epidemiologists and medical researchers for its ease of use and its flexibility. In a single environment the user has access to a wide range of commands from simple tables to complex models, in any sequential order. The software has a large library of contributed functions written by users. The documentation for these functions is published in the STATA Technical Bulletin (STB) and the code made available on the STATA website. For example, our optimal sampling programs have been published in STB-58, November 2000. A powerful capability in STATA is the web compatibility; the 'search' command inside STATA allows the user to quickly and easily identify contributed functions and the 'net install' command allows one to directly install from the web, as part of STATA, any function they request. Our optimal sampling functions can be installed directly from the Internet in this way if you already have STATA software installed in your computer.

We developed our optimal sampling package using STATA 6, although STATA 7 and STATA 8 have been released since then. However all our programs can be run in the newer versions of Stata as we have used the `'version 6.0'` command to inform STATA of the version used to write the code.

4.8 Appendix 2: The Optimal Sampling Package

The *optimal sampling* functions in R, S-PLUS and STATA are very similar, with the exception of some slight variations in the naming conventions required by the different statistical packages. In this section we provide an introduction to the features that are common to all three systems, before presenting more details about installing and running the software in the different environments. Readers intending to download and run any version of the package are advised to first read this section carefully, as common features will not be elaborated on further under the headings dealing with the specific statistical packages.

The study designs derived by the *optimal* package are optimal with respect to a specified coefficient in the logistic regression model, in the same way that the usual power calculations for a simple hypothesis aim to minimise the standard error of a chosen estimate. Hence the user must always specify the predictor variable to be optimised when calling any of the functions. The package contains three different functions for deriving optimal designs under the three different scenarios mentioned earlier:

1. *Fixed sample size*: calculates optimal second-stage sampling fractions subject to fixed total sample sizes at the first and second stages. The 'fixed *n*' function computes the optimal second-stage sampling fractions (and sample sizes) for each stratum defined by the different levels of response variable and first-stage variables. Users need to provide the first-stage sample size for each stratum, the second-stage sample size required, and the name of the predictor variable to be optimised. In order to understand the order in which the vector of first-stage sample sizes must be provided, users are encouraged to first run a 'coding' function (see below).

2. *Fixed budget*: calculates the total number of study observations and the second-stage sampling fractions that will maximise precision subject to an available budget. In addition to specifying the budget, the user must also supply: (i) the unit cost of observations at the first and second stage, (ii) the vector of prevalences of each of the strata defined by the different levels of response variable and first-stage variables, and (iii) the name of the variable to be optimised. If there is any doubt about the order in which one must supply the vector of prevalences, the 'coding' function should be run first.

3. *Fixed precision*: calculates the total number of study observations and the second-stage sampling fractions that will minimise the study cost subject to a fixed variance for a specified coefficient. In addition to specifying the required variance the user must also supply: (i) the unit cost of observations at the first and second stage, (ii) the vector of prevalences of each of the strata defined by different levels of the response variable and first-stage covariates and (iii) the name of the variable to be optimised. Again, the 'coding' function should run to advise on the order in which one must supply the vector of prevalences.

The *coding* function is an important auxiliary function that advises users on the order in which certain inputs are to be provided to the optimal sampling functions. The coding function orders the strata formed by different levels of the response variable and first-stage covariates. Within the coding function a new variable is created which contains the distinct groups formed by different levels of the dependent variable (Y) and first-stage covariates (Z). A list is printed indicating the definition of each stratum and this is the order in which the vector of first-stage sample sizes or prevalences must be provided when calling any of the optimal sampling functions described above. This function will be included in our illustrations of the different software tools below.

4.8.1 Illustrative data sets

Two illustrative data sets are provided with the optimal package. The first of these is a pilot data set taken from the 8096 patients in the CASS study (Vliestra *et al.*, 1980) who underwent bypass surgery. The breakdown by sex and operative mortality of these 8096 patients was given in Table 4.1. Our illustration assumes that only this 'first-stage' information is available and that one wishes to ascertain the ages of a second-stage subsample of subjects in order to estimate the age-adjusted effect of sex on mortality. We manufactured a simple pilot data set by selecting 25 observations at random from each of the strata formed by the different levels of mortality and sex, so that the pilot data consists of 200 observations on the three variables, mortality sex and age. This is the same pilot data set used in Reilly (1996).

The second illustrative data set which is also used in Reilly (1996) is taken from the same source and reflects the set-up discussed by Cain and Breslow (1988). In addition to mortality and sex, we now have a three-level categorical weight variable available at the first stage, but the true weight is to be ascertained, together with five other risk factors (angina, congestive heart failure, left ventricular end diastolic blood pressure, urgency of surgery, and age) at the second stage. The numbers of subjects in each of the first-stage strata were given in Table 4.2. The pilot data set was manufactured by taking all subjects from the smallest stratum and randomly sampling 10 observations from each of the other strata, so that the sample consists of 118 observations.

4.9 Appendix 3: Using the Optimal Package in R

For users of R, the optimal functions are available as a package named *twostage* from the following sites:

- http://www.r-project.org
- http://www.meb.ki.se/~marrei/software/

The zip file contains a 'README.packages' file where one can find the procedures on how to install the package. Once the package has been installed the functions can be used by issuing the command '*library(twostage)*' in the R session window. The command '*help(package = twostage)*' will open a window where more details are given.

For our illustration below, we will use only the 'budget' function and so we provide the syntax and features here. The other functions are similar, and interested readers who download the package will find all details provided in the help files and user documentation.

4.9.1 Syntax and features of optimal sampling command 'budget' in R

budget	**Optimal sampling design for 2-stage studies with fixed budget**

Usage:

```
budget     (x=x,y=y,z=z,prev=prev,factor=NULL,var=NULL,b=b,
           c1= c1,c2=c2)
```

Arguments:

REQUIRED ARGUMENTS

y	: response variable (binary 0-1)
x	: matrix of predictor variables
z	: matrix of the surrogate or auxiliary variables (can be more than one column)
prev	: the prevalence of each (y,z) stratum, where (y,z) is the different levels of y and z
var	: The name of the predictor variable whose coefficient is to be optimised. If this is a factor variable, see **DETAILS** at the end of this section.
b	: the total budget available
c1	: the cost per first stage observation
c2	: the cost per second stage observation

OPTIONAL ARGUMENTS

factor	: the names of any factor variables in the predictor matrix

Value:

The following lists will be returned:

n	: the optimal number of observations (first stage sample size)

```
design    : a list consisting of the following items:
ylevel    : the different levels of the response variable
zlevel    : the different levels of first stage covariates z.
prev      : the prevalence of each (ylevel,zlevel) stratum
n2        : the sample size of pilot observations for each (yle-
            vel,zlevel) stratum
prop      : optimal 2nd stage sampling proportion for each (yle-
            vel,zlevel) stratum
samp.2nd : optimal 2nd stage sample size for each (ylevel,zlevel)
            stratum
se        : the standard error of estimates achieved by the opti-
            mal design
```

4.9.2 Example

We give an example using the second pilot sample from the CASS data described above. The data are in the cass2 matrix, which can be loaded using data(cass2) and a description of the data set can be seen using help(cass2). Our model is:

$$\log\frac{P(Y_i=1)}{1-P(Y_i=1)} = \beta_0 + \beta_1 SEX_i + \beta_2 weight_i + \beta_3 age_i + \beta_4 CHF_i + \beta_5 ANGINA_i + \beta_6 LVE_i + \beta_7 Surgery_i$$

and the response variable is operative mortality.

Following the same scenario as Cain and Breslow (1988) we use sex and categorical weight as auxiliary variables. Given an available budget of £10 000, a first-stage cost of £1/unit and second-stage cost of £0.5/unit, we illustrate here the steps that calculate the sampling strategy to optimise the precision of the coefficient for surgery (**surg**), and the exact output obtained.

```
data(cass2)

 y_cass2[,1]         #response variable
 z_cass2[,c(2,3)]    #auxiliary variable
 x_cass2[,c(2,4:9)] #predictor variables

# run CODING function to see in which order we should enter pre-
valences

 coding(x=x,y=y,z=z)

[1] ''For calls requiring n1 or prev as input, use the following
order''
```

```
      ylevel sex wtcat new.z n2
 [1,]      0   0     1     1 10
 [2,]      0   1     1     2 10
 [3,]      0   0     2     3 10
 [4,]      0   1     2     4 10
 [5,]      0   0     3     5 10
 [6,]      0   1     3     6 10
 [7,]      1   0     1     1  8
 [8,]      1   1     1     2 10
 [9,]      1   0     2     3 10
[10,]      1   1     2     4 10
[11,]      1   0     3     5 10
[12,]      1   1     3     6 10
```

```
# supplying the prevalence (from Table 5, Reilly 1996)
prev_c(0.0197823937,0.0544015826,0.1339020772,0.0503214639,
0.6698813056,0.0467359050,0.0009891197,0.0022255193,
0.0040801187,0.0032146390, 0.0127349159,0.0017309594)

# optimise surg coefficient

budget(x=x,y=y,z=z,var="surg",prev=prev,b=10000,c1=1,
c2=0.5)

[1] ''please run coding function to see the order in which you''
[1] ''must supply the first-stage sample sizes or prevalences''
[1] ''Type ?coding for details!''
[1] ''For calls requiring n1 or prev as input, use the following
order''

      ylevel sex wtcat new.z n2
 [1,]      0   0     1     1 10
 [2,]      0   1     1     2 10
 [3,]      0   0     2     3 10
```

```
 [4,]   0    1    2        4 10
 [5,]   0    0    3        5 10
 [6,]   0    1    3        6 10
 [7,]   1    0    1        1  8
 [8,]   1    1    1        2 10
 [9,]   1    0    2        3 10
[10,]   1    1    2        4 10
[11,]   1    0    3        5 10
[12,]   1    1    3        6 10
[1] ''Check sample sizes/prevalences''

$n

[1] 8752

$design

      ylevel zlevel            prev  n2   prop  samp.2nd

 [1,]      0      1 0.0197823937  10 0.6181      107
 [2,]      0      2 0.0544015826  10 1.0000      476
 [3,]      0      3 0.1339020772  10 0.5107      598
 [4,]      0      4 0.0503214639  10 0.1465       65
 [5,]      0      5 0.6698813056  10 0.1061      622
 [6,]      0      6 0.0467359050  10 1.0000      409
 [7,]      1      1 0.0009891197   8 1.0000        9
 [8,]      1      2 0.0022255193  10 1.0000       19
 [9,]      1      3 0.0040801187  10 1.0000       36
[10,]      1      4 0.0032146390  10 1.0000       28
[11,]      1      5 0.0127349159  10 1.0000      111
[12,]      1      6 0.0017309594  10 1.0000       15

$se
                     [,1]
(Intercept)   1.181028836
sex           0.219738725
weight        0.006671856
```

```
age                 0.014483114
angina              0.241016357
chf                 0.074636552
lve                 0.009852953
surg                0.175539553
```

If we examine the recommended optimal sampling fractions in the context of the original data, we see that the optimal strategy is to sample all cases, all female controls in the lowest and highest category of weight, and varying fractions of the other controls. Hence where sampling costs differ for first- and second-stage subjects, the balanced design recommended by Cain and Breslow (1988) is not necessarily optimal.

We see from the output that the achieved standard error of the surgery coefficient is 0.1755. If our objective had been to achieve this precision at minimum cost, then the *precision* function would have arrived at the same design as above and the same total cost of £10 000 (within rounding error).

4.10 Appendix 4: Using the Optimal Package in S-Plus

Our *optimal* package has been tested under S-PLUS 4 for windows, so some modifications may be needed for other versions. Users who have S-PLUS installed on their computer can download our program from the following sites:

- http://lib.stat.cmu.edu/DOS/S/ [STATLIB website]
- http://www.meb.ki.se/~marrei/software/

The zip file contains a README file with the instructions on how to install the package. Once the package has been installed the functions are made available by issuing the command *'library(optimal)'* in the S-PLUS session window. Since the syntax and features of the package are so similar in Splus and in R, we refer readers to the previous section for a description and illustrative example.

4.11 Appendix 5: Using the Optimal Package in STATA

We have written the *optimal* package in STATA version 6, and it runs in STATA version 6 or later. The package can be installed directly from the STATA website. However, since we submitted the program to the STB we have improved the calling syntax and the output format and the most recent version is available at http://www.meb.ki.se/~marrei/software/

Note that the program provided at this website differs slightly from the STB version. The most noticeable change is that the functions now call the name of the variable to be optimised instead of the position of the variable in the predictor matrix. Since the syntax and features section below are written based on the version in our website, there may be some minor differences from the STB help files.

When the *optimal* package is installed one can test if all the components are installed by typing help optfixn or help optbud or help optprec, and trying some of the examples.

Again we use the fixed budget optimisation (function named **optbud**) to demonstrate the features. The other two functions (**optfixn** and **optprec**) are very similar and will present no difficulty to the user.

4.11.1 Syntax and features of 'optbud' function in STATA

optbud **Optimal sampling design for 2-stage studies with fixed budget**

Command line:

```
optbud depvar [indepvars] [if exp] [in range] [, first[varlist]
prev[vecname] b(#) c1(#) c2(#) optvar(varname) coding(#)
```

Options:

first[varlist]	specifies the first stage variables
prev[vecname]	vector of prevalences for each stratum formed by different levels of dependent variable and first stage covariates.
b(#)	available budget
c1(#)	cost per study subject at the first stage
c2(#)	cost per study subject at the second stage
optvar(varname)	the covariate whose variance estimate is to be minimised (i.e. optimised). If the covariate is a factor variable you need to specify the level whose coefficient is to be optimised (see ANALYSIS WITH CATEGORICAL VARIABLES).
coding(#)	a logical flag: default of 0 (FALSE) means that prior to calling the optbud function you have run the ''coding'' function (help coding for details) to create the vector grp_yz, containing the distinct groups (strata) formed by the different levels of response (Y) and first stage covariates (Z). If you have not run ''coding'' and you call the "optbud" function with coding=1, the grp_yz vector will be created within the optbud function, but

it is imperative that the vector [vecname] is provided to optbud in the correct order! For this reason, we strongly suggest that any call to optbud is preceded by a call to coding.

4.11.2 Analysis with categorical variables

When we have categorical predictor variables we usually fit separate coefficients for each category. The **xi** command prefix is a standard STATA command that can be used with **optfixn**, **optbud** and **optprec** to accommodate this need. STATA will create some new variables with names I'varname'_'level'. For example variable Isex_1 is a categorical variable for variable SEX with level = 1. If you want to optimise the variance estimates for this variable you should set optvar(Isex_1) in the command syntax as follows:

```
xi : optfixn mort i.sex age, first(sex) n1(fstsamp) n2(1000) optvar(Isex_1)
```

4.11.3 Illustrative example

We illustrate the same example from CASS as used above for R. We assume, as before, that we have a £10 000 budget and the first- and second-stage costs per observation are £1 and £ 0.5 respectively. The first-stage variables are sex and categorical weight. Again we will suppose that we would like to optimise the precision of the urgency of the surgery (**surg**) coefficient. The following steps will run the analysis in STATA, where the lines preceded by ** are explanatory comments and not part of the STATA session:

```
use wtpilot

coding mort sex wtcat


** enter the prevalences in the order suggested by coding function

** NOTE: the transpose operator ′ is essential
matrix prev=(0.02,.134,.670,.054,.05,.047,.001,.004,.013,
.002,.003,.002)′

** optimise the surg coefficient

optbud mort sex-surg,first(sex wtcat) prev(prev) optvar(surg)
b(10000) c1(1) c2(0.5)
```

OUTPUT:

```
the second stage sample sizes

--------+--------
group           |
(mort sex       |
wtcat)          |        Freq.
--------+--------
            1   |           10
            2   |           10
            3   |           10
            4   |           10
            5   |           10
            6   |           10
            7   |            8
            8   |           10
            9   |           10
           10   |           10
           11   |           10
           12   |           10
--------+--------

please check the sample sizes!

grp_yz   mort      sex   wtcat   grp_z    prev   n2_pilot
     1      0        0       1       1     .02         10
     2      0        0       2       2    .134         10
     3      0        0       3       3     .67         10
     4      0        1       1       4    .054         10
     5      0        1       2       5     .05         10
     6      0        1       3       6    .047         10
     7      1        0       1       1    .001          8
     8      1        0       2       2    .004         10
```

```
 9    1    0    3    3    .013    10
10    1    1    1    4    .002    10
11    1    1    2    5    .003    10
12    1    1    3    6    .002    10

the optimal sampling fraction (sample size) for grp_yz 1 = .604 (106)
the optimal sampling fraction (sample size) for grp_yz 2 = .504 (591)
the optimal sampling fraction (sample size) for grp_yz 3 = .106 (624)
the optimal sampling fraction (sample size) for grp_yz 4 = 1 (473)
the optimal sampling fraction (sample size) for grp_yz 5 = .146 (64)
the optimal sampling fraction (sample size) for grp_yz 6 = 1 (412)
the optimal sampling fraction (sample size) for grp_yz 7 = 1 (9)
the optimal sampling fraction (sample size) for grp_yz 8 = 1 (35)
the optimal sampling fraction (sample size) for grp_yz 9 = 1 (114)
the optimal sampling fraction (sample size) for grp_yz 10 = 1 (18)
the optimal sampling fraction (sample size) for grp_yz 11 = 1 (26)
the optimal sampling fraction (sample size) for grp_yz 12 = 1 (18)
the optimal number of obs = 8756
the minimum variance for surg: .0311946
total budget spent: 10001
```

Note that the optimal study size and second-stage sampling fractions are slightly different from those obtained using R and S-PLUS, but this is simply because we rounded the prevalence vector. We see from the output that the minimum variance achieved for the coefficient of interest (surg) is 0.0311946. If our objective had been to achieve this precision at minimum cost, then the **optprec** function would have arrived at the same design as above and the same total cost of £10 000.

References

Breslow, N. E. and Cain, K. C. (1988). Logistic regression for two-stage case-control data, *Biometrika*, **75**: 11–20.

Breslow, N. E. and Holubkov R. (1997). Maximum likelihood estimation of logistic regression parameters under two-phase, outcome-dependent sampling. *Journal of the Royal Statistical Society B*, **59**: 447–461.

Buonaccorsi, J. P. (1990). Double sampling for exact values in some multivariate measurement error problems, *Journal of the American Statistical Association*, **85**: 1075–1082.

Cain, K. C. and Breslow, N. E. (1988). Logistic regression analysis and efficient design for two-stage studies, *American Journal of Epidemiology*, **128**(6): 1198–1206.

Chambers, J. M. (1998). *Programming with Data*, Springer, New York (http://cm.bell-labs.com/cm/ms/departments/sia/Sbook/).

Cochran, W. G. (1963). *Sampling Techniques*, John Wiley & sons, Ltd, New York.

Dempster, A. P., Laird, N. M. and Rubin, D. B. (1977). Maximum likelihood from incomplete data via the EM algorithm. *Journal of the Royal Statistical Society B*, **39**: 1–38.

Flanders, W. D. and Greenland, S. (1991). Analytic methods for two-stage case-control studies and other stratified designs, *Statistics in Medicine*, **10**: 739–747.

Nam, J. (1973). Optimum sample sizes for the comparison of the control and treatment, *Biometrics*: **29**, 101–108.

Prentice, R. L. and Pyke, R. (1979). Logistic disease incidence models and case-control studies, *Biometrics*, **66**(3): 403–411.

Reilly, M. (1996). Optimal sampling strategies for two-stage studies, *American Journal of Epidemiology*, **143**(1): 92–100.

Reilly, M. and Pepe, M. S. (1995). A Meanscore method for missing and auxiliary covariate data in regression models, *Biometrika*, **82**(2): 299–314.

Reilly, M. and Salim, A. (2001). http://www.mep.ki.se/~marrei/software/

R Project for Statistical Computing. http://www.r-project.org

Rubin, D. B. (1987). *Multiple Imputation for Non-response in Surveys*, John Wiley & Sons, Ltd, New York.

S-PLUS. http://www.mathsoft.com/splus

STATA. http://www.stata.com/

Tosteson, T. D. and Ware, J. H. (1990). Designing a logistic regression study using surrogate measures for exposure and outcome, *Biometrika*, **77**: 11–21.

Venables, W. N. and Ripley, B. D. (1999). *Modern Applied Statistics with S-PLUS*, 3rd edn, Springer Verlag, Berlin and New York.

Vliestra, R. E., Frye, R. L. and Kromnal, R. A. *et al.* (1980). Risk factors and angiographic coronary artery disease: a report from the Coronary Artery Surgery Study (CASS). *Circulation*, **62**: 254–261.

5

Response-Driven Designs in Drug Development

Valerii V. Fedorov and Sergei L. Leonov

GlaxoSmithKline, 1250 So. Collegeville Road, Collegeville, PA 19426-0989, USA

5.1 Introduction

We present an overview of optimal design methods, together with some new findings that can be applied in drug development. In Section 5.2, we introduce optimal design concepts via binary dose response models. In Section 5.3, examples of continuous models are given, and the optimal design problem for a general nonlinear regression model is formulated. In Section 5.4, the classic results are extended to multi-response problems and to models with a variance function that depends on unknown parameters. In this case direct substitution of the estimated variance into the information matrix obtained from the simpler regression framework leads, in general, to non-optimal designs. We show how to properly take the information contained in the variance component into account. In Section 5.5 a model with experimental costs is introduced which is studied within the normalized design paradigm and can be applied, for example, to the analysis of clinical trials with multiple endpoints. In Section 5.6, we describe the use of adaptive designs.

Applied Optimal Designs Edited by M.P.F. Berger and W.K. Wong
 ISBN: 0-470-85697-1 (HB)

5.2 Motivating Example: Quantal Models for Dose Response

Dose response studies arise in various pharmaceutical applications and are an important counterpart of efficient drug development. Often dose response studies can be considered within the framework of binary response experiment such as success–failure, or dead–alive in toxicology studies; see Fedorov and Leonov (2001). For a quantal model, a binary variable Y is introduced which depends on dose x,

$$Y = Y(x) = \begin{cases} 1, & \text{a response is present on dose } x, \\ 0, & \text{no response,} \end{cases} \tag{5.1}$$

and the probability of response is modelled as

$$P\{Y = 1|x\} = \eta(x, \theta),$$

where $0 \leq \eta(x, \theta) \leq 1$ is a given function, and $\theta = (\theta_0, \theta_1, \ldots, \theta_{m-1})^T$ is a vector of unknown parameters. It is often assumed that

$$\eta(x, \theta) = \pi(z), \quad \text{where} \quad z = \theta^T f(x), \quad f(x) = (f_0(x), f_1(x), \ldots, f_{m-1}(x))^T, \tag{5.2}$$

and $f_i(x)$ are given functions of dose x. Here we resort to a two-parameter case:

$$z = \theta_0 + \theta_1 x, \qquad f(x) = (1, x)^T. \tag{5.3}$$

Various generalizations are straightforward.

In many applications, $\pi(z)$ is selected as a probability distribution function. However, there are cases when $\pi(z)$ is not monotonic with respect to z (for example, in some oncology studies). Among the most popular choices of functions π are:

- Logistic (logit):

$$\pi_L(z) = \mathrm{e}^z/(1 + \mathrm{e}^z). \tag{5.4}$$

- Probit: $\pi(z)$ is a standard normal cumulative distribution function,

$$\pi_P(z) = \frac{1}{\sqrt{2\pi}} \int_{-\infty}^{z} \mathrm{e}^{-u^2/2} du.$$

In practice, when properly normalized, both models lead to almost identical results, as was noted in a number of publications; see Cramer (1991, Ch. 2.3) and Finney (1971, p. 98).

For any parametric model, it is one of the main goals to obtain the most precise estimates of unknown parameters θ. So if a practitioner can select various doses, in some admissible range, and number of patients on each dose, then, given a sample size N, the question is how to allocate those doses and patients to obtain the 'best' estimates of unknown parameters. The quality of estimators is traditionally measured by their variance–covariance matrix.

5.2.1 Optimality criteria

Let $\mu(x, \theta)$ be an $(m \times m)$ information matrix of a single observation at point x,

$$\mu(x, \theta) = E\left[\frac{\partial \log L(x, \theta)}{\partial \theta} \frac{\partial \log L(x, \theta)}{\partial \theta^T}\right],$$

where θ is an m-dimensional vector of model parameters, $\theta \in \Omega \in R^m$, and $L(x, \theta)$ is a likelihood function. For model (5.2)

$$L(x, \theta) = \eta(x, \theta)^Y [1 - \eta(x, \theta)]^{1-Y},$$

where Y is introduced in (5.1).

If n_i independent observations are taken at point x_i, $\Sigma_i n_i = N$, then the information matrix of N experiments equals

$$M_N(\theta) = \sum_{i=1}^{n} n_i \mu(x_i, \theta). \tag{5.5}$$

Let $M(\xi, \theta)$ be a normalized information matrix,

$$M(\xi, \theta) = \frac{1}{N} M_N(\theta) = \sum_i w_i \mu(x_i, \theta), \quad w_i = \frac{n_i}{N},$$

where $\xi = \{x_i, w_i\}$ is a normalized (continuous, or approximate) design; weights w_i satisfy $\Sigma_i w_i = 1$. In this setting N may be viewed as a resource available to a practitioner; see Section 5.5 for a different normalization. The variance–covariance matrix of maximum likelihood estimator (MLE) $\hat{\theta}$ for large samples is approximated by (e.g. Rao, 1973, Ch. 5)

$$\mathrm{Var}(\hat{\theta}) \approx M_N^{-1}(\theta) = \frac{1}{N} M^{-1}(\xi, \theta). \tag{5.6}$$

In convex design theory, it is standard to minimize various functionals depending on the normalized variance matrix $M^{-1}(\xi, \theta)$,

$$\xi^* = \arg\min_{\xi} \Psi[M^{-1}(\xi, \theta)], \tag{5.7}$$

where Ψ is a selected functional (criterion of optimality). The optimization is performed with respect to designs ξ,

$$\xi = \{w_i, 0 \leq w_i \leq 1, \sum_i w_i = 1; x_i \in \mathbf{X}\},$$

where $\mathbf{X}$ is an admissible range of doses, and weights w_i are continuous variables. This leads to the concept of approximate optimal designs since $n_i = Nw_i$ can be non-integer. For comparatively large N, this is not a problem if one rounds n_i to the nearest integer subject to $\Sigma_i n_i = N$.

Among popular optimality criteria are (see, for example, Fedorov, 1972, Silvey, 1980 or Pukelsheim, 1993):

- $\Psi = \log|M^{-1}(\xi, \theta)|$ – D-criterion which is often called a generalized variance and is related to the volume of the confidence ellipsoid;
- $\Psi = \mathrm{tr}AM^{-1}(\xi, \theta)$–A-criterion (linear), where A is an $m \times m$ non-negative definite matrix. For example, if $A = (1/m)I_m$, where I_m is an $m \times m$ identity matrix, then this criterion reduces to the average variance of parameter estimates: $\Psi = (1/m)\Sigma_{i=0}^{m-1}\mathrm{Var}(\hat{\theta}_i)$.
- If the prediction of some given function $\phi(x, \theta)$ is of interest, then criteria like

$$\mathrm{Var}[\hat{\phi}(x_0, \theta)], \qquad \max_{x \in X_0} \mathrm{Var}[\hat{\phi}(x, \theta)], \qquad \int_{X_0} \mathrm{Var}[\hat{\phi}(x, \theta)]\mathrm{d}x$$

are widely used; here x_0 and X_0 are particular point and region of interest, respectively. Note that $\phi(x, \theta)$ may not coincide with response $\eta(x, \theta)$ which was used in the motivation example.

Optimal designs for quantal models

If function $\pi(z)$ in model (5.2) has a derivative $\pi'(z)$, then the information matrix of a single observation equals (e.g. Wu, 1988, Torsney and Musrati, 1993)

$$\mu(x, \theta) = \frac{[\pi'(z)]^2}{\pi(z)[1 - \pi(z)]} f(x) f^T(x).$$

For many functions $\pi(z)$, under model (5.3) the D-optimal designs are two-point designs, with half the measurements at each dose. For logistic model (5.4), this result was established by White (1975); see also Kalish and Rosenberger (1978), Silvey (1980, Ch. 6.6), Sitter (1992), Abdelbasit and Plackett (1983) and Minkin (1987). We remark that for a sufficiently large range of doses, D-optimal designs are uniquely defined in the z-space. Moreover, for each of the models discussed in the

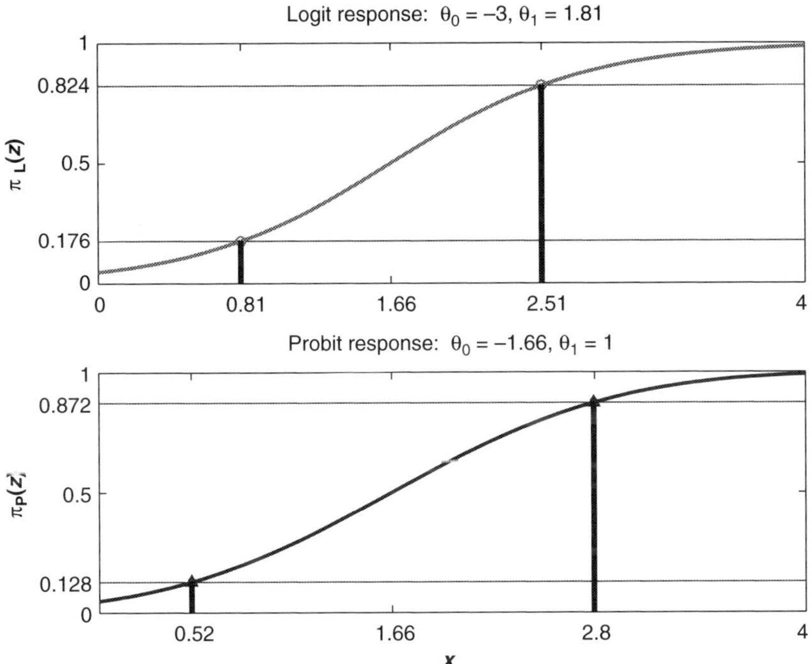

Figure 5.1 D-optimal designs for logistic and probit models, $x = (z - \theta_0)/\theta_1$. Upper panel: logistic model, $z = -3 + 1.81x$. Lower panel: probit model, $z = -1.66 + x$

beginning of this section, optimal points correspond to certain probabilities of response. For the logistic model (5.4),

$$z_{\text{opt}} \approx \pm 1.543 \quad \text{and} \quad \pi_{\text{L}}(-1.543) = 0.176, \quad \pi_{\text{L}}(1.543) = 0.824;$$

see Figure 5.1, top. Thus, if x_1 and x_2 are the two optimal doses corresponding to D-optimal design, then

$$\theta_0 + \theta_1 x_1 = -1.543, \qquad \theta_0 + \theta_1 x_2 = 1.543.$$

For the probit model,

$$z_{\text{opt}} \approx \pm 1.138, \qquad \pi_{\text{P}}(-1.138) = 0.128, \qquad \pi_{\text{P}}(1.138) = 0.872;$$

see Figure 5.1, bottom. In Figure 5.1, parameters θ of logit and probit distributions $[(-3, 1.81)^T$ and $(-1.66, 1)^T$, respectively] are chosen such that the mean and the variance of the two distributions are the same. For other examples, see Ford *et al.* (1992) and Torsney and Musrati (1993).

Note that if x_p is a dose level that causes a particular response in 100 $p\%$ of subjects, i.e. $\eta(x_p, \theta) = p$, then for models (5.2) and (5.3) the normalized variance of the MLE $\hat{x}_p$ is a special case of A-criterion (c-criterion) and can be written as

$$\Psi = \text{Var}(\hat{x}_p) = \text{tr}\, A_p M^{-1}(\xi, \theta), \qquad \text{with } A_p = \frac{f(x_p) f^T(x_p)}{\theta_1^2};$$

see Wu (1988) and Sitter (1992). For a discussion on A-optimal designs for binary models, see Sitter and Wu (1993), Zhu and Wong (2000) and Mathew and Sinha (2001). Various biomedical applications of optimal designs in quantal models are discussed in Begg and Kalish (1984), Kalish (1990) and Minkin and Kundhal (1999).

In general, optimal designs depend on the choice of function $\eta(x, \theta)$, and on parameter values θ, which is a standard situation for nonlinear parameterization. This leads to the concept of locally optimal designs, as in Chernoff (1953) and Fedorov (1972). However, as soon as the model is chosen and a preliminary parameter estimate $\tilde{\theta}$ is selected, the next steps are straightforward:

1. Draw a corresponding curve $y = \eta(x, \tilde{\theta})$ (logistic, probit, etc).
2. In the case of logistic model, draw two lines $y_1 = 0.176$ and $y_2 = 0.824$.
3. Find the points on the x-axis where those two lines intersect with the curve $y = \eta(x, \tilde{\theta})$.

Obviously, the quality of the above designs may be poor if the preliminary estimate $\tilde{\theta}$ differs significantly from the true value of unknown parameters. This problem can be tackled by using various techniques which include minimax designs (Fedorov and Hackl, 1997), Bayesian designs (Chaloner and Verdinelli, 1995) and adaptive designs (Box and Hunter, 1965); Fedorov, 1972, Ch. 4; Zacks, 1996; Flournoy *et al.*, 1998). In oncology dose toxicity studies, one of the most popular formal adaptive procedures for finding dose x_p in one-parameter models is the continual reassessment method; see O'Quigley *et al.* (1990) and O'Quigley (2001). We postpone a detailed description of adaptive designs to Section 5.6.

5.3 Continuous Models

In this section, we start with the examples of continuous models which are used in pharmaceutical applications, and then formulate optimal design problems for general nonlinear regression models. For a discussion on linear dose response models, see Wong and Lachenbruch (1996).

5.3.1 Example 3.1

The four-parameter logistic model,or $E_{\max}$ model, is used in many pharmaceutical applications including bioassays, of which ELISA (enzyme-linked immunosorbent assay) and cell-based assays are popular examples; see Finney (1976), Karpinski (1990), Hedayat *et al.* (1997) and Källén and Larsson (1999):

$$E(y|x;\theta) = \eta(x,\theta) = \theta_3 + \frac{\theta_4 - \theta_3}{1 + (x/\theta_1)^{\theta_2}}, \tag{5.8}$$

where for a cell-based assay, y is the number of cells at concentration x; parameter θ_1 is often denoted as ED_{50}; θ_2 is a slope parameter; θ_3 and θ_4 are lower and upper asymptotes.

5.3.2 Example 3.2

The compartmental model with first-order absorption is popular in pharmacokinetic applications; see Lindsey *et al.* (1999):

$$\eta(x,\theta) = \frac{D\theta_1}{\theta_3(\theta_1 - \theta_2)}(\mathrm{e}^{-\theta_2 x} - \mathrm{e}^{-\theta_1 x}),$$

where D is dose; x is time after drug administration (independent variable); $\eta(x,\theta)$ is drug concentration at time x; θ_1 and θ_2 are absorption and elimination rates respectively; θ_3 is volume of distribution.

Both examples can be put into the general framework of nonlinear regression models:

$$E(y_{ij}|x_i;\theta) = \eta(x_i,\theta), \quad i = 1,\ldots,n;\ j = 1,\ldots,n_i; \sum_{i=1}^{n} n_i = N,$$

where y_{ij} are often assumed to be independent random variables with $\mathrm{Var}(y_{ij}|x_i;\theta) = \sigma^2(x_i)$ (i.e. in dose response studies variance of response σ_i^2 may vary from dose to dose), x is a control (independent) variable which can be chosen by a practitioner: $x_i \in \mathbf{X} \subset R^k$ ($\mathbf{X}$ is the admissible range of doses and $k = 1$ in our examples) and θ is a vector of unknown parameters, $\theta \subset \Omega \subset R^m$.

Under the normality assumption for the response variable y, the variance matrix of the MLE can be approximated by (5.6), where the normalized information matrix is defined by

$$M(\xi,\theta) = \sum_{i=1}^{n} w_i \mu(x_i,\theta), \quad \text{with} \quad w_i = \frac{n_i}{N},\ \mu(x_i,\theta) = \sigma^{-2}(x_i) f(x_i,\theta) f^T(x_i,\theta). \tag{5.9}$$

$f(x,\theta)$ is a vector of partial derivatives of the response function $\eta(x,\theta)$,

$$f(x,\theta) = \left[\frac{\partial\eta(x,\theta)}{\partial\theta_1}, \frac{\partial\eta(x,\theta)}{\partial\theta_2}, \ldots, \frac{\partial\eta(x,\theta)}{\partial\theta_m}\right]^T.$$

Now, as far as optimal experimental design is concerned, the sequence of steps that follow is exactly the same as for quantal models:

1. Select a preliminary parameter estimate $\tilde{\theta}$;
2. Select a criterion of optimality $\Psi = \Psi[M^{-1}(\xi, \tilde{\theta})]$;
3. Minimize $\Psi[M^{-1}(\xi, \tilde{\theta})]$ with respect to designs ξ, i.e. construct the locally optimal design.

There exists a variety of theoretical results and efficient numerical algorithms to solve the above minimization problem; see Fedorov (1972), Silvey (1980), Atkinson and Donev (1992), Pukelsheim (1993) and Fedorov and Hackl (1997).

Sometimes the functional structure of the nonlinear model makes optimal designs independent of some unknown parameters which reduces the complexity of computational algorithms. Logistic model (5.8) provides an example of a *partially* nonlinear model where parameters θ_3, θ_4 enter the response function in a linear fashion. This, in turn, implies that D-optimal designs do not depend on values of θ_3 and θ_4; see Hill (1980).

5.4 Variance Depending on Unknown Parameters and Multi-response Models

It is often assumed that variance σ^2 depends not only on control variables x, but on unknown parameters θ as well: $\sigma_i^2 = s(x_i, \theta)$. For instance, variance may depend on the mean, as in the power model (e.g. see Finney, 1976):

$$S(x, \theta) \sim \eta^{\delta}(x, \theta), \quad \delta > 0.$$

In cell-based assay studies, since cell counts are measured, one can assume that response has Poisson distribution, and consequently $\delta = 1$. In pharmacokinetic studies, as in Example 3.2, parameter δ has often been set to 2; see Lindsey *et al.* (1999). Another possible generalization is the introduction of a model with multiple correlated responses, such as several endpoints in clinical trials, or simultaneous investigation of efficacy and toxicity in dose response studies; see Heise and Myers (1996) and Fan and Chaloner (2001).

Let the observed responses be normally distributed $(k \times 1)$ vectors,

$$E[y|x] = \eta(x, \theta), \qquad \text{Var}[y|x] = S(x, \theta) \geq S_0 > 0. \tag{5.10}$$

where $\eta(x,\theta) = (\eta_1(x,\theta), \ldots, \eta_k(x,\theta))^T$, and $S(x,\theta)$ is a $(k \times k)$ matrix. Then the Fisher information matrix of a single observation is given by

$$\mu(x,\theta) = [\mu_{\alpha\beta}(x,\theta)]_{\alpha,\beta=1}^{m} = \mu^R(x,\theta) + \mu^V(x,\theta),$$

$$\mu_{\alpha\beta}^R(x,\theta) = \frac{\partial \eta^T(x,\theta)}{\partial \theta_\alpha} S^{-1}(x,\theta) \frac{\partial \eta(x,\theta)}{\partial \theta_\beta}; \tag{5.11}$$

$$\mu_{\alpha\beta}^V(x,\theta) = \frac{1}{2} \mathrm{tr} \left[S^{-1}(x,\theta) \frac{\partial S(x,\theta)}{\partial \theta_\alpha} S^{-1}(x,\theta) \frac{\partial S(x,\theta)}{\partial \theta_\beta} \right],$$

cf. Magnus and Neudecker (1988, Ch. 6) or Muirhead (1982, Ch. 1). In a single-response case, the second term on the right-hand side of (5.11) reduces to

$$\mu^V(x,\theta) = \frac{1}{2S^2(x,\theta)} \frac{\partial S(x,\theta)}{\partial \theta} \frac{\partial S(x,\theta)}{\partial \theta^T} = \frac{1}{2} \frac{\partial \ln S(x,\theta)}{\partial \theta} \frac{\partial \ln S(x,\theta)}{\partial \theta^T}.$$

The latter presentation is helpful for the power model of variance, see (5.22) below. In general, the dimension and structure of y, η, and S can vary for different x; see examples in Section 5.5.

Before proceeding to the discussion of the optimal design issues we shall make a few remarks on the estimation algorithms. If the variance function S does not depend on unknown parameters, then the generalized least squares estimator (LSE) $\tilde{\theta}_N^{(1)}$,

$$\tilde{\theta}_N^{(1)} = \arg\min_\theta \sum_{i=1}^{N} [y_i - \eta(x_i,\theta)]^T S^{-1}(x_i) [y_i - \eta(x_i,\theta)], \tag{5.12}$$

possesses optimal properties and coincides with the MLE in the case of Gaussian errors. However, when S depends on θ, then the estimator $\tilde{\theta}_N^{(2)}$,

$$\tilde{\theta}_N^{(2)} = \arg\min_\theta \sum_{i=1}^{N} [y_i - \eta(x_i,\theta)]^T S^{-1}(x_i,\theta) [y_i - \eta(x_i,\theta)], \tag{5.13}$$

is, in general, not consistent; see Fedorov (1974), Malyutov (1982), Beal and Sheiner (1988) or Vonesh and Chinchilli (1997, Chapter 9.2.4). Instead, modifications of the generalized LSE must be used for the consistent estimation. For example, an iteratively reweighted least squares (IRLS) estimator $\hat{\theta}^{(1)}$,

$$\hat{\theta}_N^{(1)} = \lim_{t\to\infty} \theta_t, \tag{5.14}$$

$$\theta_t = \arg\min_\theta \sum_{i=1}^{N} [y_i - \eta(x_i,\theta)]^T S^{-1}(x_i,\theta_{t-1}) [y_i - \eta(x_i,\theta)],$$

is strongly consistent with the asymptotic variance–covariance matrix

$$D_1 = \text{Var}(\hat{\theta}_N^{(1)}) \simeq \frac{1}{N}\left[\sum_i w_i \frac{\partial \eta(x_i,\theta)}{\partial \theta} S^{-1}(x_i,\theta) \frac{\partial \eta(x_i,\theta)}{\partial \theta^T}\right]^{-1}\Bigg|_{\theta=\tilde{\theta}_N}. \quad (5.15)$$

However, the IRLS estimator does not take the information contained in the variance component into account and, thus, is inferior to the MLE in terms of non-negative matrix ordering.

Fedorov *et al.* (2001) and Fedorov and Leonov (2004) introduced a combined IRLS estimator (CIRLS) which included not only the weighted squared deviations of predicted responses from observations, as in (5.14), but also the squared deviations of the predicted dispersion matrices from observed residual matrices,

$$\hat{\theta}_N^{(2)} = \lim_{t\to\infty} \theta_t, \quad \text{where } \theta_t = \arg\min_\theta R_N(\theta, \theta_{t-1}), \quad (5.16)$$

$$R_N(\theta,\theta') = \sum_{i=1}^N [y_i - \eta(x_i,\theta)]^T S^{-1}(x_i,\theta')[y_i - \eta(x_i,\theta)]$$
$$+\frac{1}{2}\sum_{i=1}^N \text{tr}\left[\{[y_i - \eta(x_i,\theta')][y_i - \eta(x_i,\theta')]^T\right.$$
$$\left. - \Delta\eta(x_i;\theta,\theta') - S(x_i,\theta)\}S^{-1}(x_i,\theta')\right]^2 \quad (5.17)$$

and

$$\Delta\eta(x_i;\theta,\theta') = [\eta(x_i,\theta) - \eta(x_i,\theta')][\eta(x_i,\theta) - \eta(x_i,\theta')]^T.$$

It was shown that under mild regularity conditions, the CIRLS is asymptotically equivalent to the MLE, and therefore its variance–covariance matrix D_2 is 'better' than that of the IRLS, $D_2 \leq D_1$.

Recall now that the information matrix $M(\xi,\theta)$ in (5.9) was defined in the case of the variance depending on control variables x and not depending on θ. When the variance also depends on θ, as in the power model, it was common to substitute $S^{-1}(x_i,\theta)$ for $\sigma^{-2}(x_i)$ in (5.9) and to use

$$\tilde{M}(\xi,\theta) = \sum_{i=1}^n w_i S^{-1}(x_i,\theta) f(x_i,\theta) f^T(x_i,\theta),$$

for future optimization and search for optimal designs; see Bezeau and Endrenyi (1986). As remarked in (5.15), matrix $[N\tilde{M}(\xi,\theta)]^{-1}$ approximates the variance of the IRLS. If the second term on the right-hand side of (5.11) is small compared to the first one, then designs obtained by optimizing $\tilde{M}(\xi,\theta)$ will be close to the optimal ones. If the second term cannot be neglected, then the naive substitution of the estimated variance into (5.9) will lead, in general, to non-optimal designs since the matrix $\tilde{M}(\xi, \theta)$ does not take the information contained in the variance component into account.

The interesting situation occurs for a single response, $k = 1$, when the variance is proportional to the square of the mean, i.e. $S(x, \theta) = a\eta^2(x, \theta)$. If a is a given constant, then the simple calculations show that the Fisher information matrix of a single normally distributed observation is

$$\mu(x, \theta) = \frac{f(x, \theta) f^T(x, \theta)}{\eta^2 (x, \theta)} \left(\frac{1}{a} + 2 \right),$$

which differs from $S^{-1}(x, \theta) f(x, \theta) f^T(x, \theta)$ only by a constant multiplier. Thus in this particular case the 'naive' substitution leads to the same optimal designs, even though the variance–covariance matrix of the LSE is inferior to that of the MLE.

Many theoretical results and numerical algorithms of optimal experimental design theory rely on the generalized equivalence theorem; see Kiefer and Wolfowitz (1960), Fedorov (1972) and White (1973). In particular, if $\sigma_i^2 = \sigma^2(x_i)$ and consequently the normalized information matrix is defined by (5.9), then the selection of D-optimal design is governed by a so-called sensitivity function

$$\psi(x, \xi, \theta) = \frac{\mathrm{Var}[\tilde{\eta}(x, \theta)]}{\sigma^2(x)} = \frac{f^T(x, \theta) M^{-1}(\xi, \theta) f(x, \theta)}{\sigma^2(x)},$$

which plays a major role in convex design theory. The function $\psi(x, \xi, \theta)$ indicates points where taking an observation provides most information with respect to the chosen optimality criterion; see Fedorov and Hackl (1997, Ch. 2.4). For instance, in dose response studies on each step of the iterative procedure a dose $x*$, which maximizes the sensitivity function $\psi(x, \xi, \theta)$ over the admissible range of doses, has to be added to the current design ξ.

Using classic optimal design techniques, analogues of the equivalence theorem can be established for models with multiple responses and variance that depends on unknown parameters; see Fedorov *et al.* (2001, 2002). A necessary and sufficient condition for a design ξ^* to be locally optimal is the inequality

$$\psi(x, \xi, \theta) = \mathrm{tr}[\mu(x, \theta) M^{-1}(\xi^*, \theta)] \leq m, \quad m = \dim \theta, \tag{5.18}$$

in the case of the D-criterion, and

$$\psi(x, \xi, \theta) = \mathrm{tr}[\mu(x, \theta) M^{-1}(\xi^*, \theta) A M^{-1}(\xi^*, \theta)] \leq \mathrm{tr}\, [A M^{-1}(\xi^*, \theta)] \tag{5.19}$$

in the case of the linear criterion. Equality in (5.18), or (5.19), is attained at the support points of the optimal designs ξ^*.

For model (5.10) with a single response, $k = 1$, the sensitivity function of the D-criterion can be written as

$$\psi(x, \xi, \theta) = \frac{d_1(x, \xi, \theta)}{S(x, \theta)} + \frac{d_2(x, \xi, \theta)}{2S^2(x, \theta)}, \tag{5.20}$$

where

$$\begin{aligned} d_1(x, \xi, \theta) &= \frac{\partial \eta(x, \theta)}{\partial \theta^T} M^{-1}(\xi, \theta) \frac{\partial \eta(x, \theta)}{\partial \theta}, \\ d_2(x, \xi, \theta) &= \frac{\partial S(x, \theta)}{\partial \theta^T} M^{-1}(\xi, \theta) \frac{\partial S(x, \theta)}{\partial \theta}; \end{aligned} \tag{5.21}$$

see Downing *et al.* (2001). The scalar case was extensively discussed by Atkinson and Cook (1995) for various partitioning schemes of parameter vector θ, including separate and overlapping parameters in the variance function.

5.4.1 Example 4.1

To illustrate D-optimal designs for the logistic model (5.8) with the power model of variance,

$$S(x, \theta) = \delta_1 \eta^{\delta_2}(x, \gamma), \tag{5.22}$$

where $\gamma = (\gamma_1, \gamma_2, \gamma_3, \gamma_4)^T$, $\theta = (\gamma_1, \gamma_2, \gamma_3, \gamma_4, \delta_1, \delta_2)^T$, we use the data from a study on the ability of a compound to inhibit the proliferation of bone marrow erythroleukemia cells in a cell-based assay; see Downing *et al.* (2001). The vector of unknown parameters was estimated by fitting the data collected from 96-well plate assay, with 6 repeated measurements at each of the 10 concentrations, $\{500, 250, 125, \ldots, 0.98\ \text{ng/ml}\}$, which represents a twofold serial dilution design with $N = 60$ and weights $w_i = 1/10$, i.e.

$$\xi_2 = \{w_i = 1/10, x_i = 500/2^i, \quad i = 0, 1, \ldots, 9\}.$$

Thus the design region was set to $\mathbf{X} = [-0.02, 6.21]$ on the log scale. Figure 5.2 presents the locally optimal design and the variance of prediction for the model defined by (5.8) and (5.22), where $\theta = (75.2, 1.34, 616, 1646, 0.33, 0.90)^T$.

The upper and lower right panels of Figure 5.2 present the normalized variance of prediction $\psi(x, \xi, \theta)$ defined in (5.20), for the twofold serial dilution design ξ_2 (upper right) and D-optimal design ξ^* (lower right). The solid lines show the function $\psi(x, \xi, \theta)$ while the dashed and dotted lines display the first and second terms on the right-hand side of (5.20), respectively. The unnormalized variance of

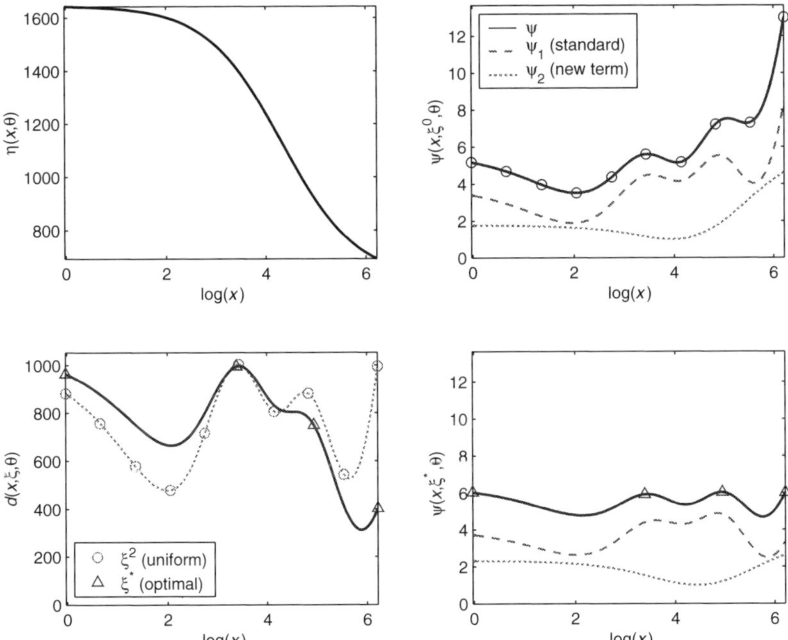

Figure 5.2 Plots for Example 4.1. Upper left panel: response fuction $\eta(x,\theta)$. Upper right panel: normalized variance $\psi(x,\xi,\theta)$ for twofold serial dilution design ξ_2. Lower right: normalized variance $\psi(x,\xi,\theta)$ for optimal design ξ^*. Lower left: unnormalized variance $d_1(x,\xi,\theta)$ (triangles-optimal design, circles-serial dilution). Function $\psi(x,\xi,\theta)$ is defined in (5.20), function $d_1(x,\xi,\theta)$ is defined in (5.21). Dashed line in the right panels is $\psi_1 = d_1(x,\xi,\theta)/S(x,\theta)$, see (5.20). Dotted line in the right panels is $\psi_2 = d_2(x,\xi,\theta)/[2S^2(x,\theta)]$

prediction $d_1(x,\xi,\theta)$ defined in (5.21), is given on the lower left panel. Note that the weights of the support points are not equal. In our example $p = \{0.28, 0.22, 0.22, 0.28\}$.

It is worthy to note that the optimal design in our example is supported at just four points, which is less than the number of estimated parameters $m = 6$. This happens because the information matrix $\mu(x,\theta)$ of a single observation at point x is defined as the sum of the two terms; see (5.11). So even in the case of a single response, $k = 1$, the rank of the information matrix $\mu(x,\theta)$ may be greater than one. On the other hand, in the single-response regression models where the variance function does not depend on unknown parameters, the second term on the right-hand side of (5.11) is missing. Therefore, in that case for the information matrix of any design $\tilde{\xi}$ to be non-degenerate, it is necessary to have at least m support points in the design.

5.4.2 Optimal designs as a reference point

It is worthwhile to remark that the direct use of locally optimal designs is not always realistic. As mentioned in Section 5.2, if the logistic model (5.4) is used in the dose response experiment, then optimal design theory suggests putting half of the patients on each of the two doses, $ED_{17.6}$ and $ED_{82.4}$. Note that dose $ED_{17.6}$ may be too low and not efficacious while dose $ED_{82.4}$ may be quite toxic and not well tolerated by many patients. Recall also that, though the optimal designs are unique in z-space, one needs preliminary estimates of parameters θ to find optimal doses in the original x-space. Hence, appealing to D-optimality of doses in a real-life study is very likely not to be a strong argument.

Nevertheless, the importance of locally optimal designs is in that they provide a reference point for the sensitivity analysis and allow us to calculate relative efficiency of alternative designs.

For instance, in the study described in Example 4.1, the scientists were traditionally using the twofold serial dilution design ξ_2, as shown in Figure 5.2. In practice, serial dilution designs can be easily implemented with an arbitrary dilution factor. So the main practical goal was not in constructing optimal designs per se, but to answer the following questions:

- How far is the twofold dilution design from the optimal one?
- Are there 'good' serial dilution designs ξ_a with factors a other than 2?

Since there existed a special interest in the estimation of parameter $\theta_1 = ED_{50}$, we compared relative efficiency of various serial dilution designs with respect to D-optimality and ED_{50}-optimality:

$$\mathrm{Eff}_D(\xi_a) = \left[\frac{|M(\xi_a,\theta)|}{|M(\xi_*,\theta)|}\right]^{1/m}, \qquad \mathrm{Eff}_{ED_{50}}(\xi_a) = \frac{[M^{-1}(\xi^*,\theta)]_{11}}{[M^{-1}(\xi_a,\theta)]_{11}},$$

where by $[A]_{11}$ we denote the first diagonal element of matrix A. Serial dilution designs are invariant with respect to model parameters, so they serve as a robust alternative to optimal designs. If it could be shown that the efficiency of serial dilutions was close to one, it would provide a solid support for the use of such designs in practice.

We fitted the data for over 20 different assays with twofold serial dilution design ξ_2 (concentrations from 500 to 0.98 ng/ml), constructed locally optimal designs, and calculated the relative efficiency of dilution designs. The results could be split into two groups:

(G1) In most cases design ξ_2 covered the whole range of the logistic curve, and the loss of efficiency was within 10–15%. This means that 10–15% more observations are required to get the same efficiency as in the optimal design.

(G2) However, in some examples either the left or right end of the curve was not properly covered (as the right end in Example 4.1, see Figure 5.2, top left), and the efficiency of designs ξ_2 was less than 0.6, both with respect to D-criterion and variance of $\hat{\theta}_1$.

Therefore we suggested trying designs $\xi_{2.5}$ or ξ_3, to start with higher concentrations, and to have 10 different concentrations on the same 96-well plate as before, to preserve the logistics of the assays. For example, one of the options with design $\xi_{2.5}$ is to start from 2000 ng/ml, and go down to $800, 320, \ldots, 0.52\,\text{ng/ml}$.

The comparison of designs $\xi_{2.5}$ with the original twofold designs showed that for the cases in group (G1), designs $\xi_{2.5}$ performed similarly to designs ξ_2, with essentially the same efficiency. On the other hand, for cases in group (G2) designs $\xi_{2.5}$ were significantly superior, with ratio $\text{Eff}(\xi_{2.5})/\text{Eff}(\xi_2)$ often greater than 1.5 for both optimality criteria. Moreover, designs $\xi_{2.5}$ were relatively close to the optimal ones, with the loss of efficiency not more than 15%. Thus, it was suggested switching to 2.5-fold serial dilution designs for future studies.

For more discussion on robustness issues, see Sitter (1992).

5.4.3 Remark 4.1

In the study discussed above the estimated values of slope parameter θ_2 were always within the interval [1.2–1.5], so the logistic curve was not very steep. This was one of the reasons for the efficiency of 2.5- or 3-fold designs. When parameter θ_2 is significantly bigger than 1 (say 4 and above), then the quality of designs with large dilution factors deteriorates, especially with respect to the estimation of θ_1. It happens because when the logistic curve is rather steep, then designs ξ_a with larger a will more likely miss points in the middle of the curve, near values $x = ED_{50}$, which in turn will greatly reduce the precision of the parameter estimation.

5.5 Optimal Designs with Cost Constraints

Traditionally when normalized designs are discussed, the normalization factor is equal to the number of experiments N; see Section 5.2. Now let each measurement at point x_i be associated with a cost $c(x_i)$, and there exists a restriction on the total cost,

$$\sum_{i=1}^{n} n_i c(x_i) \leq C. \tag{5.23}$$

In this case it is quite natural to normalize the information matrix by the total cost C and introduce

$$M_C(\xi,\theta) = \frac{M_N(\theta)}{C} = \sum_i w_i \tilde{\mu}(x_i,\theta), \quad \text{with } w_i = \frac{n_i c(x_i)}{C}, \ \tilde{\mu}(x,\theta) = \frac{\mu(x,\theta)}{c(x)}. \tag{5.24}$$

Note that the considered case should not be confused with the case when additionally to (5.24) one also imposes that $\Sigma_i n_i \leq N$. The latter design problem is more complicated and must be addressed as discussed in Cook and Fedorov (1995).

Once cost function $c(x)$ is defined, one can introduce a cost-based design $\xi_C = \{w_i, x_i\}$ and use the well-elaborated techniques of constructing continuous designs for various criteria of optimality

$$\xi^* = \arg\min_{\xi} \Psi[M_C^{-1}(\xi,\theta)], \quad \text{where } \ \xi = \{w_i, x_i\}.$$

As usual, to obtain frequencies n_i, values $\tilde{n}_i = w_i C/c(x_i)$ have to be rounded to the nearest integers n_i subject to $\Sigma_i n_i c(x_i) \leq C$.

We think that the introduction of cost constraints and the alternative normalization (5.24) makes the construction of optimal designs for models with multiple responses more sound. It allows for a meaningful comparison of 'points' x with distinct number of responses.

To illustrate the potential of this approach, we start this section with an example of the two-dimensional response function $\eta(x,\theta) = [\eta_1(x,\theta), \eta_2(x,\theta)]^T$, with the variance matrix

$$S(x,\theta) = \begin{pmatrix} S_{11}(x,\theta) & S_{12}(x,\theta) \\ S_{12}(x,\theta) & S_{22}(x,\theta) \end{pmatrix}.$$

Let a single measurement of function $\eta_i(x,\theta)$ cost $c_i(x)$, $i = 1, 2$. Additionally, we impose a cost $c_v(x)$ on any single or pair of measurements. The rationale behind this model comes from considering a hypothetical visit of a patient to the clinic to participate in a clinical trial. It is assumed that each visit costs $c_v(x)$, where x denotes a patient (or more appropriately, some patient's characteristics). There are three options for each patient:

1. Take test t_1 which by itself costs $c_1(x)$; the total cost of this option is $C_1(x) = c_v(x) + c_1(x)$.
2. Take test t_2 which costs $c_2(x)$; the total cost is $C_2(x) = c_v(x) + c_2(x)$.
3. Take both tests t_1 and t_2; in this case the cost is $C_3(x) = c_v(x) + c_1(x) + c_2(x)$.

Another interpretation could be to measure pharmacokinetic profile (blood concentration) at one or two time points.

To consider this example within the traditional framework, introduce binary variables x_1 and x_2, $x_i = \{0 \text{ or } 1\}$, $i = 1, 2$. Let $X = (x, x_1, x_2)$ where x belongs to a 'traditional' design region $\mathbf{X}$, and pair (x_1, x_2) belongs to

$$\mathbf{X}_{12} = \{(x_1, x_2) : x_i = 0 \text{ or } 1, \quad \max(x_1, x_2) = 1\}.$$

Define

$$\eta(X, \theta) = I_{x_1,x_2}\, \eta(x, \theta), \qquad S(X, \theta) = I_{x_1,x_2}\, S(x, \theta)\, I_{x_1,x_2}, \tag{5.25}$$

where

$$I_{x_1,x_2} = \begin{pmatrix} x_1 & 0 \\ 0 & x_2 \end{pmatrix}.$$

Now introduce the 'extended' design region $\tilde{\mathbf{X}}$,

$$\tilde{\mathbf{X}} = \mathbf{X} \times \mathbf{X}_{12} = Z_1 \bigcup Z_2 \bigcup Z_3, \tag{5.26}$$

where

$$\begin{aligned} Z_1 &= \{(x, x_1, x_2) : x \in \mathbf{X}, x_1 = 1, x_2 = 0\}, \\ Z_2 &= \{(x, x_1, x_2) : x \in \mathbf{X}, x_1 = 0, x_2 = 1\}, \\ Z_3 &= \{(x, x_1, x_2) : x \in \mathbf{X}, x_1 = x_2 = 1\}, \quad \text{and} \quad Z_1 \bigcap Z_2 \bigcap Z_3 = \emptyset. \end{aligned}$$

The normalized information matrix $M_C(\xi, \theta)$ and design ξ are defined as

$$M_C(\xi, \theta) = \sum_{i=1}^{n} w_i \tilde{\mu}(X_i, \theta), \quad \sum_{i=1}^{n} w_i = 1; \xi = \{X_i, w_i\},$$

where $\tilde{\mu}(X, \theta) = \mu(X, \theta)/C_i(x)$ if $X \in Z_i$, $i = 1, 2, 3$; with $\mu(X, \theta)$ defined in (5.11) and $\eta(X, \theta)$, $S(X, \theta)$ introduced in (5.25).

Note that the generalization to $k > 2$ is straightforward; one has to introduce k binary variables x_i and matrix $I_{x_1,\ldots,x_k} = \text{diag}(x_i)$. The number of subregions Z_i in this case is equal to $2^k - 1$.

The presentation (5.26) is used in the example below to demonstrate the performance of various designs with respect to the sensitivity function $\psi(X, \theta, \xi)$, $X \in Z_i$.

5.5.1 Example 5.1

We take

$$\eta_1(x,\theta) = \gamma_1 + \gamma_2 x + \gamma_3 x^2 + \gamma_4 x^3 = F_1^T(x)\gamma;$$
$$\eta_2(x,\theta) = \gamma_1 + \gamma_5 x + \gamma_6 x_+ = F_2^T(x)\gamma,$$

where $F_1(x) = (1, x, x^2, x^3, 0, 0)^T$; $F_2(x) = (1, 0, 0, 0, x, x_+)^T$; $x \in \chi = [-1, 1]$, and $x_+ = \{x \text{ if } x \geq 0, \text{ and } 0 \text{ otherwise}\}$. Cost functions are selected as constants $c_v, c_1, c_2 \in [0, 1]$ and do not depend on x. Similarly, the variance matrix $S(x,\theta)$ is constant,

$$S(x,\theta) = \begin{pmatrix} S_{11} & S_{12} \\ S_{12} & S_{22} \end{pmatrix}.$$

In our computations, we take $S_{11} = S_{22} = 1$, $S_{12} = \rho$, $0 \leq \rho \leq 1$, thus changing the value of S_{12} only. Note that $\theta = (\gamma_1, \gamma_2, \ldots, \gamma_6; S_{11}, S_{12}, S_{22})^T$.

The functions η_1, η_2 are linear with respect to unknown parameters γ. Thus optimal designs do not depend on their values. On the contrary, in this example optimal designs do depend on the values of the variance parameters S_{ij}, i.e. we construct locally optimal designs with respect to their values. We considered a rather simple example to illustrate the approach. Nevertheless, it allows us to demonstrate how the change in cost functions and variance parameter ρ affects the selection of design points.

For the first run, we choose $c_v = 1$, $c_1 = c_2 = 0$, $\rho = 0$; see Figure 5.3 which shows the sensitivity function $\psi(X,\theta,\xi) = \text{tr}[\tilde{\mu}(X,\theta)M_C^{-1}(\xi^*,\theta)]$ for $X \in Z_j$, $j = 1, 2, 3$. Not surprisingly, in this case the selected design points are in subregion Z_3. Indeed, since individual measurements cost nothing, it is beneficial to take two measurements instead of a single one to gain more information and to decrease the variability of parameter estimates. The weights of the support points are shown in the plot which illustrates the generalized equivalence theorem: the sensitivity function ψ hits the reference line $m = 9$ at the support points of D-optimal design; recall that $\dim(\theta) = 9$.

The next case deals with positive correlation $\rho = 0.3$ and $c_v = 1$, $c_1 = c_2 = 0.5$; see Figure 5.4. Now there are just four support points in the design: two of them are at the boundaries of subregion Z_3 with weights $w_1 = w_2 = 0.33$, and the other two are in the middle of subregion Z_1, $x_{3,4} = \pm 0.45$ with weights $w_3 = w_4 = 0.17$.

Unlike the traditional normalization by the total sample size, in case of cost constraints the randomization ratio $n_1 : n_2 : \ldots$ is *not* defined simply by the ratio of weights $w_1 : w_2 : \ldots$. Recall that with cost constraints values n_i are given by $n_i = w_i C / c(X_i)$. Therefore, in the last example

$$\frac{n_1}{n_3} = \frac{w_1}{w_3}\frac{c(X_3)}{c(X_1)} = \frac{0.33 \times 1.5}{0.17 \times 2} \approx 1.45,$$

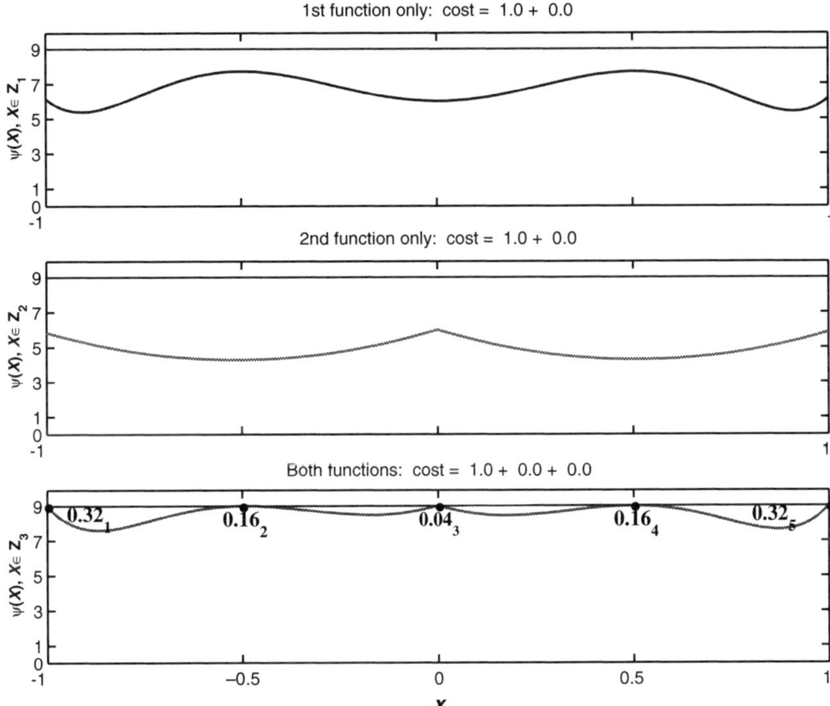

Figure 5.3 Sensitivity function $\psi(X, \theta, \xi)$ for three subregions Z_1 (top), Z_2 (middle) and Z_3 (bottom), and D-optimal design for $c_v = 1$, $c_1 = c_2 = 0$; $\rho = 0$ for Example 5.1. All optimal points, marked by dots, are in subregion Z_3. The value of sensitivity function ψ is equal to $m = 9$ at the optimal points

and therefore $n_1 = n_2 \approx 0.29 \cdot N$, $n_3 = n_4 \approx 0.21 \cdot N$, where N is the total sample size.

In the next example we demonstrate how to incorporate cost constraints in the random coefficients regression model.

5.5.2 Example 5.2 Pharmacokinetic model, serial sampling

At the early stages of clinical trials, patients often undergo serial blood sampling to determine activity level of various factors. Compartmental models are an important tool in mathematical modelling of how drugs circulate through the body over time, even though the compartments usually have no physiological equivalents; cf. Gibaldi and Perrier (1982). It is often assumed that the rate of transfer between compartments and the rate of elimination from compartments follow first-order (linear) kinetics which leads to systems of linear differential equations. For

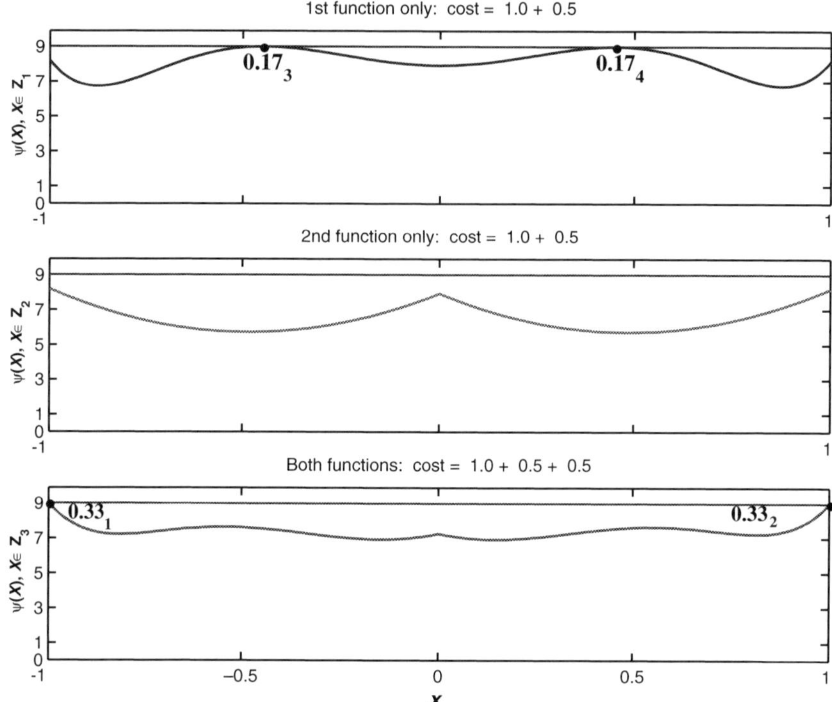

Figure 5.4 Sensitivity functions and optimal design for $c_v = 1$, $c_1 = c_2 = 0.5$, $\rho = 0.3$ for Example 5.1. Two optimal points are left in subregion Z_3, the other two are in subregion Z_1

example, if a dose D is administered orally at time $x = 0$, a two-compartment model with first-order absorption and elimination from central compartment can be used,

$$\begin{cases} \dot{g}_0(x) = -K_a g_0(x) \\ \dot{g}_1(x) = K_a g_0(x) - (K_{12} + K_{10}) g_1(x) + K_{21} g_2(x) \\ \dot{g}_2(x) = K_{12} g_1(x) - K_{21} g_2(x) \end{cases} \tag{5.27}$$

with initial condition $g_0(0) = D$, $g_1(0) = g_2(0) = 0$. Here K_a is an absorption rate constant, K_{10} is an elimination rate constant, and K_{12} and K_{21} are transfer rate constants between compartments. The central, or compartment 1, corresponds to plasma (serum); the peripheral, or compartment 2, is associated with tissue. Measurements $\{y\}$ are taken on the central compartment,

$$E(y|\gamma, x) = \eta(x, \gamma) = g_1(x)/V_1, \quad \text{Var}(y|\gamma, x) = \sigma^2,$$

where V_1 is the apparent volume of distribution of the central compartment, and by γ we denote the vector of unknown parameters, $\gamma = (V_1, K_a, K_{10}, K_{12}, K_{21})$ of dimension m_γ. Solving system (5.27) leads to

$$\eta(x, \gamma) = \frac{K_a D}{V_1}\left[\frac{(K_{21} - K_\mathrm{a})\mathrm{e}^{-K_\mathrm{a}x}}{(u_1 - K_\mathrm{a})(u_2 - K_\mathrm{a})} + \frac{(K_{21} - u_1)\mathrm{e}^{-u_1 x}}{(K_\mathrm{a} - u_1)(u_2 - u_1)} + \frac{(K_{21} - u_2)\mathrm{e}^{-u_2 x}}{(K_a - u_2)(u_1 - u_2)}\right], \tag{5.28}$$

with

$$u_{1,2} = 0.5\left[K_{12} + K_{21} + K_{10} \pm \sqrt{(K_{12} + K_{21} + K_{10})^2 - 4K_{21}K_{10}}\right];$$

see Gibaldi and Perrier (1982, Appendix B), or Seber and Wild (1989, Example 8.5).

We assume that given γ, all measurements are independent, and that parameters γ_j of patient j are independently sampled from the normal distribution with

$$E(\gamma_j) = \gamma^0, \quad \mathrm{Var}(\gamma_j) = \Lambda, \tag{5.29}$$

where γ^0 and Λ are usually referred to as 'population', or 'global' parameters. In practice, different number of samples k_j can be obtained for different j's (patients). Therefore, it seems reasonable to introduce costs similar to Example 5.1. In particular, one of the options is to impose a cost c_v for each visit, and a cost c_s for each of the individual samples. Then if k samples are taken, the total cost per patient will be described as

$$C_k = c_\mathrm{v} + kc_\mathrm{s}, \quad k = 1, \ldots, q,$$

with the restriction (5.23) on the total cost.

Let $X_j = (x_{1j}, x_{2j}, \ldots, x_{k_j j})$ and $Y_j = (y_{1j}, y_{2j}, \ldots, y_{k_j j})$ be sampling times and measured concentrations, respectively, for patient j. If $\eta(X_j, \gamma^0) = [\eta(x_{1j}, \gamma^0), \ldots, \eta(x_{k_j j}, \gamma^0)]^T$ and $F(X_j, \gamma^0)$ is a $(k_j \times m_\gamma)$ matrix of partial derivatives,

$$F(X_j, \gamma^0) = \left[\frac{\partial \eta(X_j, \gamma)}{\partial \gamma_\alpha}\right]\bigg|_{\gamma = \gamma^0},$$

then using (5.29), one gets the following approximation of the variance–covariance matrix S,

$$S(X_j, \theta) = \mathrm{Var}(Y_j | X_j) = F(X_j, \gamma^0)\Lambda F^T(X_j, \gamma^0) + \sigma^2 I_{k_j}, \tag{5.30}$$

where $\theta = (\gamma^0, \Lambda, \sigma^2)$ is a combined vector of model parameters.

To formally introduce the design region **X**, we note that there is a natural ordering corresponding to the timing of the samples. The design subregion for a patient depends upon the number and timing of samples taken from that patient. For example, if a single sample is drawn, the design subregion Z_1 will consist of a single point x, $0 \leq x \leq T$. For a patient having two samples taken, the design subregion Z_2 will consist of vectors $X = (x_1, x_2)$. Moreover, the two consecutive samples have to be separated in time by at least some positive Δ, therefore the design subregion Z_2 is defined as

$$Z_2 = \{X = (x_1, x_2): \quad 0 \leq x_1,\ x_1 + \Delta \leq x_2 \leq T\},$$

etc. If q samples are taken from a patient, then

$$Z_q = \{X = (x_1, \ldots, x_q): \quad 0 \leq x_1,\ x_1 + \Delta \leq x_2, \ldots, x_{q-1} + \Delta \leq x_q \leq T\}.$$

Finally, $X = Z_1 \bigcup Z_2 \bigcup \ldots Z_q$, and the normalized information matrix can be defined as

$$M_C(\xi, \theta) = \sum_{i=1}^{n} w_i \tilde{\mu}(X_i), \quad \sum_{i=1}^{n} w_i = 1,$$

where $\tilde{\mu}(X, \theta) = \mu(X, \theta)/C_k$, if $\dim(X) = k$ and $X \in Z_k$; the information matrix $\mu(X, \theta)$ is defined in (5.11), with $S(X, \theta)$ introduced in (5.30).

5.5.3 Remark 5.1

If one uses the Taylor expansion for $\eta(x, \gamma)$ up to the second-order terms,

$$\eta(x, \gamma) = \eta(x, \gamma^0) + \left[\frac{\partial \eta(x, \gamma^0)}{\partial \gamma_\alpha}\right]^T [\gamma - \gamma^0] + \frac{1}{2}[\gamma - \gamma^0]^T H(\gamma^0)[\gamma - \gamma^0] + \cdots$$

where

$$H(\gamma^0) = \left[\frac{\partial^2 \eta(x, \gamma)}{\partial \gamma_\alpha \partial \gamma_\beta}\right]\Bigg|_{\gamma = \gamma^0},$$

then it follows (5.29) that the expectation of $\eta(x, \gamma_j)$ with respect to the distribution of parameters γ_j can be approximated as

$$E_\gamma \eta(x, \gamma_j) \approx \eta(x, \gamma^0) + \frac{1}{2}\mathrm{tr}[H(\gamma^0)\Lambda]. \tag{5.31}$$

If the second term on the right-hand side of (5.31) is not 'small' compared to $\eta(x, \gamma^0)$, then to estimate global parameters γ^0, some adjustments need to be made in the estimation procedures (such as iteratively reweighted algorithms (5.14)–(5.17)), to incorporate the additional term in the mean function. However, if the second term is relatively small, then disregarding this term should not significantly affect the estimation algorithm. This is exactly the case in our numerical example below.

As mentioned at the beginning of the section, the introduction of costs allows for a meaningful comparison of patients with distinct number of blood samples taken.

For this example, we used the data from the study where patients received oral drug and had one or two blood samples drawn. The pharmacokinetic model was parameterized in terms of oral clearance CL, distributional clearance Q, and volumes of distribution V_1, V_2 of central and peripheral compartments, respectively. This leads to the following expressions for the compartment rate constants in (5.27):

$$K_{10} = CL/V_1, \qquad K_{12} = Q/V_1, \qquad K_{21} = Q/V_2.$$

The absorption rate constant was fixed to $1.70\,\mathrm{h}^{-1}$, and the parameters of response function $\gamma = (CL, Q, V_1, V_2)$, covariance parameters Λ and residual variance σ^2 were estimated from the data. Those estimates were used to construct locally optimal designs. For the examples below, we take the diagonal covariance matrix $\Lambda = \mathrm{diag}(\lambda_i)$, so that the combined vector of unknown parameters is

$$\theta = (CL, Q, V_1, V_2; \lambda_{CL}, \lambda_Q, \lambda_{V_1}, \lambda_{V_2}; \sigma^2)^T, \quad \dim(\theta) = 9.$$

We assume that the drug is administered at $x = 0$; the sampling, one or two samples per patient, is allowed every hour from $x = 1$ to $x = 20$, with samples separated by at least $\Delta = 4$ hours. The design region $\mathbf{X}$ is presented in Figure 5.5, top panel. D-optimal design and sensitivity function under the assumption of equal costs for single samples and pairs, i.e. $C_1 = C_2 = 1$, are shown in Figure 5.5, bottom panel. Similar to the first case in Example 5.1, it is not surprising that the pairs of samples are selected,

$$X_1 = (1, 5), \qquad X_2 = (3, 11), \qquad X_3 = (10, 20)$$

with weights 0.39, 0.27 and 0.34, respectively. Note that for subregion Z_2, the sensitivity function hits the reference plane $\psi = 9$ at the D-optimal points. In contrast, the sensitivity function for subregion Z_1 is well below reference line $m = 9$ and is not presented here.

If the cost of drawing two blood samples is increased to $C_2 = 10$, then four single points appear in the D-optimal design, see Figure 5.6:

$$X_{1-4} = \{1, 3, 8, 20\} \quad \text{with weights } w_{1-4} = \{0.2, 0.21, 0.2, 0.19\}.$$

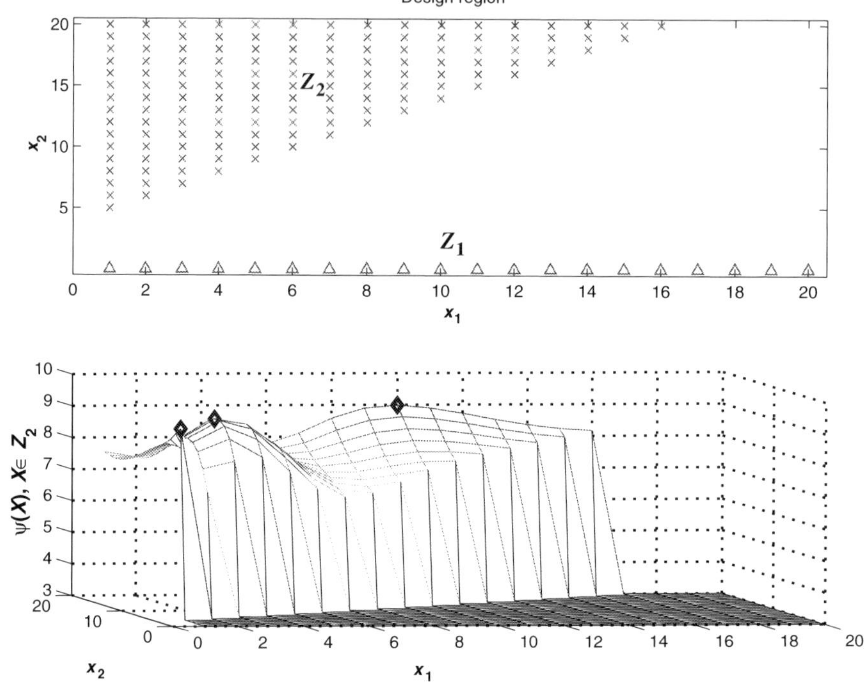

Figure 5.5 Design region and sensitivity function for Example 5.2. Top panel: design region, singles Z_1 (triangles) and pairs Z_2 (crosses). Bottom panel: sensitivity function ψ for equal costs. Note that $\psi = 9$ at the support points of optimal design

Still, there are two optimal pairs of samples, $X_5 = (1, 6)$ and $X_6 = (14, 18)$ with weights $w_5 = 0.09$ and $w_6 = 0.11$.

As in Example 5.1, to get actual randomization ratio, one needs to take cost function $c(X)$ into account. For example,

$$\frac{n_1}{n_5} = \frac{0.2 \times 10}{0.09 \times 1} \approx 22.2, \text{etc.}$$

and finally,

$$n_1 = n_3 = 0.244 \times N, \qquad n_2 = 0.256 \times N,$$
$$n_4 = 0.232 \times N; \qquad n_5 = 0.013 \times N,$$
$$n_6 = 0.011 \times N.$$

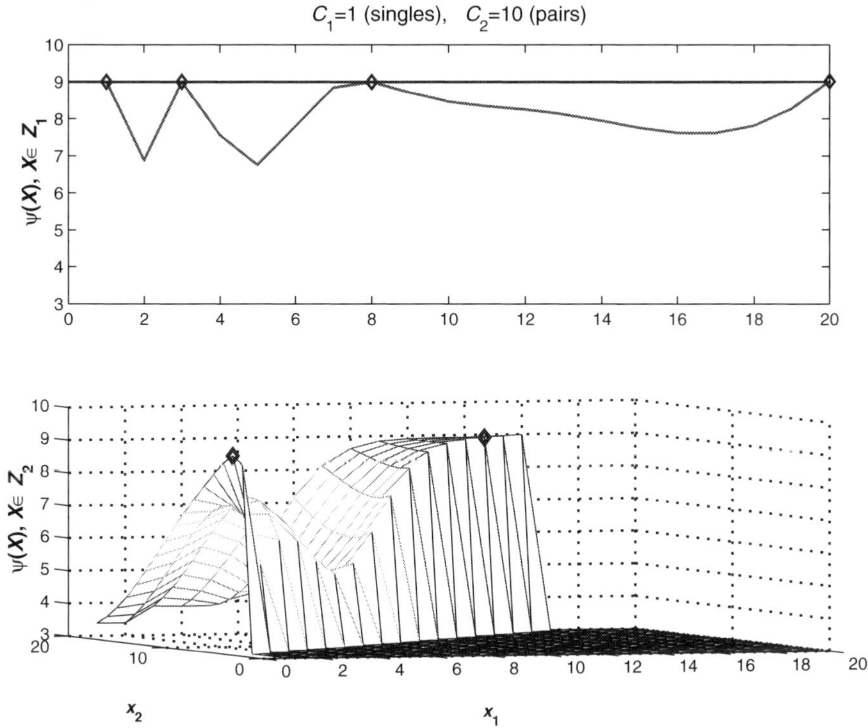

Figure 5.6 Sensitivity functions for Example 5.2 with unequal costs: cost(singles) = 1, cost(pairs) = 10. Singles Z_1 (top panel) and pairs Z_2 (bottom panel)

The interpretation of the optimal design is as follows:

- Take one sample at times $\{1, 3, 8, 20\}$ using 24.4, 25.6, 24.4 and 23.2% of the patients, respectively.
- Take two samples at times (1, 6) and (14, 18) using 1.3 and 1.1% of the patients, respectively.

For other methods of generating optimal designs with constraints in random effects models, see Mentré *et al.* (1997), where the authors allow for repeated measurements at the same design point. This, however, does not seem appropriate in our applications of interest.

5.6 Adaptive Designs

As mentioned earlier, for nonlinear models the information matrix depends, in general, on the values of unknown parameters which leads to the concept of locally optimal designs. The misspecification of the preliminary estimates may result in the poor performance of locally optimal designs. Therefore the use of an adaptive, or sequential, approach in this situation can be beneficial; see Box and Hunter (1965), Fedorov (1972, Ch. 4), Wu (1985a,b), Atkinson and Donev (1992, Ch. 11) and Zacks (1996). In the rest of this section, we shall illustrate the adaptive approach in the context of dose response experiments.

The adaptive designs are performed in stages in order to correct dose levels at each stage and approach the optimal doses as the number of stages increases. The adaptive approach relies on the reiteration of two steps: estimation and selection of new doses. We first briefly outline the adaptive algorithm and then give a more formal description:

(A.0) Initial non-singular design (doses) is chosen and preliminary parameter estimates are obtained.

(A.1) Additional dose(s) is selected from the available range of doses which provides the biggest improvement of the design with respect to selected criterion of optimality and current parameter estimates.

(A.2) The estimates of unknown parameters are refined given additional observation(s).

Steps (A.1) and (A.2) are repeated given available resources (for instance, maximal number of patients to be enrolled in a study).

For a formal introduction of adaptive designs, it is convenient to revert to unnormalized information matrix $M_N(\theta) = M_N(\theta, \xi)$ introduced in (5.5). In this section by $\xi = \{x_i, n_i\}$ we denote an unnormalized design, if not stated otherwise.

Step 0. A preliminary experiment is performed according to some initial non-singular design ξ_0 with N_0 observations, i.e. $Ey_i^0 = \eta(x_i^0, \theta)$, $i = 1, 2, \ldots, N_0$. Then an initial parameter estimate θ^0 is obtained from model fitting.

Step 1. Let $\xi_1 = \xi_0 \cup x_1$, where

$$x_1 = \arg\min_{\mathrm{x}} [\mathrm{M}_{\mathrm{N_0+1}}^{-1}(\theta^0, \xi_{0,\mathrm{x}})],$$

Ψ is a chosen criterion of optimality, and $\xi_{0,x} = \xi_0 \cup x$, which means that one additional observation at dose level x is added to design ξ_0. The minimization is performed with respect to all admissible doses. Next, a new experiment at dose level x_1 is carried out, $Ey_1 = \eta(x_1, \theta)$, and an updated parameter estimate θ_1 is obtained from model fitting. . . .

Step $s + 1$. Let $\xi_{s+1} = \xi_s \cup x_{s+1}$, where

$$x_{s+1} = \arg\min_{x} \Psi[M_{N_0+s+1}^{-1}(\theta^s, \xi_{s,x})],$$

and $\xi_{s,x} = \xi_s \cup x$. Then an updated parameter estimate θ^{s+1} is obtained, etc.

For D-criterion, at step $s+1$ one has to select a dose with the largest variance of prediction for current variance matrix $\mathrm{Var}(\theta^s)$; see Fedorov (1972, Ch. 4).

Since the selection of design points is based on optimizing a function that involves previous observations, random variables y_s become dependent, and therefore matrix $M_{N_0+s}(\theta^{s-1}, \xi_s)$ is no longer the Fisher information matrix in the traditional sense; see Ford and Silvey (1980) and Rosenberger and Hughes-Oliver (1999). However, our experience with adaptive algorithms shows that these algorithms perform quite well in applications. For more discussion on the asymptotic properties of adaptive designs, see Fedorov and Uspensky (1975, Ch. 2), where for similar models it was shown that under mild conditions (continuity and differentiability of the response function, compactness of the design region) the sequence of adaptive designs $\{\xi_s\}$ converges weakly to the optimal design,

$$\lim_{s\to\infty} s\Psi[M^{-1}_{N_0+s}(\theta^{s-1}, \xi_s)] = \min_{\xi} \Psi[M^{-1}(\xi, \theta)],$$

where the minimization on the right-handside is performed with respect to normalized designs ξ, and θ is the true vector of the unknown parameters; see also Malyutov (1988).

In practice, the selection of new doses and re-estimation of unknown parameters at each step of the algorithm is not realistic. A more practical approach is based on 'batch' processing, i.e. when not one, but several doses are selected at once, and/or when the estimation and parameter refinement are performed after every kth step of the algorithm.

5.6.1 Example 6.1

To illustrate the performance of the adaptive algorithm, we use the four-parameter logistic model (5.8). We observed that the power model of variance (5.22) often leads to extremely high correlation ρ_{56} of parameters θ_5 and θ_6, as big as $\rho_{56} = -0.98$ in some runs. So in our simulations we used another variance function with additive and multiplicative error terms,

$$S(x, \theta) = \delta_1 + \delta_2 \eta(x, \gamma), \qquad \theta = (\gamma_1, \gamma_2, \gamma_3, \gamma_4; \delta_1, \delta_2). \tag{5.32}$$

Though the correlation coefficient ρ_{56} is not very small in (5.32) as well ($\rho_{56} = -0.71$), this model makes more physical sense in that it allows us to take into account the background, or media noise, by including the additive term δ_1.

We utilize batch processing, with parameter estimation being performed after every 20 steps of the algorithm. The design region was set to $\mathbf{X} = [-0.02, 7.60]$ on the log scale, and the data were generated using $\theta = (74.7, 1.37, 100, 1000, 10, 0.1)^T$, so that $\{y_i\}$ are i.i.d. normal variables with mean $\eta(x_i, \theta)$ and variance $S(x_i, \theta)$. To estimate the vector of unknown parameters θ, we use the CIRLS algorithm which is introduced in (5.16) and (5.17), and which is asymptotically equivalent to the MLE; see Downing *et al.* (2001), Fedorov *et al.* (2001) and Fedorov and Leonov (2004).

Often in practice certain restrictions exist on the available range of doses. For instance, for quantal models discussed in Section 5.2 high doses may be toxic and doses exceeding a particular level ED_α have to be excluded from the study. On the other hand, for cell-based assays described by the logistic model (5.8), very small doses may lead to a large number of the 'bad' cells and need to be dropped.

Figures 5.7–5.9 illustrate the adaptive algorithm for the logistic model (5.8) and the variance function (5.32) with the restricted range of doses. Here we exclude small doses from the study; the cut-off point x_α is determined from

$$\eta(x_\alpha, \theta^s) = \theta_3^s + \alpha(\theta_4^s - \theta_3^s). \tag{5.33}$$

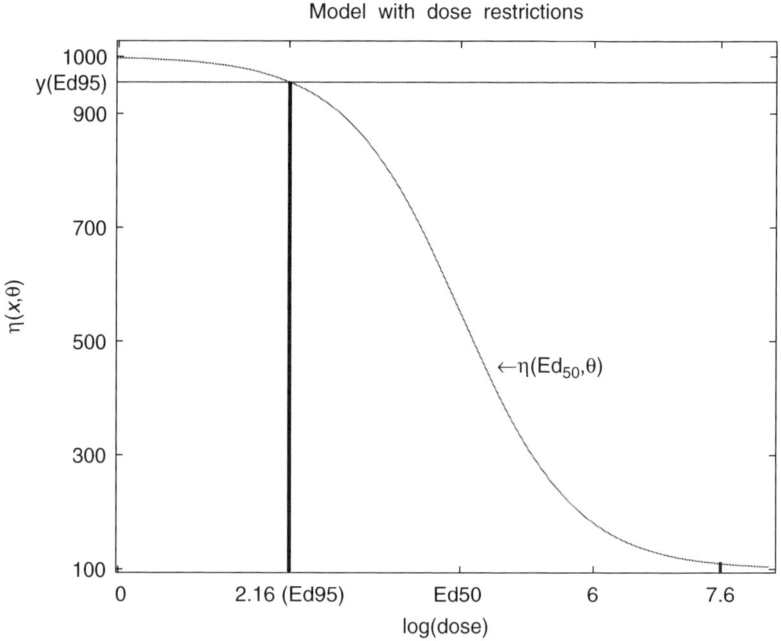

Figure 5.7 Logistic model with a restricted dose range for Example 6.1

Therefore on step s, given estimate θ^s, one has to find $x_\alpha = x_\alpha(\theta^s)$ from (5.33) and restrict the search of new dose levels to the region $\{x \geq x_\alpha(\theta^s)\}$. In this example level $\alpha = 0.95$ was selected, so that $\log x_\alpha(\theta) = 2.16$, where $x_\alpha(\theta)$ is a cut-off point corresponding to the true parameter values; see Figure 5.8.

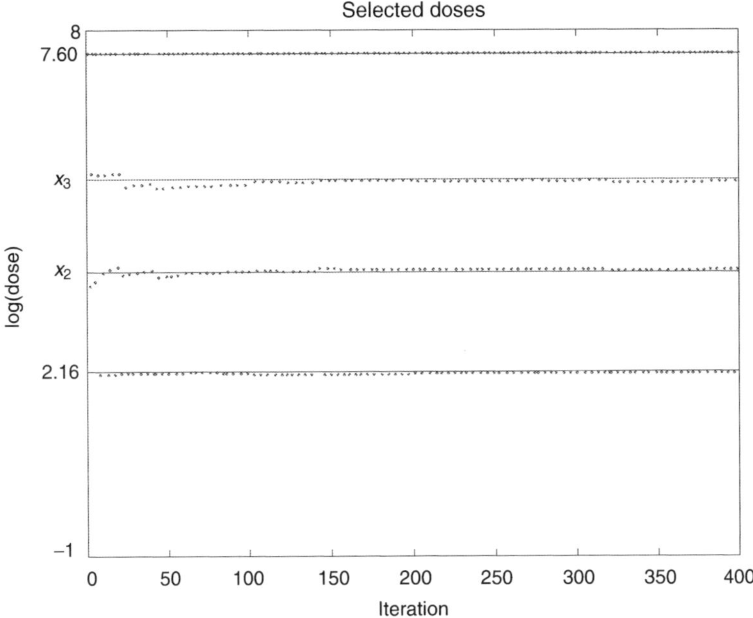

Figure 5.8 Selected doses, restricted dose range for Example 6.1. Horizontal lines correspond to the support points of D-optimal design

Figure 5.8 shows the selected doses $\{x_i\}$ and Figure 5.9 presents parameter estimates after 400 iterations. The locally D-optimal design, which corresponds to the true parameter values and design region $\mathbf{X} = [2.16, 7.60]$, is supported on four doses $\{2.16, 3.94, 5.44, 7.60\}$. The reference lines in Figure 5.8 correspond to those dose levels. The reference lines in Figure 5.9 correspond to the true values θ_i of unknown parameters, $i = 1, \ldots, 6$. On the first few iterations, the selected dose levels deviate slightly from optimal levels x_2, x_3 due to some discrepancy of parameter estimates from the true values. However, once the parameter estimates are refined, i.e. more and more information is collected, the selected dose levels essentially coincide with the optimal doses. It follows from (5.33) that cut-off points $x_\alpha(\theta^s)$ depend on current estimates θ^s, thus the lowest available dose does not necessarily coincide with $x_\alpha(\theta)$; see Figure 5.8.

5.7 Discussion

Optimal experimental design is a critical aspect of research in various biomedical applications, such as dose response studies, analysis of bioassays, pharmacokinetic/pharmacodynamic modelling. The optimal design theory and numerical algorithms

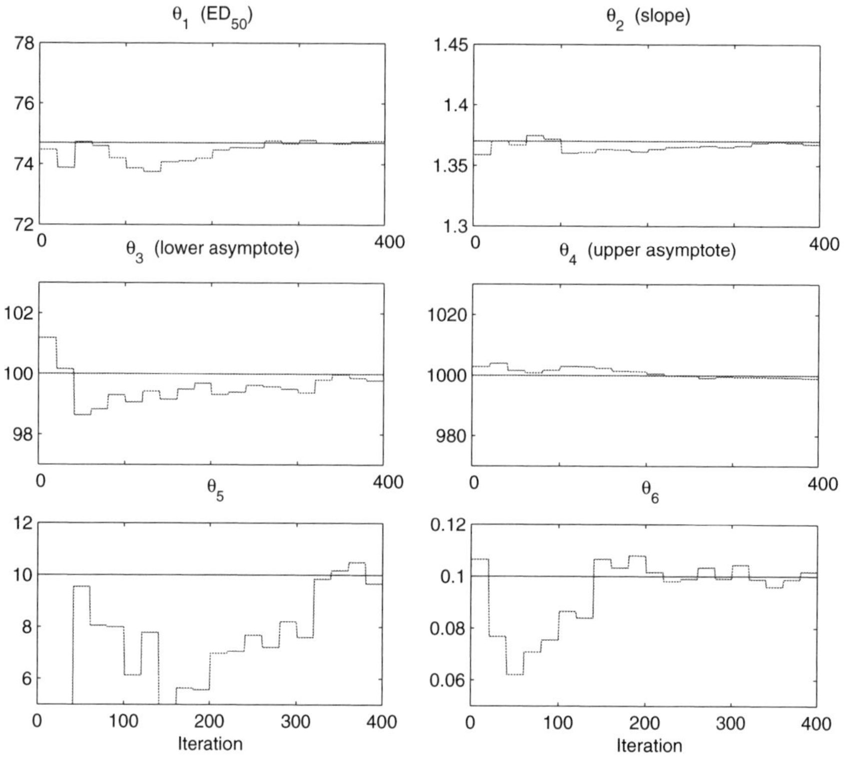

Figure 5.9 Parameter estimates for Example 6.1 with a restricted dose range

are well developed and may be implemented in various stages of drug development. In dose response studies, optimal design methods may be utilized under constraints on the admissible range of doses and responses. For instance, one may introduce a penalty function which reflects concerns associated with administering higher doses in safety studies. This leads to the introduction of constraints in the optimization problem which is defined in (5.7). The results which were developed for standard regression models admit straightforward generalization for this class of problems; see Cook and Fedorov (1995) and Torsney and Mandal (2001). Sometimes the number of doses may be fixed, or certain dose levels may be forced into designs. Such problems can be considered within the framework of regularized designs; see Fedorov and Hackl (1997, Ch. 2.6). In pharmacokinetic/pharmacodynamic (PK/PD) studies with multiple samples per patient, certain sampling times may be forced in the designs which mimic forced variables in stepwise regression algorithms (e.g. trough values in PK studies).

For a continuous response, logistic models with several parameters are recommended. They belong to the class of partially nonlinear models which simplifies the construction of optimal designs. Special attention should be given to the variance structure of observations, especially in the case when the variance depends on unknown parameters of the model.

For nonlinear models, optimal designs, in general, depend on parameter values which lead to the introduction of locally optimal designs. A practitioner should be cautious when directly applying locally optimal designs because the misspecification of a preliminary estimate can lead to a relatively poor performance of the locally optimal design. In this situation, the use of adaptive, Bayesian or minimax designs is strongly recommended. Still, if the dimension of the parameter space is low, then constructing locally optimal designs for various θ often leads to a reasonable compromise; see Atkinson and Fedorov (1988).

Sometimes the direct use of optimal designs is not realistic, for instance the use of two-point designs in quantal models. Moreover, alternative designs, such as serial dilution designs in bioassay studies, often perform reasonably well and are sufficiently robust to the misspecification of parameter values. Nevertheless, optimal designs can serve as a reference point for the computation of the relative efficiency of the alternative designs.

The introduction of cost functions and the application of cost constraints to normalized designs provide experimenters with a basis for incorporating a total cost. Such a design offers several benefits, not the least of which is to enable the experimenter to conduct the study within budget while obtaining reliable parameter estimates. Moreover, it allows for a meaningful comparison of 'points' with a distinct number of responses, such as patients with a distinct number of blood samples taken in PK/PD studies. The introduction of costs seems promising in considering cross-over trials which we plan to address in our future studies.

Acknowledgements

The authors wish to thank the referees for many valuable comments on an earlier version of the chapter which helped to improve presentation of the results.

References

Abdelbasit, K. M. and Plackett, R. L. (1983). Experimental design for binary data. *J. Amer. Stat. Assoc.*, **78**, 90–98.

Atkinson, A. C. and Cook, R. D. (1995). D-optimum designs for heteroscedastic linear models. *J. Amer. Stat. Assoc.*, **90**(429), 204–212.

Atkinson, A. C. and Donev, A. (1992). *Optimum Experimental Design*. Clarendon Press, Oxford.

Atkinson, A. C. and Fedorov, V. V. (1988). The optimum design of experiments in the presence of uncontrolled variability and prior information. In: Dodge, Y., Fedorov, V. V. and Wynn, H. (eds). *Optimal Design and Analysis of Experiments*, North-Holland, Amsterdam, pp. 327–344.

Beal, S. L. and Sheiner, L. B. (1988). Heteroscedastic nonlinear regression. *Technometrics*, **30**(3). 327–338.

Begg, C. B. and Kalish, L. A. (1984). Treatment allocation for nonlinear models in clinical trials: the logistic model. *Biometrics*, **40**, 409–420.

Bezeau, M. and Endrenyi, L. (1986). Design of experiments for the precise estimation of dose-response parameters: the Hill equation. *J. Theoret. Biol.*, **123**, 415–430.

Box, G. E. P. and Hunter, W. G. (1965). Sequential design of experiments for nonlinear models. In: Korth, J. J. (ed.), *Proceedings of IBM Scientific Computing Symposium*, IBM, White Plains, New York, pp. 113–137.

Chaloner, K. and Verdinelli, I. (1995). Bayesian experimental design: a review. *Stat. Sci.*, **10**, 273–304.

Chernoff, H. (1953). Locally optimal designs for estimating parameters. *Ann. Math. Stat.*, **24**, 586–602.

Cook, R. D. and Fedorov, V. V. (1995). Constrained optimization of experimental design, *Statistics*, **26**, 129–178.

Cramer, J. S. (1991). *The Logit Model: An Introduction for Economists*. Edward Arnold, London.

Downing, D. J., Fedorov, V. V. and Leonov, S. L. (2001). Extracting information from the variance function: optimal design. In: Atkinson, A. C., Hackl, P. and Müller, W. G. (eds). *MODA6–Advances in Model-Oriented Design and Analysis*, Heidelberg, Physica-Verlag, pp. 45–52.

Fan, S. K. and Chaloner, K. (2001). Optimal design for a continuation-ratio model. In: Atkinson, A. C., Hackl, P. and Müller, W. G. (eds). *MODA6–Advances in Model-Oriented Design and Analysis*, Heidelberg, Physica-Verlag, pp. 77–85.

Fedorov, V. V. (1972). *Theory of Optimal Experiment*. Academic Press, New York.

Fedorov, V. V. (1974). Regression problems with controllable variables subject to error. *Biometrika*, **61**, 49–56.

Fedorov, V. V., Gagnon, R. C. and Leonov, S. L. (2001). *Optimal Design for Multiple Responses with Variance Depending on Unknown Parameters*. GlaxoSmithKline BDS Technical Report 2001-03, Collegeville, Pa.

Fedorov, V. V., Gagnon, R. C. and Leonov, S. L. (2002). Design of experiments with unknown parameters in variance. *Appl. Stoch. Models Bus. Ind.*, **18**, 207–218.

Fedorov, V. V. and Hackl, P. (1997). *Model-Oriented Design of Experiments*. Springer-Verlag, New York.

Fedorov, V. V. and Leonov, S. L. (2001). Optimal design of dose-response experiments: a model-oriented approach. *Drug Inform. J.*, **35**(4). 1373–1383.

Fedorov, V. V. and Leonov, S. L. (2004). On parameter estimation for models with unknown parameters in variance. *Comm. Statist., Theory Meth.*, **33**(11) (to appear).

Fedorov, V. V. and Uspensky, A. B. (1975). *Numerical Aspects of Design and Analysis of Regression Experiments*. Moscow State Univ. Press, Moscow (in Russian).

Finney, D. J. (1971). *Probit Analysis*. Cambridge University Press, Cambridge.

Finney, D. J. (1976). Radioligand assay. *Biometrics*, **32**, 721–740.

Flournoy, N., Rosenberger, W. F. and Wong, W. K. (eds) (1998). *New Developments and Applications in Experimental Design*. Institute of Mathematical Statistics, Lecture Notes, vol. 34, Hayward, Calif.

Ford, I. and Silvey, S. D. (1980). A sequentially constructed design for estimating a nonlinear parametric function. *Biometrika*, **67**(2). 381–388.

Ford, I., Torsney, B. and Wu, C. F. J. (1992). The use of a canonical form in the construction of locally-optimal designs for nonlinear problems. *J. Roy. Statist. Soc.*, Ser. B, **54**, 569–583.

Gibaldi, M. and Perrier, D. (1982). *Pharmacokinetics*, 2nd edn. Marcel Dekker, New York.

Hedayat, A. S., Yan, B. and Pezutto, J. M. (1997). Modeling and identifying optimum designs for fitting dose response curves based on raw optical data. *J. Amer. Stat. Assoc.*, **92**, 1132–1140.

Heise, M. A. and Myers, R. H. (1996). Optimal designs for bivariate logistic regression. *Biometrics*, **52**, 613–624.

Hill, P. D. H. (1980). D-optimal designs for partially nonlinear regression models. *Technometrics*, **22**, 275–276.

Kalish, L. A. (1990). Efficient design for estimation of median lethal dose and quantal dose-response curves. *Biometrics*, **46**, 737–748.

Kalish, L. A. and Rosenberger, J. L. (1978). *Optimal Designs for the Estimation of the Logistic Function*. Tech. Report 33, Penn. State University, University Park, Pa.

Källén, A. and Larsson, P. (1999). Dose response studies: how do we make them conclusive? *Stat. Med.*, **18**, 629–641.

Karpinski, K. F. (1990). Optimality assessment in the enzyme-linked immunosorbent assay (ELISA). *Biometrics*, **46**, 381–390.

Kiefer, J. and Wolfowitz, J. (1960). The equivalence of two extremum problems. *Canad. J. Math.*, **12**, 363–366.

Lindsey, J. K., Wang, J., Byrom, W. D. and Jones, B. (1999). Modeling the covariance structure in pharmacokinetic crossover trials. *J. Biopharm. Stat.*, **9**(3). 439–450.

Magnus, J. R. and Neudecker, H. (1988). *Matrix Differential Calculus with Applications in Statistics and Econometrics*. John Wiley & Sons, Ltd, New York.

Malyutov, M. B. (1982). On asymptotic properties and application of IRGNA-estimates for parameters of generalized regression models. In: *Stochastic Processes and Applications*, Moscow, pp. 144–165 (in Russian).

Malyutov, M. B. (1988). Design and analysis in generalized regression model F. In: Fedorov, V. V. and Läuter, H. (eds). *Model-Oriented Data Analysis*, Springer-Verlag, Berlin, pp. 72–76.

Mathew, T. and Sinha, B. K. (2001). Optimal designs for binary data under logistic regression. *J. Stat. Plan. Inf.*, **93**, 295–307.

Mentré, F., Mallet, A. and Baccar, D. (1997). Optimal design in random-effects regression models, *Biometrika*, **84**(2). 429–442.

Minkin, S. (1987). Optimal designs for binary data. *J. Amer. Stat. Assoc.*, **82**, 1098–1103.

Minkin, S. and Kundhal, K. (1999). Likelihood-based experimental design for estimation of ED_{50}. *Biometrics*, **55**, 1030–1037.

Muirhead, R. (1982). *Aspects of Multivariate Statistical Theory*. John Wiley & Sons, Ltd, New York.

O'Quigley, J., Pepe, M. and Fisher, L. (1990). Continual reassessment method: a practical design for Phase I clinical trial in cancer. *Biometrics*, **46**, 33–48.

O'Quigley, J. (2001). Dose-finding designs using continual reassessment method. In: Crowley, J. (ed.), *Handbook of Statistics in Clinical Oncology*, Marcel Dekker, New York, pp. 35–72.

Pukelsheim, F. (1993). *Optimal Design of Experiments*, John Wiley & Sons, Ltd, New York.

Rao, C. R. (1973). *Linear Statistical Inference and Its Applications*, 2nd edn. John Wiley & Sons, Ltd, New York.

Rosenberger, W. F. and Hughes-Oliver, J. M. (1999). Inference from a sequential design: proof of conjecture by Ford and Silvey. *Stat. Prob. Lett.*, **44**, 177–180.

Seber, G. A. F. and Wild, C. I. (1989). *Nonlinear Regression*, John Wiley & Sons, New York.

Silvey, S. D. (1980). *Optimal Design*. Chapman & Hall, London.

Sitter, R. R. (1992). Robust designs for binary data. *Biometrics*, **48**, 1145–1155.

Sitter, R. R. and Wu, C. F. J. (1993). Optimal designs for binary response experiments: Fieller, D, and A criteria. *Scand. J. Statist.*, **20**, 329–341.

Torsney, B. and Mandal, S. (2001). Construction of constrained optimal designs. In: Atkinson, A., Bogacka, B. and Zhigljavsky, A. (eds). *Optimum Design 2000*, Kluwer, Dordrecht, pp. 141–152.

Torsney, B. and Musrati, A. K. (1993). On the construction of optimal designs with applications to binary response and to weighted regression models. In: Müller, W. G., Wynn, H. P. and Zhigljavsky, A. A. (eds). *Model-Oriented Data Analysis*, Physica-Verlag, Heidelberg, pp. 37–52.

Vonesh, E. F. and Chinchilli, V. M. (1997). *Linear and Nonlinear Models for the Analysis of Repeated Measurements*. Marcel Dekker, New York.

White, L. V. (1973). An extension of the general equivalence theorem to nonlinear models. *Biometrika*, **60**(2). 345–348.

White, L. V. (1975). The optimal design of experiments for estimation in nonlinear model. Ph. D. Thesis, University of London.

Wong, W. K. and Lachenbruch, P. A. (1996). Tutorial in biostatistics. Designing studies for dose response. *Stat. Med.*, **15**, 343–359.

Wu, C. F. J. (1985a). Asymptotic inference from sequential design in a nonlinear situation. *Biometrika*, **72** (3), 553–558.

Wu, C. F. J. (1985b). Efficient sequential designs with binary data. *J. Amer. Stat. Assoc.*, **80**, 974–984.

Wu, C. F. J. (1988). Optimal design for percentile estimation of a quantal-response curve. In: Dodge, Y., Fedorov, V. V. and Wynn, H. P. (eds), *Optimal Design and Analysis of Experiments*, Elsevier, North-Holland, pp. 213–223.

Zacks, S. (1996). Adaptive designs for parametric models. In: Ghosh, S. and Rao, C. R. (eds), *Design and Analysis of Experiments. Handbook of Statistics*, Vol. 13. Elsevier, Amsterdam, pp. 151–180.

Zhu, W. and Wong, W. K. (2000). Multiple-objective designs in a dose-response experiment. *J. Biopharm. Stat.*, **10**(1), 1–14.

6

Design of Experiments for Microbiological Models

Holger Dette,[1] Viatcheslav B. Melas[2] and Nikolay Strigul[3]

[1] *Ruhr-Universität Bochum, Fakultät für Mathematik, 44780 Bochum, Germany*
[2] *St Petersburg State University, Department of Mathematics, St Petersburg, Russia*
[3] *Princeton University, Department of Ecology and Evolutionary Biology, Princeton, NJ, USA*

6.1 Introduction

Mathematical modelling is the usual method for describing quantitative data and predicting outcomes of microbiological processes (Pirt, 1975; Baranyi and Roberts, 1995). Several types of empirical regression models are used in applications. Most of the models that describe microbial growth are nonlinear first-order differential equations in several variables. As a consequence the commonly used models in microbiology are regression models, where the unknown parameters enter the model nonlinearly. In the statistical literature these models are called nonlinear regression models. An appropriate choice of the experimental conditions can improve the quality of the experiment substantially. The present chapter is concerned with the optimal design of experiments for estimating parameters in nonlinear regression models. This problem has received much attention in the literature and the available references can be divided into the following three groups:

Applied Optimal Designs Edited by M.P.F. Berger and W.K. Wong
 ISBN: 0-470-85697-1 (HB)

(i) papers containing theoretical results on optimal experimental designs;

(ii) papers devoted to numerical construction of optimal designs or/and to stochastic modelling of biological data;

(iii) papers devoted to the application of optimal designs in real experiments.

Note that some of the references cited in this chapter will fit in both groups (i) and (ii). However most papers can be put in group (ii) and only a few of the cited references belong to group (iii). The reason for the scarcity of papers in group (iii) may be that practitioners are not well enough aware of the basic methods and advantages of optimal design.

The main purpose of this review is to explain some of the basic ideas of design of experiments and its applications in a form accessible to readers with a microbiological education. For this purpose we will avoid many technical details, which would require a deep mathematical background. Although nonlinear regression models appear in many applications, such as econometrics, genetics and agriculture, we will mainly concentrate on nonlinear models used in microbiology and will give a review of results from the literature on this issue.

The Chapter is organized in the following way. In Section 6.2 we will introduce some basic concepts and results relating to estimation and experimental designs for nonlinear regression models, used in microbiology. We discuss the properties of the least squares estimates, design optimality criteria and basic approaches to optimal design for nonlinear models. In Section 6.3 we discuss these concepts in more detail for the model given by the Monod differential equation. This model is taken as a typical example from microbiology to explain the practical benefits and difficulties in the implementation of optimal designs. The same section also contains a review of optimal designs for other typical nonlinear regression models in microbiology. A Bayesian approach to the construction of optimal designs is considered separately in Section 6.4, while Section 6.5 is devoted to some conclusions.

6.2 Experimental Design for Nonlinear Models

In microbiological problems the quantity to be measured is usually dependent on one scalar control variable and a few parameters. The control variable is either the time of observation or a concentration of a substrate. For the sake of universality we will denote the control variable by the symbol t, despite the fact that in some microbiological applications it should be replaced by c, where c is a concentration value. In many cases the variable t is also called the explanatory variable and due to natural biological conditions it is reasonable to assume that t varies in an interval, say $[0, T]$, where T is a maximal value for t. From a mathematical point of view observations taken under particular experimental conditions are random values. In microbiological applications these values can usually be considered as independent random variables. Supposing that our model describes a microbiological phenomenon adequately we can assume that each observation consists of a deterministic part, which describes the microbiological phenomenon, and a random error, which is modelled by a random variable with expectation equal to 0. However, the variance of a particular observation (or of the corresponding error) can depend on the experimental condition t under which the experiments have been performed. Generally speaking, this dependence is unknown but we can hypothesize a mathematical model for it and then study the influence of the assumed model on the procedures for choosing experimental design and estimation of parameters.

Thus we will assume that the experimental results at experimental conditions t_i, $i = 1, 2, \ldots, N$ (where N is the total number of observations) are described by the following nonlinear regression model

$$y_i = \eta(t_j, \theta) + g(t_j, \theta, \gamma)\varepsilon_j, \quad j = 1, \ldots, N \tag{6.1}$$

where $y_1, \ldots, y_N$, are experimental observations taken under the experimental conditions $t_1, \ldots, t_N$. η and g are known functions, describing the microbiological phenomenon in the absence of observation errors, but the parameters $\theta = (\theta_0, \ldots, \theta_m)^T$ and $\gamma = (\gamma_1, \ldots, \gamma_r)^T$ are unknown parameters and have to be estimated from the available data $y_1, \ldots, y_N$. In many cases the main interest of the experimenter lies in a precise estimation of the parameters $\theta_0, \ldots, \theta_m$ while the parameters $\gamma_1, \ldots, \gamma_r$ can be considered as nuisance parameters (Silvey, 1980). Finally the random errors $\varepsilon_1, \ldots, \varepsilon_N$ are assumed to be independent and identically distributed random values with zero mean and variance 1. For the sake of a transparent notation we will explain the main ideas of parameter estimation and design of experiments for nonlinear regression models under the assumption that the function g is constant, say $g \equiv \sigma > 0$, which corresponds to the case of a homoscedastic nonlinear regression model. The main concepts in the heteroscedastic case, where the function g is not constant, are exactly the same with an additional amount of notation (see e.g. Atkinson and Cook, 1995).

Generally speaking, in microbiological problems the function $\eta(t, \theta)$ is given implicitly as a solution of an ordinary differential equation of the form

$$\eta'(t, \theta) = F(t, \theta, \eta(t, \theta)) \tag{6.2}$$

with an initial condition $\eta(0, \theta) = \theta_0$, where $\theta = (\theta_0, \ldots, \theta_m)^T$ and the function F is explicitly given. Moreover, the function $F(t, \theta, \eta)$ is usually assumed to be representable in the form

$$F(t, \theta, \eta) = \psi(t, \theta_{(1)})\varphi(\eta, \theta_{(2)}), \quad \eta = \eta(t, \theta), \tag{6.3}$$

where $(\theta_{(1)}^T, \theta_{(2)}^T) = \theta^T$ and for fixed θ the function ψ depends only on t, whereas the function φ depends only on η. A few examples will be given in this section and some more examples are considered in Section 6.3.

Using (6.3), since $\eta'(t,\theta) = \mathrm{d}\eta/\mathrm{d}t$, we obtain from (6.2), by multiplication of both sides by $\mathrm{d}t$ and then dividing by φ, the following equation:

$$[1/\varphi(\eta,\theta_{(2)})]\mathrm{d}\eta = \psi(t,\theta_{(1)})\mathrm{d}t.$$

Integrating both sides we obtain

$$\Phi(\eta) = \Psi(t), \tag{6.4}$$

where

$$\Phi(\eta) := \Phi(\eta,\theta_{(1)}) = \int_{\eta_0}^{\eta} [1/\varphi(\zeta,\theta_{(2)})]\mathrm{d}\zeta, \quad \eta = \eta(t,\theta),$$

$$\Psi(t) := \Psi(t,\theta_{(1)}) = \int_0^t \psi(u,\theta_{(1)})\mathrm{d}u.$$

Equation (6.4) determines the function $\eta = \eta(t,\theta)$ implicitly and also allows an efficient calculation of its values for any fixed value of t and θ. In a few important cases the function η can be found explicitly. However, in most cases the function η is only defined implicitly.

6.2.1 Example 2.1 The exponential regression model

Let $\psi(t,\theta_{(1)}) \equiv 1, \varphi(\eta,\theta_{(2)}) = -\theta_1\eta$, then $\theta_{(2)} = \theta_1, \theta = (\theta_0,\theta_1)^T$ (note that $\theta_{(1)}$ is not needed here). It can be easily verified that

$$\eta(t,\theta) = \theta_0 \mathrm{e}^{-\theta_1 t}$$

is the unique solution of Equation (6.4).

6.2.2 Example 2.2 Three-parameter logistic distribution

Assume that $\psi(t,\theta_{(1)}) \equiv 1, \varphi(\eta,\theta_{(2)}) = \theta_2\eta(1-\eta/\theta_0)$, $\theta_2 > 0$. Then we have

$$\eta(t,\theta) = \frac{\theta_0 \mathrm{e}^{\theta_2 t+\theta_1}}{1+\mathrm{e}^{\theta_2 t+\theta_1}}$$

where θ_1 is arbitrary. This function is called the three-parameter logistic distribution.

6.2.3 Example 2.3 The Monod differential equation

Consider the equation (Monod, 1949; Pirt, 1975)

$$\eta'(t,\theta) = \mu(t)\eta(t,\theta),$$

where $\theta = (\theta_0, \theta_1, \theta_2, \theta_3)^T$ denotes the unknown parameter

$$\mu(t) = \mu(t,\theta) = \theta_1 \frac{s(t)}{s(t)+\theta_2},$$

$$s(t) - s_0 = \frac{\eta_0 - \eta(t)}{\theta_3},$$

and $s_0 = s(\theta)$, $\eta_0 = \eta(\theta) = \theta_0$ are given initial condition and $\theta = (\theta_1, \theta_2, \theta_3)^T$. This model can be rewritten in the form

$$\eta'(t,\theta) = \varphi(\eta(t,\theta),\theta),$$

where $\eta = \eta(t,\theta)$,

$$\varphi(\eta,\theta) = \theta_1 \frac{s_0\theta_3 + \eta_0 - \eta}{s_0\theta_3 + \eta_0 - \eta + \theta_2\theta_3}\eta.$$

By an integration we obtain Equation (6.4) in the form

$$t = \frac{1}{\theta_1}\left[(1+b)\ln(\eta/\eta_0) - b\ln\frac{c-\eta}{c-\eta_0}\right],$$

where the constant b is defined by $b = \theta_2\theta_3/[s_0\theta_3 + \eta_0]$.

Assume that under the experimental condition t_j $(j = 1, 2, \ldots, n)$ an experiment is repeated r_j times, while errors of different experiments under the same conditions are independent and let $\sum_{j=1}^{n} r_j = N$ denote the total number of observations in the experiment. The specification of the experimental conditions $t_1, \ldots, t_n$ and the relative proportions $\omega_n = r_n/N, \ldots, \omega_1 = r_1/N$ of total observations taken under these conditions is called experimental design and denoted by a matrix

$$\xi = \begin{pmatrix} t_1, \ldots, t_n \\ \omega_1, \ldots, \omega_n \end{pmatrix}, \tag{6.5}$$

where $\sum_{j=1}^{n} \omega_j = 1$, $\omega_j > 0$, $t_j \in [0, T]$, $j = 1, 2, \ldots, n$. In other words, a design of the form (6.5) advises the experimenter to take at each point t_j exactly $r_j = \omega_j N$ observations y_{ij} (if the numbers $\omega_j N$ are not integers a rounding procedure should be applied such that the resulting values r_j satisfy $\sum r_j = N$). The results of the experiments from such type of design can be conveniently written in the form

$$y_{ji} = \eta(t_j, \theta) + \varepsilon_{ji}, \quad i = 1, \ldots, r_j, \quad j = 1, \ldots, n, \tag{6.6}$$

where ε_{ji} denote independent random variables with expectation 0 and constant variance. Note that the index j in (6.6) corresponds to the different experimental conditions t_j under which observations are obtained, while the index i corresponds to the r_j observations obtained under a particular fixed condition.

The most popular technique for estimating parameters is the least squares technique. The estimate $\hat{\theta}_{(N)}$ is obtained as the value θ for which the sum of squares

$$\sum_{j=1}^{n} \sum_{i=1}^{r_j} (y_{ij} - \eta(t_j, \theta))^2 \tag{6.7}$$

attains its minimal value. Note that there may exist several values for θ minimizing this sum and this property depends on the particular regression function η under consideration. As a consequence there may exist nonlinear regression models where the least squares estimator is not uniquely determined. However, in many models used in microbiology it can be proved by mathematical arguments that the least squares estimator is uniquely determined. For example, it was proved for the Monod model (Dette *et al.*, 2003e) that for every θ^0 with positive coordinates the least squares estimate is uniquely determined (note that in the Monod model the regression function cannot be represented in an explicit form). The technique of that paper can be applied to most of the models used in microbiology.

Throughout this chapter let θ^* denote the 'true' but unknown value of the parameter θ. In other words, Equation (6.6) holds for $\theta = \theta^*$. We assume that the least squares estimate $\hat{\theta}_{(N)}$ obtained from the r_j experiments under the experimental conditions t_j, $j = 1, \ldots, n$ is unique (note that $\sum_{j=1}^{n} r_j = N$ is the total number of observations). Under some assumptions of regularity it was shown by Jennrich (1969) that for a sufficiently large sample size N the vector $(\hat{\theta}_{(N)} - \theta^*)$ has approximately a normal distribution with zero mean and covariance matrix

$$\frac{\sigma^2}{N} M^{-1}(\xi, \theta^*)$$

where σ denotes the standard deviation of the errors in model (6.6) and the matrix $M(\xi, \theta^*)$ is defined by

$$M(\xi, \theta) = \left(\sum_{k=1}^{n} \omega_k \frac{\partial \eta(t_k, \theta)}{\partial \theta_i} \frac{\partial \eta(t_k, \theta)}{\partial \theta_j} \right)_{i,j=0}^{m} \tag{6.8}$$

This matrix is called the Fisher information matrix in the literature and we assume that it is non-degenerate throughout this chapter. In principle the covariance matrix is a measure for the precision of the least squares estimator for the unknown parameter θ^* and a 'smaller' matrix yields more precise estimates. For example, the ith diagonal element of the matrix $(\sigma^2/N)M^{-1}(\xi,\theta^*)$ will be denoted by $((\sigma^2/N)M^{-1}(\xi,\theta^*))_{ii}$ and is an approximation of the variance or mean squared error for the ith component $\hat{\theta}_{i(N)}$ of the least squares estimator $\hat{\theta}_{(N)}$. An approximate confidence interval for the ith component θ_i of the vector θ is given by

$$\left[\hat{\theta}_{i(N)} - \frac{\hat{\sigma}u_{1-\frac{\alpha}{2}}}{\sqrt{N}}\sqrt{(M^{-1}(\xi,\hat{\theta}_{(N)}))_{ii}}, \hat{\theta}_{i(N)} + \frac{\hat{\sigma}u_{1-\frac{\alpha}{2}}}{\sqrt{N}}\sqrt{(M^{-1}(\xi,\hat{\theta}_{(N)}))_{ii}}\right]$$

where $u_{1-\alpha/2}$ denotes the $1-\alpha/2$ quantile of the standard normal distribution and $\hat{\sigma}^2$ is an estimate of the unknown variance of the error. For a concrete nonlinear regression model closeness of the estimator $\hat{\theta}_{(N)}$ to θ^* and the variance $\mathcal{D}(\hat{\theta}_{i(N)})$ to the variance $(\sigma^2/N)(M^{-1}(\xi,\hat{\theta}_{(N)}))_{ii}$ can and should be verified by stochastic simulation techniques. Such a verification was done in cases for the commonly used models in microbiology. For most cases it was shown that for moderate sample sizes N the sampling variances of the parameter estimates are well approximated.

Note that the precision of the estimates can always be decreased by increasing the sample size N, which yields a 'smaller' covariance matrix and smaller variances of the least squares estimates. However, in practice the sample size is usually fixed, due to cost considerations of each additional experiment. To improve the quality of the estimates or, from a different point of view, to reduce the number of experiments needed to obtain the estimates with a given accuracy, we note that the variances of the estimates $\hat{\theta}_{i(N)}$ and the covariance matrix of the vector $\hat{\theta}_{(N)}$ also depend on the given design ξ, which determines the relative proportion of total observations to be taken at the experimental conditions $t_1, \ldots, t_n$ and its location. Therefore the question arises whether one can find a design in some optimal way.

6.2.4 Example 2.4

Consider a special case of the exponential regression model in Example 2.1, that is

$$\eta(t,\theta) = \mathrm{e}^{-\theta t}$$

where $t \in [0, 10]$ and assume that the unknown parameter is given by $\theta^* = 0.05$ and that $\sigma^2 = 1$. The length of the confidence interval is given by

$$\frac{2u_{1-\frac{\alpha}{2}}}{\sqrt{N}}\sqrt{(M^{-1}(\xi,\theta^*))_{11}}$$

where the Fisher information is given by

$$M(\xi,\theta) = \sum_{i=1}^{n} \omega_i \left(\frac{\partial}{\partial\theta} \eta(t_i,\theta) \right)^2 = \sum_{i=1}^{n} \omega_i t_i^2 \mathrm{e}^{-2t_i\theta}.$$

Assume that a 95% coverage probability is desired and that $N = 100$ observations are available. If observations are taken according to the uniform design (i.e. $t_i = i/10; \omega_i = 1/100,\ i = 1, \ldots, n = 100$) the length of the confidence interval is approximately

$$2 \cdot 0.196 \cdot 0.2481 \approx 0.0972.$$

On the other hand, assume that the experimenter takes the 100 observation only at $n = 5$ different experimental conditions uniformly, that is $t_i = 2i, \omega_i = 1/5,\ i = 1, \ldots, 5$, then the length of the confidence interval is

$$2 \cdot 0.196 \cdot 0.2244 \approx 0.0879.$$

Thus we obtain a reduction of the length of the tolerance region simply by sampling at different experimental conditions. A further reduction of the length could be obtained by sampling at only two different experimental conditions, e.g. $t_1 = 10, t_2 = 20, \omega_1 = \omega_2 = 1/2$. In this case the length of the interval would be

$$2 \cdot 0.196 \cdot 0.1483 \approx 0.0581$$

and the question arises whether the experimental conditions can be chosen such that the length of the confidence interval is minimal. This is precisely the problem of finding an optimal experimental design, which will be described and discussed next.

In general an optimal design maximizes or minimizes the value of a given function of the Fisher information matrix (see, for example, Silvey, 1980). However, this matrix depends on the vector θ^*, which is unknown. To overcome this difficulty a simple and often rather efficient approach is to substitute an initial guess θ^0 for the unknown value θ^*. The corresponding designs are called locally optimal designs (Chernoff, 1953). The most popular criteria of optimality are *D*-, *c*- and *E*-criteria (Pukelsheim, 1993). These criteria have appeared in the literature on optimal design for microbiological nonlinear models and we review some of them for the sake of completeness.

A design is called locally *D*-optimal if it maximizes the quantity

$$\det M(\xi, \theta^0),$$

where θ^0 is a given initial value for the true parameter vector. It can be shown that an optimal design with respect to this criterion yields approximately a minimal value for the volume of the tolerance ellipsoid of the estimates (see, for example, Karlin and Studden, 1966, Ch. X). In addition, due to the famous equivalence theorem introduced in Kiefer and Wolfowitz (1960), this design minimizes an approximation of the worst variance of the predicted response $\eta(t, \hat{\theta}_N)$ over the interval $[0, T]$, that is

$$\max_{t \in [0,T]} \mathcal{D}(\eta(t, \hat{\theta}_N)).$$

For a given vector $c \in \mathbb{R}^{m+1}$ a design is called locally c-optimal if it minimizes the variance

$$\mathcal{D}(c^t \hat{\theta}_N) = \frac{\sigma^2}{N} c^T M^{-1}(\xi, \theta^0) c$$

of the estimate of a given linear combination of the parameters $\theta^T c = \sum_{i=0}^{m} \theta_i c_i$. The important choices for the vector c are contrasts, where the sum of the components of the vector c is 0 and unit vectors, i.e.

$$c = e_i = (\underbrace{0, \ldots, 0}_{i}, 1, 0, \ldots, 0)^T.$$

In the latter case a c-optimal design minimizes the variance of the least squares estimate for the ith parameter (under the condition that $\theta^* = \theta^0$). Such a criterion was used in Example 2.4, where the interest was the length of a particular confidence interval. If the experimenter has particular interest in several linear combinations of the parameters a locally E-optimal design seems to be appropriate. This design maximizes the minimum eigenvalue

$$\lambda_{\min}(M(\xi, \theta^0))$$

of the Fisher information matrix. A standard argument from linear algebra shows that this maximization is equivalent to the minimization of the maximum eigenvalue of the inverse Fisher information matrix $M^{-1}(\xi, \theta^0)$ or to the minimization of the worst variance among all estimates $c^T \hat{\theta}_N$ for the linear combinations $c^T \theta$ with a given norm $c^T c = 1$.

Another important criterion used in microbiological studies is the modified E-criterion

$$\frac{\lambda_{\min}(M(\xi, \theta^0))}{\lambda_{\max}(M(\xi, \theta^0))},$$

and an optimal experimental design with respect to this criterion makes this quantity as small as possible.

For many models used in microbiology it has been proved either by rigorous mathematical arguments or by intensive numerical experiments that the number n of different experimental conditions for a locally optimal design is equal to the number of parameters in the nonlinear regression model, i.e. $n = m + 1$ (see for example Dette and Wong, 1999; Dette and Biedermann, 2003; Dette *et al.*, 2003e). It is also usually the case that the locally optimal designs advise the experimenter to take observations at the maximal possible value T for the explanatory variable. Moreover, if $n = m + 1$ the optimal relative proportions of total observations to be taken at the experimental conditions $t_0, \ldots, t_m$ can usually be easily determined as functions of the points $t_0, \ldots, t_m$ (see Pukelsheim and Torsney, 1991), where the particular function depends on the optimality criterion under consideration. For example, for the case of the D-optimality criterion these functions are constant and all equal, i.e. $\omega_i = 1/(m + 1)$, $i = 0, 1, \ldots, m$ (see, for example, Fedorov, 1972) and similar formulae are available for the c and E-optimality criterion (Pukelsheim and Torsney, 1991). These facts are very helpful for the determination of locally optimal designs, because they reduce the dimension of the optimization problem.

If we have found a locally optimal design we should study its sensitivity with respect to the choice of the initial value θ^0. This can be done by calculating the relative efficiency of a locally optimal design at the point θ_0 with respect to the locally optimal designs at other points from a given set. For example, in the case of D-optimality criterion one usually considers the quantity

$$I(\xi) = \min_{\theta \in \Omega} I(\xi, \theta), \tag{6.9}$$

where Ω is a given set of possible values for the unknown parameter θ^*,

$$I(\xi, \theta) = \left(\frac{\det M(\xi, \theta)}{\det M(\xi_\theta, \theta)} \right)^{1/(m+1)}$$

is the D-efficiency of the given design ξ with respect to the locally D-optimal design ξ_θ under the assumption that $\theta^* = \theta$ is the 'true' parameter. The value $I(\xi, \theta)$ indicates how many more observations will be needed under the design ξ to obtain a given accuracy with respect to the design $\xi_\theta, \theta = \theta^*$. If it can be shown that the accuracy of the locally optimal design for the choice θ^0 is not too sensitive with respect to the choice of the initial value θ_0 the application of locally optimal designs is well justified. For most models in microbiology it was shown that locally optimal designs are often robust with respect to the choice of the initial parameter θ^0 (see, for example, Bezeau and Endrenyi, 1986; Dette *et al.*, 2003e) and for this reason we will mainly concentrate on locally optimal designs in this chapter.

However, if the dependence of the optimal design on the nonlinear parameter is more severe, or if the experimenter has no prior knowledge about the location of the unknown parameters, some care is necessary with the application of locally optimal designs. In this case some more sophisticated optimality criteria are required for the construction of efficient and robust designs, which will be briefly mentioned here for the sake of completeness. We restrict ourselves to the D-optimality criterion, but the idea for other types of criteria is very similar.

1. If one can split the whole experiment into several stages with $N_1, \dots, N_r$ observations taken at each stage, then it can be useful to use a so-called sequential experimental design (see Fedorov, 1972; Silvey, 1980). In this approach the estimate constructed at the previous stage can be used to construct a design for a current stage. For example, for the D-criterion we start with an initial guess for the unknown parameter, say θ^0, and take N_1 observations according to the locally D-optimal design (i.e. the design, which maximizes $\det M(\xi, \theta^0)$) to obtain the least squares estimate $\hat{\theta}_{(N_1)}$. In the second stage this estimate is used as preliminary guess and a new design is determined by maximizing the determinant $\det M(\xi, \theta_{(N_1)})$ to obtain the N_2 observations for the second stage of the experiment. The least squares estimate from the first $N_1 + N_2$ observations, say $\hat{\theta}_{(N_1+N_2)}$, is then used as preliminary guess for the third stage and the next N_3 observations are taken according to the design maximizing the determinant $\det M(\xi, \hat{\theta}_{(N_1+N_2)})$. This procedure is continued for all r stages to obtain the least squares estimate $\hat{\theta}_{(N)} = \hat{\theta}_{(N_1+\cdots+N_r)}$ from the total sample. It should be noted that inference from a sequential design is not easy (Ford and Silvey, 1980; Ford *et al.*, 1985; Wu, 1985; Woodroofe and Coad, 2002). Moreover, there exist many experiments where observations cannot be taken at several stages.

2. In order to obtain a non-sequential design, which is less sensitive with respect to the choice of the parameter θ^0, several authors proposed to construct a design maximizing the value $I(\xi)$ for a given set Ω with respect to the choice of the experimental design. Such designs are called maximin efficient designs (see Müller, 1995). Maximin efficient designs for the D- and E-optimality criterion were constructed in Dette and Biedermann (2003) and Dette *et al.* (2003c) for the Michaelis–Menten model, respectively. Further examples can be found in Dette and Biedermann (2003), who considered a compartmental model, and Dette *et al.* (2004), who discussed this type of design in the Hill model. Some more general results for this type were obtained by Lopez-Fidalgo and Wong (2002) and by Dette *et al.* (2003a, b). We finally note that the determination of maximin efficient designs is a substantially more difficult problem compared to the problem of determining locally optimal designs.

3. A different robust alternative is to assume sufficient knowledge of θ to specify a prior distribution for this parameter and to average the respective optimality criteria over the plausible values of θ defined by the prior. This leads to so-called Bayesian optimality criteria (see e.g. Chaloner and Larntz, 1989; Chaloner and Verdinelli, 1995) and we will discuss this approach separately in Section 6.4.

6.3 Applications of Optimal Experimental Design in Microbiology

The parameters of models for describing quantitative data and predicting outcomes of microbiological processes often have special names and their values play an important role in the analysis of experimental results. Most of the models which describe microbial growth are nonlinear first-order differential equations in several variables. Variables are related by the stoichiometric coefficients and reffect mass-balance equations (Pirt, 1975; Howell, 1983; McMeekin and Ross, 2002). Additionally at some time point (usually at the initial point), all coefficients and variable definitions are specified. One of the important stages of the mathematical modelling application is a comparison of the model predictions with the real experimental data, and identification of the model parameters and optimal experimental designs is very useful at this stage.

The problem of identification of the model parameters concerns two important questions. The first is how well the mathematical model reflects the real microbiological process. This question is a very difficult one; the answer always depends on the level of details of the research. For example, if one considers degradation of organic pollutants, for example nitroaromatic compounds, by a community of microorganisms with complex and unknown structure (let us imagine activated sludge), then it should be sufficient to choose an unstructured growth model such as the classical Monod model or some of its modifications such as Andrews or Haldane use to describe this microbiological process (Ellis *et al.*, 1996; Knightes and Peters, 2000; Goudar and Ellis, 2001). On the other hand, if one considers a much more 'simple' microbiological process, such as utilization of glucose by *Escherichia coli*, we would find that the Monod model is not a realistic tool for describing this process and it is necessary to use models with more complicated structure (Ferenci, 1999). One of the important methods for examining whether a particular model is adequate for a given microbiological process is the comparison of the model predictions with experimental data. Unfortunately, a good fit of the model is only a necessary but not a sufficient condition to be confident that the selected model is realistic. It is not difficult to identify models which yield a good fit to experimental data but are not realistic (Hopkins and Leipold, 1996). The model selection is not a formal procedure and it requires simultaneous theoretical and experimental biological efforts.

If the first question is answered then the second problem is the identification of model parameters and the question of the efficiency of an experiment arises. This problem plays the central role in our chapter. The following questions should be answered:

1. Is it possible to determine parameters of the mathematical model which describe the microbiological process using the particular experimental procedure? And if that answer is 'no' can we find such a scheme of measurements for which the answer is 'yes'?

2. Is it possible to make a relatively small number of experiment replications and measurements without losing information?

Without a clear understanding that the model parameters are unique and their values realistic, it is not possible to use them as characteristics of a microbiological process. Ignoring these important questions leads to mistakes and unsuccessful application of mathematical models for the analysis and prediction of microbiological processes. Although the mathematics cannot guarantee that the necessary data is obtainable it can be extremely useful in the sense that it can significantly reduce the number of necessary measurements and experimental replications. A good experimental design can therefore help to economize laboratory time and expenses.

The present section consists of two parts. First we will consider ideas of optimal experimental design using the Monod model as an example. Then we will review some literature on applications of optimal design techniques to microbiological models, which are not related to the Monod model.

6.3.1 The Monod model

The Monod model can describe several important characteristics of microbial growth in a simple periodic culture of micro-organisms. This model was proposed by Nobel laureate J. Monod more than 50 years ago and it is one of the basic models for quantitative microbiology (Monod, 1949; Ferenci, 1999). After several decades of intensive studies many limitations of this model as well as restrictions of its applications are very well known (Pirt, 1975; Baranyi and Roberts, 1995; Ferenci, 1999). At the same time many modifications of this model have been proposed in some specific cases (Ellis *et al.*, 1996; Fu and Mathews, 1999; Schirmer *et al.*, 1999; Vanrolleghem *et al.*, 1999). The Monod model is still used very often without any modifications, especially in such fields as environmental and industrial microbiology. For example, it is the most common model for describing the dynamics of organic pollutant biodegradation (Blok and Struys, 1996; Knightes and Peters, 2000; Goudar and Ellis, 2001). At the same time the similarity of the Monod model and the Michaelis–Menten equation provides a very wide application of this model type throughout biological and biomedical sciences. This type of equation is very often used in biochemistry, plant physiology, biophysics and pharmacology.

The Monod model (Monod, 1949, Pirt, 1975) for periodic culture (batch) experiments may be presented as a first-order differential equation:

$$\eta'(t) = \mu(t)\eta(t)$$

where

$$\mu(t) = \theta_1 \frac{s(t)}{s(t) + \theta_2},$$

$$s(t) - s_0 = \frac{\eta_0 - \eta(t)}{\theta_3}.$$

Here $s_0 = s(0)$ and $\eta_0 = \eta(0)$ are given initial conditions, i.e. initial concentrations of the consuming substrate and microbial biomass, respectively. Three parameters characterize microbial growth. Each parameter has its own traditional notation and name: θ_1 is called the maximal specific growth rate usually denoted by $\mu_{\max}$, θ_2 the saturation (affinity) constant denoted by K_s, and θ_3 the yield coefficient usually denoted by Y. The variable t represents time, which varies in the closed interval $[0, T]$. Typical minimum values of T are several hours for optimal microbiological media, whereas the maximum is one year or more for specialized groups of micro-organisms. All three parameters, initial conditions and variables are positive because of natural biological conditions. This differential equation can be easily integrated as indicated in the previous section.

The general meaning of this model is that the rate of microbial growth (i.e. change of microbial biomass) is equal to the current biomass value multiplied by the specific growth rate. Specific growth is a function of the concentration of nutrients with two parameter characteristics for a given micro-organism (maximum specific growth rate and saturation constant or Michaelis–Menten constant). At the same time increase of microbial biomass is proportional to the decrease of nutrients where the coefficient of proportionality is the yield coefficient Y.

The parameter estimation problem for the Monod and Monod-type models has been considered many times with different mathematical and biological assumptions (Aborhey and Williamson, 1978; Holmberg, 1982; Baltes *et al.*, 1994; Dochain *et al.*, 1995; Vanrolleghem et al., 1995). Some of the results are more theoretical and others are more related to a specific experimental procedure. Usually the best techniques for parameter identification are chosen based on the experimental procedure. For Monod-type models experimental procedures vary greatly dependent on measurement technique and reactor type. There are several main variants of simple batch experiments, as indicated below:

1. The batch-type experiment where it is possible to make only measurements of the biomass value or some parameter which is related to biomass activity, for example oxygen uptake or carbon dioxide production (Blok and Struys, 1996; Vanrolleghem *et al.*, 1995). The structural identifiability of the parameters in the Monod model for this case was analysed by Chappel and Godfrey (1992).

2. Experiments where it is possible to make simultaneous measurement of the consumed substrate and the biomass concentration. An example of this type of experiment is the investigation of the biodegradation of the volatile hydrophobic organic compounds such as phenanthrene (Guha and Jaffe, 1996), where parameters in the Monod model were determined by maximum likelihood estimation from direct measurements of the phenanthrene concentrations and carbon dioxide production. Applications of least squares estimation techniques for this case were considered from a theoretical perspective in several publications (Aborhey and Williamson, 1972; Holmberg, 1982; Saez and Rittmann, 1992).

To discuss the main ideas of optimal experimental design let us consider the first type of experiment. For example, we are investigating biodegradation of some organic pollutant, which is the unique source of organic and energy for the growing micro-organisms (Sommer *et al.*, 1995), and we can only take measurements of the biomass of micro-organisms at different time points (or for example respiratory activity). Then the results of the experiment is a table $\{t_i, x_i\}$ with two rows; the first is the time of the measurement $\{t_i\}$, and the second is the value of a microbial biomass $\{x_i\}$. The first column of this table is called experimental design; it will be denoted as ξ, and to be more mathematically precise we can say that it is a discrete probabilistic measure with equal weights (see Section 6.2). It is obvious that before we start an experiment it is important to decide when we will take our measurements, but we should also leave open the possibility of changing the experimental design after the experiment has started.

To illustrate the optimal design approach artificial experimental data sets will be used. This approach is very often used with Monte Carlo simulations as a tool for the analysis of regression models (Holmberg, 1982; Ellis *et al.*, 1996; Melas and Strigul, 1999; Poschet *et al.*, 2003). Let us decide to take measurements every 2 hours up to 40 hours; then our experimental design is $\{0, 2, 4, 6 \ldots 40\}$. A typical example of data is illustrated in Figure 6.1. The experimental data will be simulated artificially using a Monod model. Let us fix some parameter set called a 'true' value θ^*, for example $\theta_1^* = 0.25$; $\theta_2^* = 0.5$; $\theta_3^* = 0.25$ (this set is the average of several experimental results, see Pirt, 1975). The Monod model solution is calculated

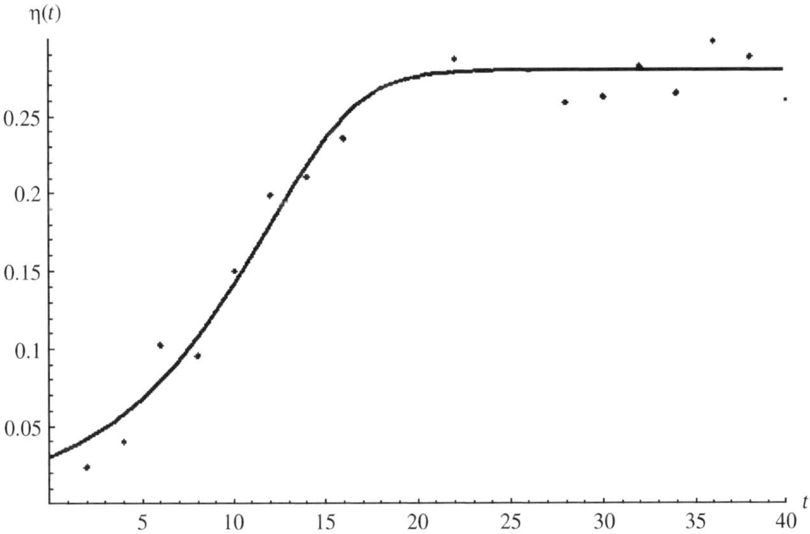

Figure 6.1 Artificial experimental data and the plot of the Monod solution (20 uniform points experimental design with normal distributed experimental errors, dispersion 0.02)

for this parameter θ^* with the initial conditions $s_0 = 1$ and $\eta_0 = 0.03$. Consider the values of the solution of the Monod model at the points of our experimental design, $\eta_i = \eta(t_i, \theta)$, $i = 1, \ldots, 20$. At this stage the ideal set of experimental data is constructed. It is such a set of experimental measurements which was obtained if there were no experimental errors in both measurements and time of measurements. But the most important property of real experimental data is the existence of random error in measurements. In real life there are never precise measurements; there are errors in the biomass measurements (for any experimental technique) and sometimes significant errors in the time of measurements. Suppose that those errors are random and they are normally distributed with some fixed values of dispersion. For example, suppose the variance of random errors equals 0.02. Then the table of the artificial experimental data using the values of the microbial biomass are the numbers randomly taken from this normal distribution with these values as averages, i.e. $\eta_i^* = \eta_i + \varepsilon_i$, where $\varepsilon_i \in N(0; 0.02)$. Assume that we can repeat the experiment several times, say, r times, and hence we will have k independent sets of the experimental data $\{\eta_i^*\}_j$, $i = 1, \ldots, 20$, $j = 1, \ldots, r$. The number of replications r is a very important characteristic of any experiment scheme, of course we would like to make as few experimental replications as possible without loss of information. The experimental optimal design approach is useful to diminish the number r of replications of the experiment.

Let us forget that we know the 'true' parameter' values θ^* and assume that after the experiment we have the results of the biomass measurements $\{\eta_i^*\}_j$. Then our aim is to estimate the Monod model parameters from these data. This situation is very close to real problems (Knightes and Peters, 2000; Schirmer *et al.*, 1999). What do we have at this stage? First, we know that the experiment is very well described by the Monod model (note that in real applications the specification of a nonlinear regression model is a non-trivial problem). Second, we have experimental data sets with some realistic replication number r, say, $r = 5$.

The problem of determining the least squares estimate in a nonlinear regression model (that is the set of parameter estimates that minimizes the sum of squared residuals in (7)) is a nonlinear optimization problem, which can be solved by standard algorithms, for example the Gauss–Newton method. In the following we denote the set of Monod parameter estimates as $\hat{\theta} = \{\hat{\theta}_1, \hat{\theta}_2, \hat{\theta}_3\}$ and the parameter estimates obtained from the r experiments by $\hat{\theta}_1, \ldots, \hat{\theta}^r$. How do we know that the obtained estimation $\hat{\theta}$ is close enough to the 'true' parameters θ^*? Actually, we can say only that the nonlinear least squares estimates are good enough asymptotically, i.e. consistent and asymptotically normally distributed if the sample size is sufficiently large (see Jennrich, 1969 and Dette *et al.*, 2003e). Unfortunately mathematics cannot guarantee that any of the estimates $\{\hat{\theta}^j\}_{j=1,\ldots k}$ will be close enough to the 'true' parameter θ^* when the sample size is not very large. As we can see from Table 6.1, the 10 independent estimates of $\hat{\theta}$ do not provide precise estimations of θ^*. The question of what is a precise estimate is also very important. If one uses at most 10% error as the criterion of good estimation, then the estimates

Table 6.1 Replications of least squares estimates of the parameters in the Monod model with a uniform design (20 equidistant distributed points in the interval [0, 40] and a locally *D*-optimal design, which uses only three different experimental conditions). The 'true' parameters are $\theta_1^* = 0.25, \theta_2^* = 0.5, \theta_2^* = 0.25$, while the experimental errors are generated from a normal distribution with variance $\sigma^2 = 0.02$

	Uniform design			*D*-optimal design		
	$\hat{\theta}_1(\mu_{max})$	$\hat{\theta}_2(K_s)$	$\hat{\theta}_3(Y)$	$\hat{\theta}_1(\mu_{max})$	$\hat{\theta}_2(K_s)$	$\hat{\theta}_3(Y)$
1	0.1645	0.0795	0.2441	0.2545	0.5097	0.2492
2	0.3087	0.9239	0.2557	0.2541	0.5123	0.2455
3	0.2520	0.5295	0.2547	0.2452	0.4814	0.2556
4	0.1497	0.0477	0.2565	0.2661	0.5725	0.2519
5	0.1639	0.0597	0.2434	0.2451	0.4533	0.2487
6	0.1851	0.2446	0.2562	0.2690	0.5430	0.2521
7	0.2007	0.2456	0.2342	0.2489	0.5224	0.2568
8	0.2937	0.7057	0.2624	0.2389	0.4631	0.2519
9	0.3251	0.9019	0.2397	0.2569	0.5313	0.2534
10	0.2255	0.3607	0.2584	0.2510	0.4848	0.2458
Average	0.2269	0.4099	0.2505	0.2530	0.5074	0.2511
SD	0.0647	0.3379	0.0094	0.0093	0.0371	0.0038

of $\hat{\theta}_1, \hat{\theta}_2$ are not close to the 'true' values θ_1^* and θ_2^* in the presented replications because the standard deviations of them are too large. At the same time the estimate $\hat{\theta}_3$ determines the component θ_3^* with sufficient accuracy practically in each replication.

For better understanding of this result let us make a correlation analysis of the results of 100 independent replications ($k = 100$). It is important that the estimators $\hat{\theta}_1$ and $\hat{\theta}_2$ are closely correlated with the correlation coefficient 0.99, at the same time the estimator $\hat{\theta}_3$ is not correlated with the other two estimates ($r = 0.28$). The two-dimensional plot (Figure 6.2) of $\hat{\theta}_1$ and $\hat{\theta}_2$ can be considered as the level surface of the nonlinear square estimation, as long as the values of the min $\sum(\eta_i^* - \eta(t_i, \theta))^2$ are very close in those experiments (Table 6.1). It is interesting to examine the three-dimensional plot of $\sum(\eta_i^* - \eta(t_i, \theta))^2$ for varying values of θ_1 and θ_2 (Figure 6.3). The lowest level of the sum of squares is located on the line at the bottom; it looks very similar to a ravine. We can conclude that linear dependence of $\hat{\theta}_1$ and $\hat{\theta}_2$ is a serious problem for a precise estimation of the parameters θ_1^* and θ_2^* for realistic sample sizes. For the Monod type of nonlinearity, i.e. for the form of the Michaelis–Menten function, this problem has been described many times by different groups of researchers (Holmberg, 1982; Magbanua *et al.*, 1998). It was shown that in the considered experimental conditions it is not possible to identify parameters of the Monod model (Holmberg and Ranta, 1982).

Since the problem of the correlation between $\hat{\theta}_1$ and $\hat{\theta}_2$ plays a central role in these considerations it is useful to obtain an elementary interpretation of this

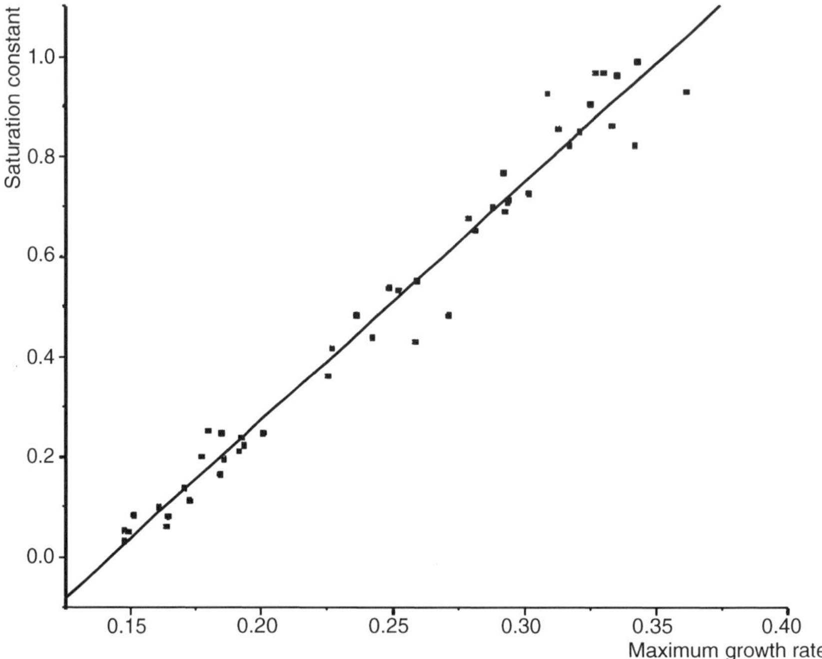

Figure 6.2 Scatter plot of the estimations of the Monod model parameters: maximum specific growth rate and stauration constant, $n = 100, r = 0.987, p < 0.0001$

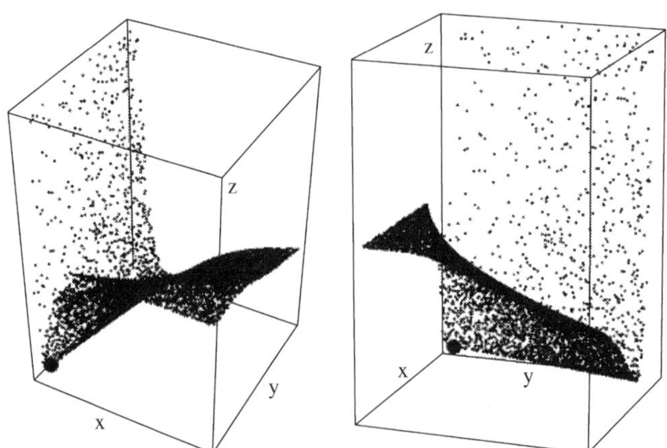

Figure 6.3 Regression surface for one artificial data set (large point is the true parameters value)

problem. In Figures 6.4–6.6 we represent results of the solution of the Monod differential equation for a variation of the model parameters. We can see that θ_3 mostly determines the value of the horizontal asymptote of the solution $\eta(t)$ and that θ_1, θ_2 determine the bottom and upper bends of the solution. The lower bend is

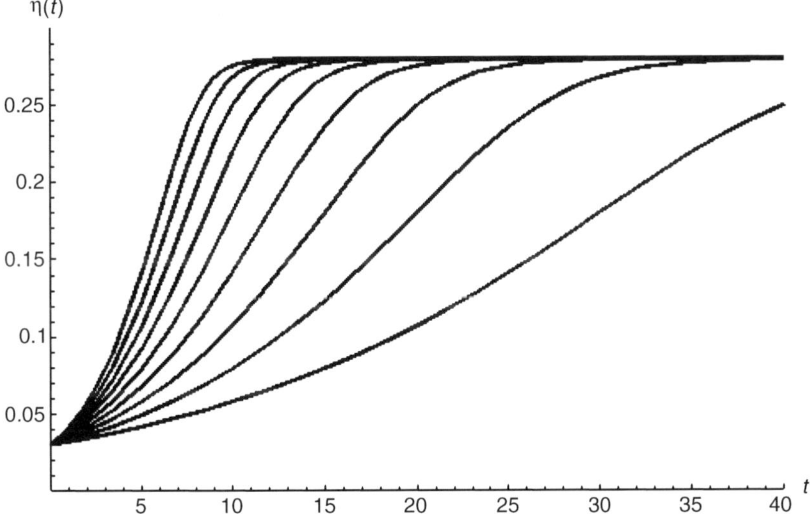

Figure 6.4 Effect of the maximum specific growth rate on the solution of the Monod equation maximal specific growth rate was changed from 0.1 to 0.5 with step 0.05, other parameters were fixed (0.5, 0.25); initial conditions $x_0 = 0.3, S_0 = 1$)

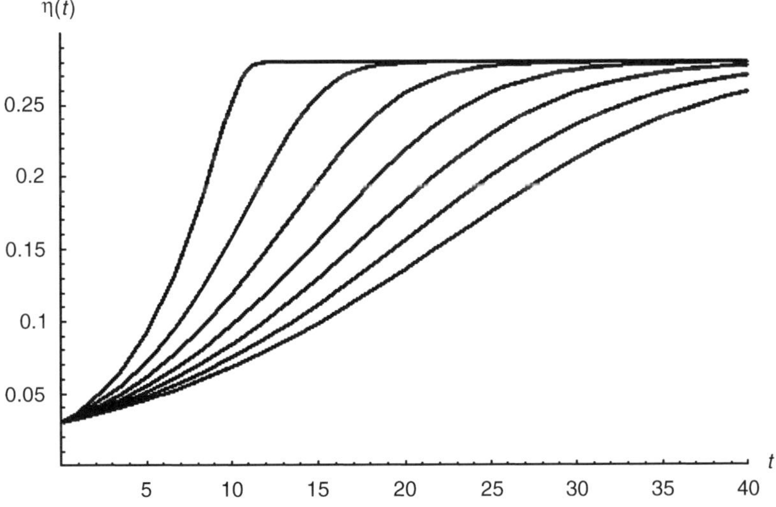

Figure 6.5 Effect of the Michaelis–Menten constant on the solution of the Monod equation (the Michaelis–Menten constant was changed from 0.1 ti 1.9 with step 0.3, other parameters were fixed (0.25, 0.25); initial conditions $x_0 = 0.3, S_0 = 1$)

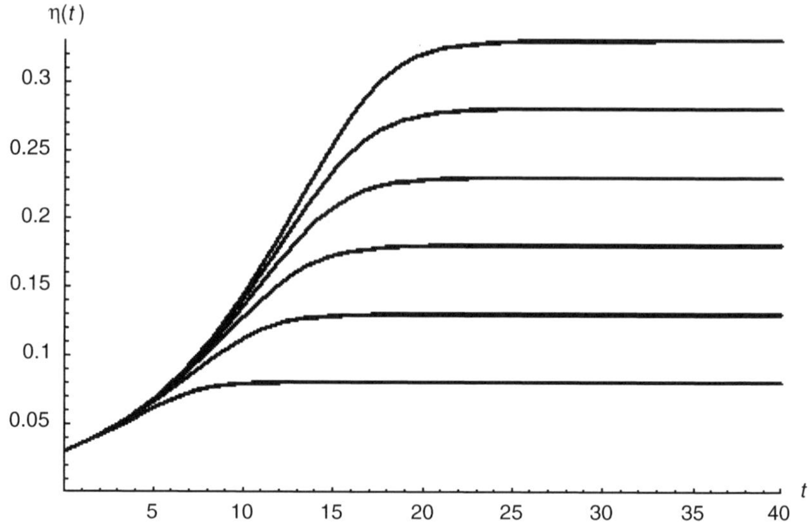

Figure 6.6 Effect of the yield coefficient on the solution of the Monod equation (the yield coefficient was changed from 0.05 to 0.3 with step 0.05, the other parameters were fixed (0.25, 0.5); initial conditions $x_0 = 0.3, S_0 = 1$)

called the exponential phase of microbial growth and the upper bend is the retardation phase (Monod, 1949). In some sense the change of the maximal specific growth rate θ_1 (for example increase) can be compensated by the opposite change of the Michaelis–Menten constant (decrease) and the graph of the solution will be very similar. Hence, the correlation between $\hat{\theta}_1$ and $\hat{\theta}_2$ means that if we plot such solutions of the Monod equation with very different values of θ_1, θ_2 then they will have the same level of nonlinear least squares estimation to the experimental points (Figure 6.7). It is possible to consider several alternative approaches to deal with this problem:

1. One could use a different mathematical model for the analysis of the data (but obviously in this case the Monod model is the correct chosen model).
2. One could increase the number r of experimental replications (but it is not known how many replications will guarantee good parameter estimates and whether the cost of replication is prohibitive).
3. One could improve the precision of the experimental measurements to reduce the dispersion of the measurement errors (this is not always possible and may require additional expensive equipment).
4. One could apply another mathematical estimation technique, for example maximum likelihood estimation, which is sometimes better than the simple nonlinear least square regression technique (Sommer *et al.*, 1995; Knightes

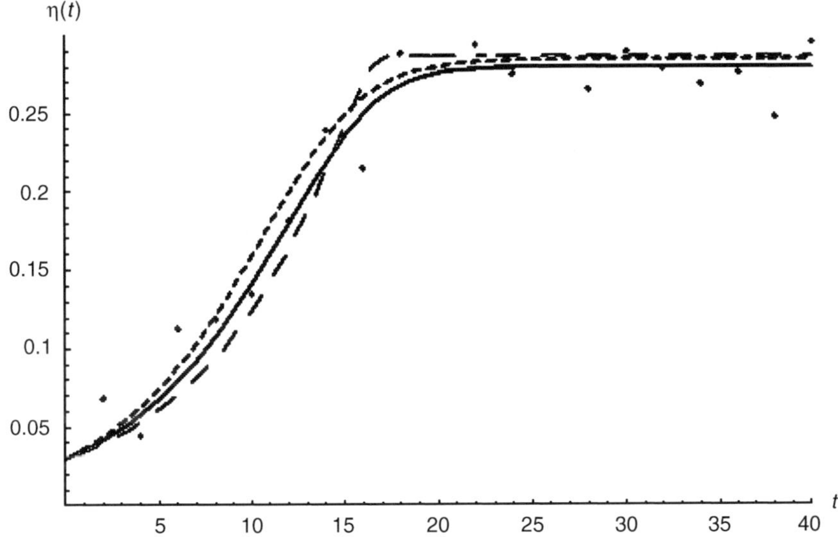

Figure 6.7 Illustration of the correlation of parameter estimation problem. Three solutions of the Monod model with different parameters but same least sequare value from the artificial experimental data set: solid line: (0.25,0.5,0.25) true parameter set; long dashed line: (0.1497, 0.047, 0.2565); short dashed line: (0.3418, 0.8224, 0.2547)

and Peters, 2000), or an estimation algorithm based on Lyapunov's stability theory (Zhang and Guay, 2002).

5. One could use an optimal experimental design to define the different experimental conditions (there is no guarantee of improvement by a design obtained from asymptotic theory, but if this approach is efficient it will have more advantages compared to the previous approaches because it does not cause additional costs).

Let us concentrate on the application of optimal experimental design and describe a way to construct an experimental design such that parameters will be identi.ed very well from a minimal number of experimental measurements n and a minimal number r of replications of the experiment. This approach for Monod-type models was developed theoretically (Munack, 1991; Baltes *et al.*, 1994; Versyck *et al.* 1998; Dette *et al.*, 2003e) and applied in several types of experiments (Vanrolleghem *et al.*, 1995; Ossenbruggen *et al.*, 1996; Merkel *et al.*, 1996). First of all it is known that the parameters of the Monod model can be identified (see Dette *et al.* 2003e; Petersen *et al.*, 2003; Holmberg and Ranta, 1982; Chappel and Godfrey, 1992). The minimum values of the least squares estimation for the regression model in a two-dimensional problem give an ellipse, and if the estimates of the parameters are closely correlated (as $\hat{\theta}_1$ and $\hat{\theta}_2$) then the ellipse is very elongated; when parameters are not correlated this regression ellipse is a circle. If we consider a problem with three parameters then the resulting object is a three-dimensional ellipsoid. But in this case parameter θ_3 is estimated very well by any experimental design which includes measurements of the stationary phase of microbial growth. Then our problem is two-dimensional, i.e. we need to estimate θ_1, θ_2. The eigenvalues of the Fisher information matrix are determined by the axes of this ellipse and their lengths are proportional to the square root of the inverse of the eigenvalues. The determinant of the Fisher information matrix is the area of this confidence ellipse. Our objective is to find an optimal experimental design in order to make this regression ellipse small in some sense.

It was proved in Dette *et al.* (2002) that the locally *D*-optimal design for the Monod model is a three-point design. Therefore to identify the Monod parameters it is enough to take measurements of the growing biomass at three optimal time points. The two points of the optimal design are on the lower and upper bends and the third point is on the horizontal asymptote. We can give a nonformal interpretation of this optimal design as follows. To obtain good estimates of our parameters we should make measurements at points of the microbial growth curve where the curve changes its behaviour determined by the parameters, i.e. the maximal specific growth rate determines the exponential phase of growth (Figure 6.4) and Michaelis–Menten constant determines the retardation phase of microbial growth (Figure 6.5). Then to obtain good estimates of those parameters it is necessary to take measurements at some critical points in the middle of those growth phases (i.e. in the middle of the lower and the upper bends). As for the yield coefficient, it is then obvious that this parameter can be determined from one measurement of the stationary growth phase (Figure 6.6). It seems truly amazing that the results obtained with advanced mathematical techniques can be explained in such a simple and obvious way.

Now having constructed the optimal designs we face several very important questions, which will determine how useful the optimal experimental design approach is for the Monod model. Some of the questions are theoretical, and some are practical:

1. How efficient is the application of the nonlinear optimal design in our experiments? Do we obtain a significant advantage from the application of those methods or maybe some naive experimental designs (as the uniform design) are nearly as good?
2. What is the relation between locally *D*-optimal design and other types of locally optimal experimental designs? Maybe, for example, locally *E*-optimal design will turn out to be more efficient in this case.
3. How precise should measurements be at optimal points? How stable is the optimal design with respect to experimental errors in the measurements?
4. The initial value of parameters θ^0 is crucial for the construction and performance of the optimal design, What will be the result of the experiment if this value deviates substantially from the 'true' value of the parameters? How will the optimal design change if the actual values of the parameters are not very close to the initial values?

Each of those questions is important in some sense. If this type of optimal design is not efficient compared to some naive experimental designs or if a much more efficient optimal design of another type exists we can reject this approach. If an optimal design is very efficient but not stable with respect to the experimental measurement error or is very sensitive to a change in the values of initial parameters, the design may also be inappropriate for practical purposes.

For optimal experimental design in the Monod model all those questions are answered in the affirmative and the answers are very positive. The improved efficiency of the obtained optimal design over the previous uniform design is obvious from the parameter estimation data (Table 6.1). The measurements in three optimal points provide much better approximations of the parameter θ^* than the measurements in 20 uniform time points! More statistically significant results relating to the efficiency of the constructed optimal design can be found in the literature (Dette *et al.*, 2003e). The comparison of different optimal designs has also been carried out (Dette *et al.*, 2002). It has been found that E- and the optimal design for estimating the individual coefficients are more efficient than D-optimal designs for estimating the parameters θ_1 and θ_2, but the D-optimal design is more efficient for the estimation of θ_3. However, if improvement of the accuracy in the estimation of the parameters θ_1 and θ_2 is considered to be more important, then the E- and e_2-optimal designs have some advantages (Dette *et al.*, 2002). The third and fourth questions have also been answered (Dette *et al.*, 2002, 2003e). It was found that optimal designs for the Monod model are robust with respect to the experimental errors and with respect to moderate misspecifications of the initial parameter.

Several theoretical considerations for microbiological models related to the Monod model have been published. Munack (1991) calculated the optimal feeding strategy for identification of Monod-type models using the modified E-criterion. Versyck *et al.* (1998) considered the fed-batch reactor and compared the modified E-criterion for Monod and Holdane kinetics. A theoretical comparison of those two models, including consideration of optimal experimental designs for parameter estimation and optimal control of the microbiological reactor, was undertaken by Smets with colleagues (Smets *et al.*, 2002), where the modified E-criterion and D-criterion were under consideration. Berkholz *et al.* (2000) applied the modified E-criterion in order to identify parameters of fed-batch hyaluronidase fermentation by *Streptococcus agalactiae*.

The application of optimal experimental design for parameter estimation of the Monod model was considered by Vanrolleghem *et al.* (1995) for activated sludge respiration. Growing microbial biomass in fed-batch experiments was characterized by their oxygen uptake with a respirograph biosensor. The single Monod model in specific form was successfully applied to the description of activated sludge kinetics:

$$\frac{dS_1}{dt} = -\frac{\mu_{\max} X}{Y_1} \frac{S_1}{K_{m1} + S_1} \tag{6.10}$$

$$\mathrm{OUR}_{\mathrm{ex}} = -(1 - Y_1) \frac{dS_1}{dt}, \tag{6.11}$$

where S_1 denotes the substrate concentration, X is the biomass concentration, Y_1 the yield coefficient, $\mu_{\max}$ the maximal specific growth rate, K_{m1} the saturation affinity constant and OUR_{ex} the exogenous oxygen uptake rate. The important assumption in this model is that the biomass concentration X is considered as a constant, because the ratio S_1/X is low and biomass growth can be neglected ($dX/dt = 0$). The efficiency of the D- and E- optimal designs for the estimation and identification of the Monod parameters was demonstrated in the experiments of organic substrate (acetate was chosen) utilization by activated sludge.

D-optimal designs have been very useful for the identification of parameters of the nitrification model for activated sludge batch experiments (Ossenbruggen *et al.*, 1996). The respiration rate of ammonium and nitrite was presented as a sum of two Michaelis–Menten functions, that is

$$q_0 = \frac{k_{\mathrm{NH}} S_{\mathrm{NH}}}{K_{\mathrm{NH}} + S_{\mathrm{NH}}} + \frac{k_{NO} S_{NO}}{K_{NO} + S_{NO}} \tag{6.12}$$

where q_0 denotes the specific respiration rate (oxygen consumption), S_{NH} is the ammonium (NH_4^+) concentration, S_{NO} the nitrite (NO_2^-) concentration, $k_{\mathrm{NH}}, k_{\mathrm{NO}}$ the maximum specific respiration rates, and $K_{\mathrm{NH}}, K_{\mathrm{NO}}$ the half-saturation constants for S_{NH} and S_{NO}, respectively. A D-optimality criterion was used to find good designs to model the two-step nitrification process with piecewise nonlinear models of ammonium and nitrite concentration. The experimental part included a series of 14 batch runs where activated sludge was spiked with ammonium chloride and sodium nitrite individually and in combination with recording of respiration activity.

6.3.2 Application of optimal experimental design in microbiological models

Numerous mathematical models other than the Monod are used in microbiology to describe kinetic processes. Practically all the famous theoretical biology growth models (such as logistic, Gompertz, exponential models, etc.) and the usual empirical regression models (linear, polynomial, exponential, different probability distributions, etc.) have been applied to study processes in microbiology. Some of them were modified and their parameters have special names and precise meanings in particular areas of microbiology. There are many analytical considerations of the optimal experiment design problem for models which are used in microbiology (Table 6.2) and several very successful examples of application of this technique in experimental practice.

Empirical regression models and growth models

Linear regression models. Linear regression models are often used in microbiology and science in general. The optimal experimental design problem for the linear

Table 6.2 Some examples of application the optimal experimental design for microbiological models

Model	Type of experiments	Design	Source
Empirical nonlinear regression models			
Exponential regression models	Application to numerous dynamic processes such as microbial growth and inactivation	Bayesian *D*-optimal design	Mukhopadhy and Haines (1995); Dette and Neugebauer (1997)
		D- and *c*-optimal designs	Han and Chaloner (2003)
Discrete and continuous logistic models	Microbial growth, risk assessments, infection processes	Bayesian design	Chaloner and Larntz (1989)
		D-optimal design	Sebastiani and Settimi (1997)
Monod model and closely related models			
Monod model	Microbial growth	*D*-, *E*-, *c*-, *A*-optimal designs	Munack (1991); Versyck *et al.* (1998); Smets *et al.* (2002); Dette *et al.* (2003e)
ASM Model No. 1	Microbial growth, biodegradation kinetics	*D*-optimal design	Hidalgo and Ayesa, 2001
Monod-type model	Activated sludge processes, respiratory measurements	*A*, D, *E* criteria	Vanrolleghem *et al.* (1995)
Monod-type model with two additive Michaelis–Menten functions	Nitrification process in the activated sludge batch experiments	*D*-optimal design	Ossenbruggen *et al.* (1996)
Monod-type model	Anaerobic degradation of the acetic acid in the batch and fed-batch experiments	*D*-optimal design	Merkel *et al.* (1996)
Haldane model	Substrate inhibited microbial growth	Modified *E*-optimal design	Versyck *et al.* (1998)
Models of predictive microbiology			
Bigelow model (Arrhenius-type model)	Inactivation of micro-organisms with temperature, determination of D,z values	*D*-optimal design	Cunha *et al.* (1997), Cunha and Oliveira (2000)
		Modified *E*-optimal design	Versyck *et al.* (1999)

Table 6.2 (*Continued*)

Model	Type of experiments	Design	Source
Baranyi model	Microbial growth model, application for the shelf-life prediction of food	*D*-optimal design	Grijspeerdta and Vanrolleghem *et al.* (1999)
Ratkowsky growth model incorporated with square root model	Inactivation of microbial activity by temperature, growth in suboptimal temperatures	Modified *E*-optimal design	Bernaerts *et al.* (2000, 2002)
Weibull probability distribution function	Microbial degradation and death kinetics, shelf-life failure, drying processes	*D*-optimal design	Cunha *et al.* (1998)
Related biochemical models			
Michaelis–Menten model	Fermentative reaction kinetics, basis for most structured model of microbial growth	*D*-, *E*-optimal design	Duggleby, (1979); Dunn (1988); Dette and Wong (1999)
		Bayesian design	Murphy *et al.* (2003)
Modified Michaelis–Menten model with enzyme deactivation	Batch enzymatic reactions where the activity of the enzyme decays with time	*D*-optimal design	Malcata (1992)
Hill equation	Enzymatic reactions, biosensors, microbial growth	*D*-optimal design	Bezeau and Endrenyi (1986)

model is one of the classical and very extensively investigated problems in the experimental design area (Fedorov, 1972; Pukelsheim, 1993). The classical example of the successful application of D-optimal experimental design for the linear model in microbiology is the design of an experiment for DNA extraction from the anaerobic rumen microbial community (Broudiscou *et al.*, 1998). The objective of the experiment is to assess the efficiency of the enzymatic pretreatment and to screen a number of detergents to maximize cell breakage and DNA recovery rate. Several potential detergents were tested: two enzymes (lysozyme and proteinase K) and five detergents (CHAPS, deoxycholic acid (SD), sodium dodecyl sulfate (SDS), sodium lauroyl sarcosine (SLaS) and triton X-100). The following linear model with seven factors was applied to optimize DNA recovery:

$$E[y] = b_0 + b_1 * (\text{proteinase K}) + b_2 * (\text{lysozyme}) + b_3 * (\text{CHAPS}) \\ + b_4 * (\text{SD}) + b_5 * (\text{SDS}) + b_6 * (\text{triton X} - 100) + b_7 * (\text{SLaS}).$$

Here $b_i, i = 0, \ldots, 7$ are the unknown coefficients to be estimated. Also the variables ‘proteinase K’ and ‘lysozyme’ were mutually excluded, as they relate to the same factor. The D-optimal design was numerically calculated by the NEMROD software (LPRAI, Université Aix-Marseille III, France). This method gave the optimal quantity of detergents for maximization of DNA recovery.

Exponential models. Exponential models are very often used for describing growth and death of micro-organisms, doseresponse analysis and risk assessment (Coleman and Marks, 1998), and kinetics of metabolite production. Exponential models are also incorporated in the numerous models in predictive microbiology for describing effects of temperature (Geeraerd *et al.*, 2000). The simplest single exponential model for microbial growth can be presented as a first-order differential equation

$$\frac{dX}{dt} = \mu_m X,$$

where X is a biomass concentration and μ_m the maximal specific growth rate. In this context the closely associated parameter, called the doubling time τ, is very often used in microbiology, i.e.

$$\tau = \frac{\ln 2}{\mu_m}.$$

The more complicated exponential models, including sums of several exponential functions, are also applied. For example, a model (Cobaleda *et al.*, 1994), which is the sum of two exponential terms with four unknown parameters, was used to fit the kinetic data of fusion between Newcastle disease virus and erythrocyte ghosts, i.e.

$$F(t) = A_1[1 - \exp(-k_1 t)] + A_2[1 - \exp(-k_2 t)],$$

where A_1, A_2, k_1, k_2 are constant (unknown) parameters and t represents time. The first exponential equation represents a fast reaction, which is the viral protein-dependent fusion process itself, and the second exponential equation represents a slow non-specific dequenching reaction (Cobaleda *et al.*, 1994). Another model, which is the sum of two exponentials with three unknown parameters, was used to describe *Escherichia coli* inactivation by pulsed electric fields (Alvarez *et al.*, 2003). The biological meaning of this model is that in one population of micro-organisms two subpopulations exist, where the first population is sensitive with respect to the inactivating factor and the second population is resistant (Pruitt and Kamau, 1993), i.e.

$$S(t) = p\exp(-k_1 t) + (1 - p)\exp(-k_2 t),$$

where $S(t)$ denotes the fraction of total survivors, t is the treatment time, p is the fraction of survivors in population 1 (sensitive population), $(1 - p)$ is the fraction of survivors in population 2 (resistant population) and k_1, k_2 are the specific death rates of subpopulations 1 and 2 respectively. The optimal designs of experiments for some of the exponential-type models have been analysed many times and their theoretical properties are well known (Melas, 1978; Mukhopadhyay and Haines, 1995; Dette and Neugebauer, 1997; Han and Chaloner, 2003; Dette *et al.*, 2003d). However, because the number of concrete practical applications is very small, several examples of applications of optimal designs for exponential models in food microbiology will be considered in Section 6.3.2.2 of this chapter.

Logistic models. The logistic-type model is traditionally very often used in biology and in particular in microbiology for the description dynamic of microbial growth, infection processes and risk assessment (Peleg, 1997; Coleman and Marks, 1998; Ratkowsky, 2002; Pouillot *et al.*, 2003). The simple continuous logistic model for microbial growth can be written as a first-order nonlinear differential equation, that is

$$\frac{\mathrm{d}X}{\mathrm{d}t} = \mu_m X(1 - kX),$$

where X is a biomass concentration, μ_m is the maximal specific growth rate, and k is a logistic growth constant. The biological meaning of the logistic growth constant k is the inverse of the maximal possible biomass concentration, i.e. $K = 1/k$.

The logistic model is one of the basic models in ecology, where it well known as the Verhilst–Pearl equation. This equation describes density-dependent growth of the biological population (Nicholson, 1954). The parameters of the logistic equation are the specific growth rate $(r = \mu_m)$ and maximal size of population $(K = 1/k)$ and have fundamental meanings in ecology as they are the basis for the concepts of r- and K-selections (May, 1981). It is interesting to note that the logistic equation can be described as the first two terms of Taylor‘s expansion of the growth equation. We will shortly present this classical consideration as developed by Lotka (1925).

Let us assume that the population (microbial culture) growth rate at any moment is a function of the population size (number or concentration of micro-organisms), i.e.

$$\frac{dX}{dt} = f(X).$$

Usually the function $f(X)$ can be expanded in a convergent Taylor series, that is

$$\frac{dX}{dt} = c_0 + c_1 X + c_2 X^2 + \cdots$$

If $X = 0$ then c_0 should equal 0 as well, because it is necessary that there is at least one micro-organism for the population to grow. Then if we set $f(X) = c_1 X$, the exponential model for the microbial growth is obtained where the coefficient c_1 is the maximal specific growth rate μ_m. The important property of this equation is that the zero growth rate may be registered only at $X = 0$ and a positive growth rate $dX/dt > 0$ is observed at any population size.

If the next term of the Taylor series is also under consideration, i.e.

$$f(X) = c_1 X + c_2 X^2,$$

then the logistic equation, with the coefficients $c_1 = \mu_m$ and $c_2 = -\mu_m k$, is obtained. The important property of the logistic equation is that it is the simplest equation where $f(X)$ has two roots, one at $X = 0$ and the second when the number of micro-organisms X reaches some positive saturation level. This property provides a typical sigmoid growth curve.

The logistic equation can be integrated by separation of variables and the solution is

$$X = \frac{1}{k(1 - e^{\alpha - \mu_m t})},$$

where α is the constant of integration, which determines the position of the solution relative to the origin. The asymptotic value $\lim_{t \to \infty} X = 1/k$ is the saturation level of the microbial culture (population) which cannot be exceeded because of the limitations of cultivation or of the natural environment.

Locally optimal experimental designs for the logistic model based on the *D*- and *E*-criteria and Bayesian designs have been extensively investigated from a theoretical prospective (Chaloner and Larntz, 1989; Ford *et al.*, 1992, Sebastiani and Settimi, 1997). We will briefly consider the Bayesian optimal design problem for the logistic model in Section 6.4.

The Hill model. The Hill equation (Bezeau and Endrenyi, 1986; Liu and Tzeng, 2000) is an empirical equation widely applied in biomedical sciences. This equation was proposed by Hill in 1910 to describe the binding of oxygen to haemoglobin (Bezeau and Endrenyi, 1986). Since then it has been widely applied in physiology, pharmacokinetic modelling and as one of the growth equations (Savageau, 1980). In microbiology this equation, for example, has been successfully applied to describe sporulation kinetics of *Bacillus thuringiensis* (Liu and Tzeng, 2000). In terms of enzyme kinetics the Hill equation describes the relationship between substrate concentration (c) and reaction velocity (v):

$$v = \frac{Vc^N}{(K + c^N)},$$

where V is the maximal velocity, K is the substrate concentration required to yield half of the maximal velocity (analogue of the Michaelis–Menten coefficient), and N is the Hill coefficient, which is the measure of deviation from hyperbolicity.

It is obvious that when $N = 1$ the Hill model gives the Michaelis–Menten model. At the same time the Hill equation is equivalent to the three-parameter logistic function (Pregibon, 1982)

$$y = A\frac{e^{\alpha+\beta x}}{1 + e^{\alpha+\beta x}},$$

where $A = V, \beta = N, \alpha = -N \ln K$, and $x = \ln c$. The D-optimal experimental design for the determination of the parameters V, K and N of the Hill equation or, what is the same, A, β and α of the logistic function, was investigated by M. Bezeau and L. Endrenyi (1986). Three-point optimal designs were constructed numerically and several important properties of the design were considered. In particular, the effect of the additive error of the observed value was investigated under the assumption that the error can be described by the power model, in which the variance of the error is proportional to the 'true' response raised to some power, i.e.

$$\sigma_i^2 = \sigma_0^2 v_i^{2\lambda},$$

where $0 \le \lambda \le 1$, and the variance σ_0^2 is the constant of proportionality. When $\lambda = 0$ it reduces to the case of homogeneous variance and when $\lambda = 1$ it reduces to the case of constant relative (or constant percentage) errors. The effect on the design efficiency of the parameter λ and initial parameter values was investigated. It was found that D-optimal designs for the Hill equation are robust with respect to moderate deviations from the assumed parameter values and efficient for different values of λ. The optimal designs were compared with the uniformly and logarithmically spaced designs. In general, the optimal design was always better than uniform design, but 4-, 5-, and 6-point logarithmical designs also performed very well. The optimal design for the case $\lambda = 0.5$ generally outperformed both uniformly and logarithmically spaced designs.

Optimal experimental design in microbial inactivation processes

Predictive microbiology (the quantitative microbial ecology of food) has received much attention during the last few decades. This relatively young research area is devoted to the analysis of microbial behaviour in food (McMeekin *et al.*, 2002). The central concept of predictive microbiology is that the growth, survival and death responses of microbes in foods should be modelled with respect to the main controlling factors, such as temperature, pH and water activity (Baranyi and Roberts, 1994; Roberts, 1995).

There is a difference between modelling approaches used in biotechnology and environmental microbiology and those used in predictive microbiology. The main differences are (Roberts, 1995):

1. The bacterial concentration of concern to food microbiologists is lower than that for fermentation microbiologists, and it may be inappropriate to use fermentation models.
2. The aim is different: in predictive microbiology it is important to prevent growth of micro-organisms rather than optimize it.
3. In food microbiology, as in environmental microbiology, the controlling factors influencing growth are generally rather heterogeneous and often poorly quantified.
4. Many foods are nutritionally rich, so nutrient limitation of microbial growth does not occur.

The classification techniques of the mathematical models used in predictive microbiology have unique features. Simple classification is based on three types of models: (1) growth models, (2) limit of growth (interface) models and (3) inactivation models (McMeekin and Ross, 2002). Evaluation of the death of micro-organisms is one of the most typical problems in food microbiology. Usually the experimental aspects of the research include a large number of experiments where the main affective factor and other factors vary. Therefore application of optimal experimental designs is a way to significantly reduce the cost of the experiments and to improve the quality of the obtained results. The typical problem is the analysis of the temperature inactivation of micro-organisms; different variations of temperature effects are most often used for growth inhibition or killing micro-organisms. The mathematical theory of this process has received much attention in recent decades.

The first-order kinetic model is the simplest and is often useful for describing inactivation of micro-organisms in one-factor experiments. For example, at a constant temperature

$$\frac{dN}{dt} = -k(T)N,$$

where N is the microbial population number (or density in a homogeneous model), and k is the specific rate of thermal inactivation, which is a constant at a given temperature. It is obvious that for this type of experiment the very thoroughly investigated optimal designs for the exponential regression model can be efficiently applied.

The Arrhenius-type model (Bigelow, 1921) and its subsequent modifications (Bernaerts *et al.*, 2000) are very often used for determining the inactivation constant k, i.e.

$$k(T) = \frac{2.303}{D_{\text{ref}}} \exp\left[\frac{2.303}{z}(T - T_{\text{ref}})\right],$$

where D_{ref} is the decimal reduction time for the reference temperature T_{ref}, and z is the number of degrees of temperature change required for a tenfold change in D value, i.e. decimal reduction time for an arbitrary temperature.

It is very common in food science to use this model and so the problem of the estimation of the parameters D and z is very important. Optimal experimental designs for this model were developed numerically using the D-criterion (Cunha *et al.*, 1997, Cunha and Oliveira, 2000) and the modified E-criterion (Versyck *et al.*, 1999). It was demonstrated theoretically that the optimal experimental designs are efficient and to illustrate the concept, a case study has been done using data from the literature (Cunha *et al.*, 1997).

Another way to inactivate micro-organisms by temperature is to keep products under suboptimal temperatures for microbial growth. Mathematical models used for describing those processes are not principally different from models of heat inactivation. Among the most popular models is the Ratkowsky model (Ratkowsky *et al.*, 1982) and its modifications (Baranyi and Roberts, 1994). Bernaerts *et al.* (2000) not only demonstrated theoretical efficiency of the E-optimal design for experiments in the Ratkowsky-type model but also conducted some practical experiments with *Escherichia coli* K12 to illustrate the application of the optimal design. The model for growth of micro-organisms as a function of temperature, proposed by Baranyi and Roberts (1994), is given by

$$\frac{\mathrm{d}n}{\mathrm{d}t} = \frac{\theta(t)}{\theta(t)+1}\mu_{\max}[T(t)][1 - \exp(n(t) - n_{\max})]$$

$$\frac{\mathrm{d}\theta(t)}{\mathrm{d}t} = \mu_{\max}[T(t)]\theta(t)$$

with an additional equation called the square root model, i.e.

$$\sqrt{\mu_{\max}[T(t)]} = b(T(t) - T_{\max}),$$

where $n(t)$ is the natural logarithm of the cell density (for homogeneous case), $n_{\max}$ is the natural logarithm of the maximal population density $N_{\max}$, and $\theta(t)$ is the variable related to the physiologic state of the microbial cells. The initial value $\theta(0)$ determines the lag phase duration which is assumed to be related to the physiologic condition of the inoculum. The maximum specific growth rate μ_{max}, as a function of the suboptimal growth temperature $T(t)$, is modelled by the two-parameter model of Ratkowsky *et al.* (1982), where b is a regression coefficient and T_{min} is termed the theoretical minimum growth temperature.

Theoretical results enabled the authors (Bernaerts *et al.*, 2000) to construct such optimal dynamic temperature inputs, where the parameters of the Ratkowsky square root model were identified from a single set of cell density measurements. In this particular case the model parameters are uncorrelated, which significantly simplifies the mathematical computations and efficiency of the optimal design technique. In a later paper they demonstrated (Bernaerts *et al.*, 2002) that a theoretical design with the modified E-criterion of the optimal temperature input was very efficient and relevant to real experimental applications. Constrained optimal temperature input guaranteed model validity and yielded accurate square root model parameters. The experimental part of the research included growth experiments with *E. coli* K12 in a computer-controlled bioreactor under optimal-valued temperature conditions and compared well with the classical static experimental design.

The probabilistic approach for the analysis of thermal inactivation considers lethal effects as probabilities (Turner, 1975; Peleg and Cole, 1998). This approach is an alternative to the deterministic consideration of microbial inactivation kinetics. The initial assumptions of the deterministic approach relate to the statements that the number of micro-organisms is so large that it is possible to consider the process as deterministic where each cell has the same probability of dying. Under those assumptions, the application of the differential equations, such as, for example, the first-order kinetics, is appropriate. In the probabilistic approach the death of each cell is considered as a single event and the 'death curve' is not a solution of a deterministic equation, but a probabilistic distribution (Turner, 1975; Peleg and Cole, 1998; Van Boekel, 2002). At the same time, there is a general theory of growth models developed in theoretical biology, where a natural connection between the probabilistic and deterministic approaches and different types of growth models is established (Turner *et al.*, 1976; Turner and Pruitt, 1978; Savageau, 1980; Chen and Christensen, 1985). In particular, under certain assumptions the Monod equation can be transformed into a special case of the logistic probability distribution (Christensen and Nyholm, 1984).

The logical consideration of the probabilistic inactivation process leads to the Weibull distribution as the most appropriate mathematical model for the interpretation of survival curves (Christensen, 1984; Peleg and Cole, 1998). In general, the Weibull distribution is widely used for reliability data, particularly the strengths of materials and failure times (Smith, 1991). In biology this distribution has been used as one of the growth models (Savageau, 1980), for dose-response function in toxicological studies (Christensen, 1984), and for the analysis of the microbial survival curves after different deactivation processes (Peleg and Cole, 1998; Peleg, 2000; Van Boekel, 2002; Collado *et al.*, 2003). Other types of probabilistic distributions are also often applied for survival process analyses (Peleg and Cole, 1998; Peleg, 2000), but the Weibull distribution has some natural advantages such as flexibility in fitting experimental data and using a minimal number of parameters (Van Boekel, 2002). The probability density function of the Weibull distribution (Smith, 1991) is given by

$$f(t) = \begin{cases} \frac{\beta}{\alpha}\left(\frac{t}{\alpha}\right)^{\beta-1}\exp\left(-\left(\frac{t}{\alpha}\right)^{\beta}\right), & t > 0 \\ 0, \quad t \leq 0, \end{cases}$$

where $\alpha, \beta > 0$. The *D*-optimal experimental design for estimation of the kinetics parameters of processes described by the Weibull probability distribution function was considered by Cunha *et al.* (1998). The optimal designs were numerically constructed for two different types of experiments. The first was the Weibull process of microbial thermal death under constant temperature. The optimal design in this case consisted of the set of two time sampling points, where fractional concentrations are irrational numbers (approximately $\eta_1 = 0.7032$ and $\eta_2 = 0.19245$) such that their product equals $1/\mathrm{e}^2$. The second experimental situation consisted of isothermal experiments conducted over a range of temperatures. In this case the rate parameter α of theWeibull distribution depended on temperature and followed the Arrhenius-type behaviour

$$\frac{1}{\alpha_i} = \frac{1}{\alpha_0}\exp\left(-\frac{Ea}{RT_i}\right).$$

At the same time, the parameter of the Weibull distribution β indicates the kinetic pattern and is independent of the temperature, within a limited range of temperatures. Those assumptions were confirmed in numerous published experiments (Cunha *et al.*, 1998). Therefore, the kinetic parameters for this practical situation are α_0, *Ea* and β. In this case the optimal design required not only the selection of sampling times but also the selection of optimal temperatures. It was found that the most informative measurements will be obtained in two isothermal experiments at the limit temperatures of the range of interest. In one experiment, two samples should be taken with the same conversions as in the first discussed experiment with constant temperature ($\eta_1 = 0.7032$ and $\eta_2 = 0.19245$), while in the other experiment one sample should be taken at the $\eta_3 = 1/\mathrm{e}$. Several case studies, based on published data, of the application of optimal experimental design for the Weibull model were presented by Cunha *et al.* (1998).

6.4 Bayesian Methods for Regression Models

The application of Bayesian methods for the the construction of good experimental designs in nonlinear regression models has been actively investigated in recent decades; it was considered in connection with the Michaelis–Menten type model in enzyme kinetics (Murphy *et al.*, 2002, 2003), with the exponential and related first-order equations (Chaloner, 1993; Mukhopadhyay and Haines, 1995; Dette and Neugebauer, 1997; Sivaganesan *et al.*, 2003), and with the logistic equation (Chaloner, 1993; Dette and Neugebauer, 1996).

The Bayesian design approach is based on the idea that a prior distribution for the unknown parameters can be specified. The posterior distribution of θ given the observations at the experimental conditions is then used for point estimations, hypothesis testing and prediction.

To be more specific, let us consider the Bayesian approach for the example of the logistic regression which has been very well investigated (Chaloner and Larntz, 1989; Chaloner, 1993; Dette and Neugebauer, 1996; Sun *et al.* 1996; Smith and Ridout, 2003). The logistic regression model can be presented as a problem of estimating the probability of success in the Bernoulli trial from a set of observations depending on an explanatory variable x. The data is given in the set $Y = \{y_1, \ldots, y_n\}$. The two unknown parameters are μ, β, i.e. the vector $\theta = \{\mu, \beta\}$, and the probability of success is given by

$$p(x, \theta) = 1/(1 + \exp(-\beta(x - \mu))).$$

The parameter μ is the value of x when the probability of success is 0.5. In microbiology and toxicology this parameter has a very important meaning, namely the 50% level of the effect, and is denoted by LD50. The second parameter β is the slope in the logit scale.

Under some realistic assumptions it can be shown that the posterior distribution of θ is approximately a multivariate normal distribution with mean equal to the maximum likelihood estimate of θ, say $\hat{\theta}$. Moreover, the covariance matrix is equal to the inverse of the observed Fisher information matrix, i.e. the matrix of the second derivatives of the log likelihood function evaluated at the maximum likelihood estimate of θ. It is also important to note that, for logistic regression, the observed and expected Fisher information matrices are identical (Chaloner and Larntz, 1989).

For an experimental design ξ on X, with n distinct design points $x_i, i = 1, \ldots n$ and proportions ω_i such that $\sum_{i=1}^{n} \omega_i = 1$, define (Chaloner and Larntz, 1989)

$$\begin{aligned}
p_i &= p(x_i, \theta)(1 - p(x_i, \theta)), \\
t &= \sum_{i=1}^{n} p_i \omega_i, \\
\bar{x} &= t^{-1} \sum_{i=1}^{n} p_i \omega_i x_i, \\
s &= \sum n_i \omega_i (x_i - \bar{x})^2,
\end{aligned}$$

here $t, \bar{x}$, and s all depend on the unknown parameter θ. The normalized Fisher information matrix $I(\theta, \eta)$ is given by

$$I(\theta, \eta) = \begin{pmatrix} \beta^2 t & -\beta t(\bar{x} - \mu) \\ -\beta t(\bar{x} - \mu) & s + t(\bar{x} - \mu)^2 \end{pmatrix}.$$

Because the matrix $I(\theta, \eta)$ is non-singular, the posterior distribution of θ using the design η is approximately multivariate normal with mean $\hat{\theta}$ and covariance matrix $(nI(\hat{\theta}, \eta)^{-1})$. Therefore preposterior expected losses can be approximated using the prior distribution of θ as the predictive distribution of $\hat{\theta}$.

Several different criteria can be used to determine an optimal experimental design (see Chaloner and Larntz, 1989; Chaloner and Verdinelli, 1995; Dette and Neugebauer, 1996, 1997). For example the Bayesian D-optimal is defined as the design which maximizes the function

$$\phi_1(\eta) = E_\theta \log \det I(\theta, \eta).$$

An alternative criterion is to maximize the function

$$\phi_2(\eta) = -E_\theta(\mathrm{tr} B(\theta) I(\theta, \eta)^{-1}),$$

where $B(\theta)$ is a given symmetric 2 by 2 matrix. In the particular case where only linear combinations of the parameters are of interest, the matrix $B(\theta)$ does not depend on θ and is a matrix with known entries. If nonlinear combinations of the parameters θ_i are of interest then $B(\theta)$ has entries which are functions of θ. Some Bayesian interpretation of these criteria in terms of minimization of expected loss can be found in Chaloner and Verdinelli (1995).

The theoretical efficiency of those criteria had been demonstrated by Chaloner and Larntz (1989). They considered the uniform and independent prior distributions of β and η. The results of these authors suggested that the Bayesian methodology for this model is preferable to the classical optimal experimental design procedure based on the best guess of parameter values, such as for example the locally D-optimal design.

The important particular case of the logistic model when the slope β is assumed to be known was considered by Dette and Neugebauer (1996). They proved that under a rather general assumption concerning the prior distribution $p_\xi(\theta)$, the Bayesian D-optimal design for the logistic regression with one unknown parameter is the one-point design at the point x^* which is given by the unique solution of the equation

$$\int_\theta \frac{2\mathrm{e}^{x-\theta}}{1 + \mathrm{e}^{x-\theta}} p_\xi(\theta) \mathrm{d}\theta - 1 = 0.$$

The numerical investigation of the same one-parametric regression model has shown that Bayesian design of experiments is an efficient tool for the statistical inference in nonlinear models (Sun *et al.*, 1996).

Unfortunately, Bayesian methods are not very often applied in microbiological practice despite their potential effectiveness and well-developed theoretical background. However, several very interesting examples of practical applications have been published recently. Bois (2001) applied a Bayesian approach to the network describing bacterial growth in the drinking-water distribution system. Authors used a mechanistic compartmental model based on differential equations to describe bacterial population behaviour in the network, affected by dissolved organic carbon and chlorine concentrations. The differential equations were based on the mass conversation laws for the chemical reaction involved. Free living biomass was increased by consumption biodegradable BOD, decreased by natural mortality and by killing by chlorine, also it was shown that biomass can attach to the cement walls and come from shearing off the walls. Consumption of BOD was assumed to follow Monod's kinetics. A Bayesian approach was used to combine prior knowledge from the scientific literature about the possible range of values for each model parameter, and data from experiments, in the context of the model parameters dynamics. A priori truncated lognormal or loguniform distributions were assigned to the parameters based on the literature. The joint posterior distribution of the parameters was determined numerically by application of Markov chain Monte Carlo techniques. This method gave an approximate convergence in about 100 000 iterations. Different realistic scenarios of the model system development were considered.

Pouillot *et al.* (2003) applied Bayesian methods for the analyses of growth of *Listeria monocytogenes* in milk under different temperatures of incubation. They developed a typical model for the predictive microbiology based on the logistic equation with time delay, which represented the lag phase of microbial growth as a primary model and a secondary model which described the effect of incubation temperature on the maximal specific growth rate. Published growth data (total 124 growth curves) of *Listeria monocytogenes* in milk was used to estimate the prior distributions of parameters. Then using Markov chain Monte Carlo techniques the posterior parameter distributions were obtained.

6.5 Conclusions

The present chapter presents some results for optimal experimental design for estimating parameters in nonlinear regression models, which are usually used in microbiology. Papers in the statistical literature devoted to this problem provide some justification for the application of the least squares method for estimating the unknown parameters in these nonlinear regression models. For a sufficiently large sample size, the least squares estimates are consistent and their distribution can be approximated by a normal distribution. Intensive simulation studies elucidated that for realistic sizes of experiments the theoretical covariance matrix of the least squares estimator is quite close to the sampling one. This observation provides a starting point for the construction and implementation of optimal experimental designs, which maximize or minimize an appropriate function of the theoretical covariance matrix. Considerable efforts were made to construct and study locally optimal designs. It is demonstrated that for many models in microbiology locally optimal designs are not too sensitive with respect to the specification of an initial value for the unknown parameter. Moreover, the application of these designs yields a substantial improvement with respect to the accuracy of the estimates and a reduction of the correlation between them.

In many cases the number of different experimental conditions defined by a locally optimal design is equal to the number of parameters in the model. It is rather difficult and not always possible to prove this merely theoretically, but it was confirmed for most of the commonly used models in microbiology models by intensive empirical studies. Elaboration of more advanced Bayesian or maximin-optimal designs seems to be promising, because these designs are more robust with respect to a substantial misspecification of the unknown parameters by the experimenter. However, the determination of optimal designs with respect to these more sophisticated optimality criteria is a substantially harder problem compared the locally optimality criteria. So far Bayesian and maximin optimal designs have been constructed only for the simplest models with one or two nonlinear parameters.

It should be stressed that the implementation of optimal designs is a potentially very important opportunity to improve efficiency of experimental studies in microbiology without increasing the costs of the experiments. However, at present, there are only a few real applications of optimal designs in microbiological practice. The authors hope that this chapter will clearly demonstrate the practical benefits of the optimal design methodology for estimating parameters of nonlinear regression models in microbiology.

Acknowledgements

The work of H. Dette and V.B. Melas was supported by the Deutsche Forschungsgemeinschaft (SFB 475: Komplexitätsreduktion in multivariaten Datenstrukturen). The authors are also grateful to Professor Charles Suffel (Stevens Institute of Technology), who read a preliminary version of this manuscript and made many extremely useful suggestions to improve the presentation. The authors would also like to thank S. Biedermann for some help with the graphical representations and Isolde Gottschlich, who carried out several corrections in a former version of the manuscript.

References

Aborhey, S. and Williamson, D. (1978). State and parameter estimation of microbial growth processes. *Automatica*, **14**(5), 493–498.

Alvarez, I., Virto, R., Raso J. and Condon, S. (2003). Comparing predicting models for the *Escherichia coli* inactivation by pulsed electric fields. *Innov. Food Sci. Emerg. Technol.*, **4**(2), 195–202.

Atkinson, A. C. and Cook, R. D. (1995). *D*-optimum designs for heteroscedastic linear models. *J. Amer. Statist. Assoc.*, **80**, 204–212.

Baltes, M., Schneider, R., Sturm, C. and Reuss, M. (1994). Optimal experimental design for parameter estimation in unstructured growth models. *Biotechnol. Prog.*, **10**(5), 480–488.

Baranyi, J. and Roberts, T. A. (1994). A dynamic approach to predicting bacterial growth in food. *Int. J. Food Microbiol.*, **23**(3–4), 277–294.

Baranyi, J. and Roberts, T. A. (1995). Mathematics of predictive food microbiology. *Int. J. Food Microbiol.*, **26**(2), 199–218.

Berkholz, R., Rohlig, D. and Guthke, R. (2000). Data and knowledge based experimental design for fermentation process optimization. *Enzyme Microb. Technol.*, **27**(10), 784–788.

Bernaerts, K., Servaes, R. D., Kooyman, S., Versyck, K. J. and Van Impe, J. F. (2002). Optimal temperature input design for estimation of the square root model parameters: parameter accuracy and model validity restrictions. *Int. J. Food Microbiol.*, **73**, 145–157.

Bernaerts, K., Versyck, K. J. and Van Impe, J. F. (2000). On the design of optimal dynamic experiments for parameter estimation of a Ratkowsky-type growth kinetics at suboptimal temperatures. *Int. J. Food Microbiol.*, **54**(1–2), 27–38.

Bezeau, M. and Endrenyi, L. (1986). Design of experiments for the precise estimation of doseresponse parameters: the Hill equation. *J. Theor. Biol.*, **123**(4), 415–430.

Biedermann, S., Dette, H. and Pepelyshev, A. (2003). Maximin optimal designs for the compartmental model. Preprint Ruhr-Universität Bochum, http://www.ruhr-unibochum.de/mathematik3/preprint.htm

Bigelow, W. D. (1921). The logarithmic nature of thermal death time curves. *J. Infect. Dis.*, **29**, 528–536.

Blok, J. and Struys, J. (1996). Measurement and validation of kinetic parameter values for prediction of biodegradation rates in sewage treatment. *Ecotoxicol. and Environ. Saf.*, **33**(3), 217–227.

Bois, F. Y. (2001). Applications of population approaches in toxicology. *Toxicol. Lett.*, **120** (1–3), 385–394.

Broudiscou, L.-P., Geissler, H. and Broudiscou, A. (1998). Estimation of the growth rate of mixed ruminal bacteria from short-term DNA radiolabeling. *Anaerobe*, **4**(3), 145–152.

Chaloner, K. (1993). A note on optimal Bayesian design for nonlinear problems. *J. Statist. Plann. Inference*, **37**(2), 229–235.

Chaloner, K. and Larntz, K. (1989). Optimal Bayesian design applied to logistic regression experiments. *J. Statist. Plann. Inference*, **21**(2), 191–208.

Chaloner, K. and Verdinelli, I. (1995). Bayesian experimental design: a review. *Statist. Sci.*, **10**, 273–304.

Chappell, M. J. and Godfrey, K. R. (1992). Structural identifiability of the parameters of a nonlinear batch reactor model. *Math. Biosci.*, **108**(2), 241–251.

Chen, Ch.-Y. and Christensen, E. R. (1985). A unified theory for microbial growth under multiple nutrient limitation. *Water Res.*, **19**(6), 791–798.

Chernoff H. (1953). Locally optimal design for estimator parameters. *Ann. Math. Stat.*, **24**, 586–602.

Christensen, E. R. (1984). Dose-response functions in aquatic toxicity testing and the Weibull model. *Water Res.*, **18**(2), 213–221.

Christensen, E. R. and Nyholm, N. (1984). Ecotoxicological assays with algae: Weibull doseresponse curves. *Envir. Sci. Technol.*, **18**(9), 713–718.

Cobaleda, C., Garcha-Sastre, A. and Villar, E. (1994). Fusion between Newcastle disease virus and erythrocyte ghosts using octadecyl Rhodamine B fluorescence assay produces dequenching curves that fit the sum of two exponentials. *Biochem. J.*, **300**(2), 347–354.

Coleman, M., Marks, H.(1998). Topics in dose-response modelling. *J. Food Prot.*, **61**(11), 1550–1559.

Collado, J., Fernandez, A., Rodrigo, M., Camats, J. and Lopez, A. M. (2003). Kinetics of deactivation of *Bacillus cereus* spores. *Food Microbiol.*, **20**(5), 545–548.

Cunha, L. M. and Oliveira, F. A. R. (2000). Optimal experimental design for estimating the kinetic parameters of processes described by the first-order Arrhenius model under linearly increasing temperature profiles. *J. Food Eng.*, **46**(1), 53–60.

Cunha, L. M., Oliveira, F. A. R. and Oliveira, J. C. (1998). Optimal experimental design for estimating the kinetic parameters of processes described by the weibull probability distribution function. *J. Food Eng.*, **37**(2), 175–191.

Cunha, L. M., Oliveira, F. A. R., Brandao, T. R. S. and Oliveira, J. C. (1997). Optimal experimental design for estimating the kinetic parameters of the bigelow model. *J. Food Eng.*, **33**(1–2), 111–128.

Dette, H. and Biedermann, S. (2003). Robust and efficient designs for the Michaelis–Menten model. *J. Amer. Statist. Assoc.*, **98**(463), 679–686.

Dette, H., Haines, L. and Imhof, L. (2003a). Maximin and Bayesian optimal designs for regression models. Preprint Ruhr–Universität Bochum, http://www.ruhr-unibochum.de/mathematik3/preprint.htm

Dette, H., Haines, L. and Imhof, L. (2003b). Maximin and Bayesian for heteroscedastic polynomial regression models. Preprint Ruhr-Universität Bochum, http://www.ruhr-unibochum.de/mathematik3/preprint.htm

Dette, H., Melas, V. B. and Pepelyshev, A. (2003c). Standardized maximin *E*-optimal designs for the Michaelis Menten model. *Stat. Sinica*, **13**, 1147–1167.

Dette, H., Melas, V. B. and Pepelyshev, A. (2003d). Locally *E*-optimal designs for exponential regression models. Preprint Ruhr-Universität Bochum, http://www.ruhr-unibochum.de/mathematik3/preprint.htm

Dette, H., Melas, V. B., Pepelyshev, A. and Strigul, N. (2002). Efficient design of experiments in the Monod model. Ruhr-Universität Bochum. http://www.ruhr-unibochum.de/mathematik3/preprint.htm (30 pp.)

Dette, H., Melas, V. B., Pepelyshev, A. and Strigul, N. (2003e). Efficient design of experiments in the Monod model. *J. R. Stat. Soc. B*, 65, Part 3, 725–742.

Dette, H., Melas, V. B. and Wong, W. K. (2004). Optimal design for goodness-of-fit of Michaelis–Menten enzyme kinetics. Preprint Ruhr-Universität Bochum, http://www.ruhruni-bochum.de/mathematik3/preprint.htm

Dette, H. and Neugebauer, H.-M. (1996). Bayesian optimal one point designs for one parameter nonlinear models. *J. Statist. Plann. Inference*, **52**(1), 17–31.

Dette, H. and Neugebauer, H.-M. (1997). Bayesian *D*-optimal designs for exponential regression models. *J. Statist. Plann. Inference*, **60**(2), 331–349.

Dette, H. and Wong, W. K. (1999). *E*-optimal designs for the Michaelis–Menten model. *Statist. Probab. Lett.*, **44**(4), 405–408.

Dochain, D., Vanrolleghem, P. A. and Van Daele, M. (1995). Structural identifiability of biokinetic models of activated sludge respiration. *Water Res.*, **29**(11), 2571–2578.

Duggleby, R. G. (1979). Experimental designs for estimating the kinetic parameters for enzyme-catalysed reactions. *J. Theoret. Biol.*, **81**, 671–684.

Dunn, G. (1988). Optimal designs for drug, neurotransmitter and hormone receptor assays. *Statist. Med.*, **7**, 805–815.

Ellis, T. G., Barbeau, D. S., Smets, B. F. and Grady Jr, C. P. L. (1996). Respirometric technique for determination of extant kinetic parameters describing biodegradation. *Water Environ. Res.*, **68**(5), 917–926.

Fedorov, V. V. (1972). *Theory of Optimal Experiments*, Academic Press, New York.

Ferenci, Th. (1999). 'Growth of bacterial cultures' 50 years on: towards an uncertainty principle instead of constants in bacterial growth kinetics. *Res. Microbiol.*, **150**(7), 431–438.

Ford, I. and Silvey, S. D. (1980). A sequentially constructed design for estimating a nonlinear parametric function. *Biometrika*, **67**, 381-388.

Ford, I., Titterington, D. M. and Wu, C.-F. J. (1985). Inference on sequential design. *Biometrika*, **72**, 545–552.

Ford, I., Torsney, B. and Wu, C.-F. J. (1992). The use of a canonical form in the construction of locally optimal designs for nonlinear problems. *J. R. Stat. Soc. B*, **54**(2), 569–583.

Fu, W. and Mathews, A. P. (1999). Lactic acid production from lactose by *Lactobacillus plantarum*: kinetic model and effects of pH, substrate, and oxygen. *Biochem. Eng. J.*, **3**(3), 163–170.

Geeraerd, A. H., Herremans, C. H. and Van Impe, J. F. (2000). Structural model requirements to describe microbial inactivation during a mild heat treatment. *Int. J. Food Microbiol.*, **59**(3), 185–209.

Goudar, C. T. and Ellis, T. G. (2001). Explicit oxygen concentration expression for estimating extant biodegradation kinetics from respirometric experiments. *Biotechnol. Bioeng.*, **75**(1), 74–81.

Guha, S. and Jaffe, P. R. (1996). Determination of Monod kinetic coefficients for volatile hydrophobic organic compounds. *Biotechnol. Bioeng.*, **50**(6), 693–699.

Han, C.and Chaloner, K. (2003). *D*- and *c*-optimal designs for exponential regression models used in viral dynamics and other applications. *J. Statist. Plann. Inference*, **115**(2), 585–601.

Holmberg, A. (1982). On the practical identifiability of microbial growth models incorporating Michaelis–Menten type nonlinearities. *Math. Biosci.*, **62**(1), 23–43.

Holmberg, A. and Ranta, J. (1982). Procedures for parameter and state estimation of microbial growth process models. *Automatica*, **18**(2), 181–193.

Hopkins, J. C. and Leipold, R. J. (1996). On the dangers of adjusting the parameters values of mechanism-based mathematical models. *J. Theor. Biol.*, **183**(4), 417–427.

Howell, J. A. (1983). Mathematical models in microbiology: mathematical tool-kit. In: *Mathematics in Microbiology*, Basin, M. (ed.), Academic Press, London, New York, 1–37.

Jennrich, R. J. (1969). Asymptotic properties of non-linear least squares estimators. *Ann. Math. Statist.*, **40**, 633–643.

Karlin, S. and Studden, W. (1966). *Tchebyshe. Systems: with Application in Analysis and Statistics*. John Wiley & Sons, Ltd, New York.

Kiefer, J. and Wolfowitz, J. (1960). The equivalence of two extremum problems. *Canad. J. Math.*, **14**, 363–366.

Knightes, C. D. and Peters, C. A. (2000). Statistical analysis of nonlinear parameter estimation for Monod biodegradation kinetics using bivariate data. *Biotechnol. Bioeng.*, **69**(2), 160–170.

Liu, B. L. and Tzeng, Y. M. (2000). Characterization study of the sporulation kinetics of *Bacillus thuringiensis. Biotechnol. Bioeng.*, **68**(1), 11–17.

Lopez-Fidalgo, J. and W. K. Wong. (2002). Design issues for the Michaelis–Menten model. *J. Theor. Biol.*, **215**(1), 1–11.

Lotka, A. J. (1925). *Elements of Physical Biology*, Williams & Wilkins, Baltimore.

McMeekin, T. A., Olley, J., Ratkowsky, D. A. and Ross, T. (2002). Predictive microbiology: towards the interface and beyond. *Int. J. Food Microbiol.*, **73**(2–3), 395–407.

McMeekin, T. A. and Ross, T. (2002). Predictive microbiology: providing a knowledge-based framework for change management. *Int. J. Food Microbiol.*, **78**(1–2), 133–153.

Magbanua, B. S., Lu, Y. T. and Grady, C. P. L. (1998). A technique for obtaining representative biokinetic parameter values from replicate sets of parameter estimates. *Water Res.*, **32**(3), 849–855.

Malcata, F. X. (1992). Starting *D*-optimal designs for batch kinetic studies of enzyme-catalyzed reactions in the presence of enzyme deactivation. *Biometrics*, **48**(3), 929–938.

May, R. M. (1981). *Theoretical Ecology: Principles and Applications*, Sinauer Associates, Sunderland, Mass.

Melas, V. B. (1978). Optimal designs for exponential regression. *Math. Operationsforsch. Stat., Series Statistics*, **9**(1), 45–59.

Melas, V. B. and Strigul, N. S. (1999). Investigation of the nonlinear regression Monod model by stochastic simulation. In: *Statistical Models with Applications to Econometrics and Related Fields*, S. M. Ermakov and Yu. N. Kashtanov (eds). Publishers of St Petersburg State University, 118–127.

Merkel, W., Schwarz, A., Fritz, S., Reuss, M. and Krauth, K. (1996). New strategies for estimating kinetic parameters in anaerobic wastewater treatment plants. *Water Sci. Technol.*, **34**(5–6), 393–401.

Monod, J. (1949). The growth of bacterial cultures. *Ann. Rev. Microbiol.*, **3**, 371–393.

Mukhopadhyay, S. and Haines, L. M. (1995). Bayesian *D*-optimal designs for the exponential growth model. *J. Statist. Plann. Inference*, **44**(3), 385–397.

Müller, Ch.H. (1995). Maximin efficient designs for estimating nonlinear aspects in linear models. *J. Statist. Plann. Inference*, **44**, 117–132.

Munack, A. (1991). Optimization of sampling. In: *Biotechnology*, Vol. 4, VCH Weinheim., 252–264.

Murphy, E. F., Gilmour, S. G. and Crabbe, M. J. C. (2002). effective experimental design: enzyme kinetics in the bioinformatics era. *Drug Discov. Today*, **7**(20), 187–191.

Murphy, E. F., Gilmour, S. G. and Crabbe, M. J. C. (2003). efficient and accurate experimental design for enzyme kinetics: Bayesian studies reveal a systematic approach. *J. Biochem. Biophys. Methods*, **55**(2), 155–178.

Nicholson, A. J. (1954). An outline of the dynamics of animal populations. *Austr. J. Zool.*, **2**, 9–65.

Ossenbruggen, P. J., Spanjers, H. and Klapwik, A. (1996). Assessment of a two-step nitrification model for activated sludge. *Water Res.*, **30**(4), 939–953.

Peleg, M. (1997). Modelling microbial populations with the original and modified versions of the continuous and discrete logistic equations. *CRC Crit. Rev. Food Sci. Nutr.*, **37**(5), 471–490.

Peleg, M. (2000). Microbial survival curves – the reality of flat 'shoulders' and absolute thermal death times. *Food Res. Intern.*, **33**(7), 531–538.

Peleg, M. and Cole, M. B. (1998). Reinterpretation of microbial survival curves. *CRC Crit. Rev. Food Sci. Nutr.*, **38**(5), 353–380.

Petersen, B., Gernaey, K., Devisscher, M., Dochain, D. and Vanrolleghem, P. A. (2003). A simplified method to assess structurally identifiable parameters in Monod-based activated sludge models. *Water Res.*, **37**(12), 2893–2904.

Pirt, S. J. (1975). *Principles of Microbe and Cell Cultivation*, John Wiley & sons, Ltd, New York.

Poschet, F., Geeraerd, A. H., Scheerlinck, N., Nicolai, B. M. and Van Impe, J. F. (2003). Monte Carlo analysis as a tool to incorporate variation on experimental data in predictive microbiology. *Food Microbiol.*, **20**(3), 285–295.

Pouillot, R., Albert, I., Cornu, M. and Denis, J. B. (2003). Estimation of uncertainty and variability in bacterial growth using Bayesian inference. Application to *Listeria monocytogenes*. *Int. J. of Food Microbiol.*, **81**(2), 87–104.

Pregibon, D. (1982). Resistant fits for some commonly used logistic models with medical application. *Biometrics*, **38**(2), 485–498.

Pruitt, K. and Kamau, D. N. (1993). Mathematical models of bacteria growth, inhibition and death under combinated stress conditions. *J. Ind. Microbiol.*, **12**, 221–231.

Pukelsheim, F. (1993). *Optimal Design of Experiments*, John Wiley & Sons, Inc., New York.

Pukelsheim, F. and Torsney, B. (1991). Optimal designs for experimental designs on linearly independent support points. *Ann. Statist.*, **19**, 1614-1625.

Ratkowsky, D. A. (2002). Some examples of, and some problems with, the use of nonlinear logistic regression in predictive food microbiology. *Int. J. Food Microbiol.*, **73**(2–3), 119–125.

Ratkowsky, D. A., Olley, J., McMeekin, T. A. and Ball, A. (1982). Relationship between temperature and growth rate of bacterial cultures. *J. Bacteriol.*, **149**(1), 1–5.

Roberts, T. A. (1995). Microbial growth and survival: developments in predictive modelling. *Int. Biodeterior. Biodegradation*, **36**(3–4), 297–309.

Saez, P. B. and Rittmann, B. E. (1992). Model-parameter estimation using least squares. *Water Res.*, **26**(6), 789–796.

Savageau, M. A. (1980). Growth equations: a general equation and a survey of special cases. *Math. Biosci.*, **48**(3–4), 267–278.

Schirmer, M., Butler, B. J., Roy, J. W., Frind, E. O. and Barker, J. F. (1999). A relative-leastsquares technique to determine unique Monod kinetic parameters of BTEX compounds using batch experiments. *J. Contam. Hydrol.*, **37**(1–2), 69–86.

Sebastiani, P. and Settimi, R. (1997). A note on *D*-optimal designs for a logistic regression model. *J. Statist. Plann. Inference*, **59**(2), 359–368.

Silvey, S. D. (1980). *Optimal Design*. Chapman and Hall, London.

Sivaganesan, M., Rice, E. W. and Marinas, B. J. (2003). A Bayesian method of estimating kinetic parameters for the inactivation of *Cryptosporidium parvum* oocysts with chlorine dioxide and ozone. *Water Res.*, **37**(18), 4533–4543.

Smets, I. Y. M., Versyck, K. J. E. and Van Impe, J. F. M. (2002). Optimal control theory: a generic tool for identification and control of (bio-)chemical reactors. *Ann. Rev. Control*, **26**(1), 57–73.

Smith, D. M. and Ridout, M. S. (2003). Optimal designs for criteria involving log(potency) in comparative binary bioassays. *J. Statist. Plann. Inference*, **113**(2), 617–632.

Smith, R. L. (1991). Weibull regression models for reliability data. *Reliability Eng. System Safety*, **34**(1), 55–76.

Sommer, H. M., Holst, H., Spliid, H. and Arvin, E. (1995). Nonlinear parameter estimation in microbiological degradation systems and statistic test for common estimation. *Environ. Int.*, **21**(5), 551–556.

Sun, D., Tsutakawa, R. K. and Lu, W.-S. (1996). Bayesian design of experiment for quantal responses: what is promised versus what is delivered. *J. Statist. Plann. Inference*, **52**(3), 289–306.

Turner, M. E. (1975). Some classes of hit-theory models. *Math. Biosci.*, **23**(3–4), 219–235.

Turner, M. E., Bradley, E. L., Kirk, K. A. and Pruitt, K. M. (1976). A theory of growth. *Math. Biosci.*, **29**(3–4), 367–373.

Turner, M. E. and Pruitt, K. M. (1978). A common basis for survival, growth, and autocatalysis. *Math. Biosci.*, **39**(1–2), 113–123.

Van Boekel, M. A. J. S. (2002). On the use of the Weibull model to describe thermal inactivation of microbial vegetative cells. *Int. J. Food Microbiol.*, **74**(1–2), 139–159.

Vanrolleghem, P. A., Van Daele, M. and Dochain, D. (1995). Practical identifiability of a biokinetic model of activated sludge respiration. *Water Res.*, **29**(11), 2561–2570.

Vanrolleghem, P. A., Spanjers, H., Petersen, B., Ginestet, Ph. and Takacs, I. (1999). Estimating (combinations of) activated sludge model no. 1 parameters and components by respirometry. *Water Sci. Technol.*, **39**(1), 195–214.

Versyck, K. J., Bernaerts, K., Geeraerd, A. H. and Van Impe, J. F. (1999). Introducing optimal experimental design in predictive modelling: A motivating example. *Int. J. Food Microbiol.*, **51**(1), 39–51.

Versyck, K. J., Claes, J. E. and Van Impe, J. F. (1998). Optimal experimental design for practical identification of unstructured growth models. *Math. Comput. Simulation*, **46**(5–6), 621–629.

Woodroofe, M. and Coad, D. S. (2002). Corrected confidence sets for sequentially designed experiments: examples. *Sequential Analysis*, **21**, 191–218.

Wu, C. F. J. (1985). Efficient sequential designs with binary data. *J. Amer. Stat. Assoc.*, **80**, 974–984.

Zhang, T. and Guay, M. (2002). Adaptive parameter estimation for microbial growth kinetics. *AIChE J.*, **48**(3), 607–616.

7

Selected Issues in the Design of Studies of Interrater Agreement

Allan Donner[1,3] and Mekibib Altaye[2]

[1]*Department of Epidemiology and Biostatistics, The University of Western Ontario, London, Canada, N6A 5C1*

[2]*Center for Epidemiology and Biostatistics, Children Hospital Medical Center, Cincinnati, Ohio 45229-3039, USA*

[3]*Robarts Clinical Trials, Robarts Research Institute, London, Ontario, Canada*

7.1 Introduction

As noted by Shrout (1998), the last decade of the twentieth century saw a proliferation of statistical advances in the measurement of interrater agreement. These advances have led to inferential procedures for indices of agreement that encompass a variety of outcome measures (continuous, dichotomous, multinomial, ordinal) in the presence of two or more raters, with and without covariate adjustment. A recent review of many of these developments is given by Banerjee *et al.* (1999).

As pointed out by Kraemer (1992), many agreement indices were originally developed and applied on an atheoretic basis. However, the application of statistical modelling to these problems has now provided investigators with new tools for both the design and analysis of agreement studies. For example, several authors (e.g. Cantor, 1996; Walter *et al.* 1998; Donner and Eliasziw, 1992; Flack *et al.* 1998, Altaye *et al.*, 2001a) have presented formulae or tables for sample size estimation that allow studies to be designed that have sufficient power to differentiate between acceptable and unacceptable levels of agreement. Yet, as pointed out by Shrout

Applied Optimal Designs Edited by M.P.F. Berger and W.K. Wong
 ISBN: 0-470-85697-1 (HB)

(1998) in the context of psychiatric research, investigators often have 'limited populations of patients and restricted number of colleagues who can serve as replicate expert raters'. Moreover, purely budgetary constraints frequently act as yet another barrier to designing a study large enough to provide the desired level of precision.

In this chapter we discuss various design issues in studies of interrater agreement, focusing on decisions concerning the outcome measure selected, the number of subjects enrolled and the number of raters recruited per subject. We are specifically interested in the extent to which various design choices differ in their implications for overall required sample size.

Donner and Eliasziw (1994) evaluated the relative efficiency (RE) associated with the choice of a dichotomous rather than a continuous outcome variable for the case of two raters, where RE was defined as the ratio of the number of subjects required to maintain a given level of statistical power. In Section 7.2 we extend these results to the case of an arbitrary number of raters m.

Many studies of interrater agreement lead naturally to an outcome variable having more than two categories. For various practical reasons these categories are often collapsed to create a dichotomous variable that is simpler to interpret. In Section 7.3 we examine the statistical implications of this policy by evaluating the relative efficiency associated with the choice of a polychomotous rather than a dichotomous trait.

A linear cost model that incorporates the cost of recruiting a subject, the cost of recruiting a rater and the cost of obtaining a measurement on a subject already recruited is introduced in Section 7.4. This model is used to determine the number of raters and subjects needed to obtain maximum precision for estimating interrater agreement with respect to a dichotomous outcome variable while adhering to a fixed budget.

Some final comments and an overall summary of our results are presented in Section 7.5.

7.2 The Choice between a Continuous or Dichotomous Outcome Variable

We assume that interest focuses on a test of $H_0 : \kappa = \kappa_0$ vs. $H_1 : \kappa \neq \kappa_0$, where κ denotes the interrater agreement coefficient in the population of interest and κ_0 is a prespecified criterion value. We denote this coefficient by κ_C for the case of a continuous outcome variable and by κ_D for the case of a dichotomous variable. As discussed by Fleiss (1981, Section 13.2), both κ_C and κ_D may be interpreted as intraclass correlation coefficients. Our aim is to compare the sample size requirements for testing H_0 in the two cases, assuming that m raters independently measure the value of the outcome variable on each of n subjects. A basic assumption underlying this comparison is that the investigators have some latitude on how to measure the primary outcome variable. For example, it may be possible to measure the progress of atherosclerosis in a planned clinical study using either a dichotomous outcome variable such as the occurrence of myocardial infarction (MI) or a more readily obtained electrographic variable measured on a continuous scale that essentially serves as a surrogate for MI.

7.2.1 Continuous outcome variable

Let Y_{ij} denote the observation taken on subject $i, i = 1, 2, \ldots, n$ by rater $j, j = 1, 2, \ldots, m$. We assume that the Y_{ij} are normally distributed with mean μ and variance σ^2 where any two observations on the same subject have correlation κ_C, but observations from different subjects are assumed to be uncorrelated.

Using one-way analysis of variance as a framework for estimation, the sample estimate of κ_C is given by

$$\hat{\kappa}_C = \frac{\text{MSA} - \text{MSW}}{\text{MSA} + (m-1)\text{MSW}} = \frac{F-1}{F+m-1}, \tag{7.1}$$

where MSA and MSW are the mean squares among and within subjects and $F = \text{MSA}/\text{MSW}$ is the usual variance ratio. The null hypothesis $H_0 : \kappa_C = \kappa_{C0}$ is rejected at significance level α if $F > cF_{1-\alpha/2}$ or $F < cF_{\alpha/2}$, where $c = [1 + (m-1)\kappa_{C0}]/(1-\kappa_{C0})$ and $F_{\alpha/2}, F_{1-\alpha/2}$, are $100(\alpha/2)$ and $100(1-\alpha/2)$ percentile points of the F-distribution with $(n-1)$ and $n(m-1)$ degrees of freedom, respectively.

The power of this test under $H_1 : \kappa_C = \kappa_{C1}$ is given (e.g. Scheffé, 1959, Section 7.2) by

$$1 - \beta = \Pr(F < C_0 F_{\alpha/2}) + \Pr(F > C_0 F_{1-\alpha/2}), \tag{7.2}$$

where

$$C_0 = (1 + m\Theta_0)/(1 + m\Theta),$$
$$\Theta_0 = \kappa_{C0}/(1 - \kappa_{C0}),$$

and

$$\Theta = \kappa_{C1}/(1 - \kappa_{C1}).$$

The sample size required to carry out the test of H_0 at level α with power $(1-\beta)$ is obtained by solving for n in formula (7.2). For example, if $\kappa_{C0} = 0.40$ and $\kappa_{C1} = 0.60$ and we wish to reject H_0 at the 5% significance level (two-sided) with 80% power, then the required number of subjects at $m = 3$ may be obtained as $n = 64$.

7.2.2 Dichotomous outcome variable

Let Y_{ij} denote the binary rating for the ith subject assigned by the jth rater, $i = 1, 2, \ldots, n$, $j = 1, 2, \ldots, m$ and let $\pi = \Pr(Y_{ij} = 1)$ denote the probability that the rating is a success, assumed constant across raters. An estimate of κ_{D} may be obtained by again using one-way analysis of variance as a framework. Following Fleiss (1981, Section 3.2), the intraclass kappa statistic may be calculated by applying formula (7.1) to the set of binary ratings.

Bahadur (1961) proposed a parsimonious model for correlated binary data that was subsequently applied by Altaye *et al.* (2001b) to modelling interrater agreement among multiple raters. Letting $Y_i = \sum_{j=1}^{m} Y_{ij}$ denote the total number of success for subject i, then the joint probability function for this model is

$$P(Y_i = y_i) = \binom{m}{y_i} \pi^{y_i}(1-\pi)^{m-y_i} \times \left[1 + \sum_{i<j} \kappa_{\mathrm{D}} Z_i Z_j + \sum_{i<j<k} \kappa_{3\mathrm{D}} Z_i Z_j Z_k + \cdots + \kappa_{m\mathrm{D}} Z_1 Z_2 \ldots Z_m\right]$$

where we define

$$y_i = \sum_{j=1}^{m} y_{ij}, \quad Z_i = \frac{(y_i - \pi)}{\{\pi(1-\pi)\}^{1/2}}, \quad \kappa_{\mathrm{D}} = E(Z_i Z_j), \ldots, \kappa_{m\mathrm{D}} = E(Z_1 Z_2 \ldots Z_m).$$

Notice that the first term of the equation arises under an ordinary binomial assumption while the remaining terms represent a correction factor that accounts for the lack of mutual independence arising from the repeated measures taken on the same subject. The parameters in this model include only π, κ_{D}, and $m - 2$ higher-order correlations. For example, at $m = 3$ this model reduces to

$$\Pr(Y_i = y_i) = \binom{3}{y_i} \pi^{y_i}(1-\pi)^{3-y_i} \times \left[1 + \kappa_{\mathrm{D}} \frac{\{(y_1-\pi)(y_2-\pi) + (y_1-\pi)(y_3-\pi) + (y_2-\pi)(y_3-\pi)\}}{\pi(1-\pi)} + \kappa_{3\mathrm{D}} \frac{(y_1-\pi)(y_2-\pi)(y_3-\pi)}{(\pi[1-\pi])^{3/2}}\right],$$

where κ_{D}, the second-order (pair-wise) correlation among the Y_{ij} is the main parameter of interest. This parameter can be shown to be equivalent to the intraclass correlation coefficient as obtained from a one-way random effects model, and is

Table 7.1 Data layout for three raters and a dichotomous outcome variable

Category	Ratings	Frequency	Probability
0	(0, 0, 0)	h_0	P_0
1	(0, 0, 1), (0, 1, 0), (1, 0, 0)	h_1	P_1
2	(1, 1, 0), (1, 0, 1), (0, 1, 1)	h_2	P_2
3	(1,1,1)	h_3	P_3
Total		n	1

also equivalent to the intraclass version of the kappa coefficient. The parameter κ_{3D} is the third-order correlation and may be thought of as measuring the degree of three-way association.

The observed ratings and corresponding probabilities can be summarized as shown in Table 7.1, where it is seen that there are four $(m+1)$ rating categories into which the three raters can place each subject. The observed category frequencies h_l in this table may be more generally defined as the number of subjects having l successes, $l = 0, 1, \ldots, m$.

Altaye *et al.* (2001b) showed how a comparison of the h_l with their expected values can be used to derive a one degree of freedom chi-square goodness-of-fit test of $H_0 : \kappa_D = \kappa_{D0}$ vs. $H_1 : \kappa_D = \kappa_{D1}$. This test can also be used to facilitate sample size requirements using standard theory of the non-central chi-square distribution. Thus for a given significance level α, power $(1-\beta)$ and prevalence π, the required sample size may be determined for specified values of κ_{D0}, κ_{D1}, and non-centrality parameter λ. Specifically, if $1-\beta(1, \lambda, \alpha)$ denotes the power of the test statistic, then the sample size needed to conduct a two-sided test of $H_0 : \kappa_D = \kappa_{D0}$ vs. $H_1 : \kappa_D = \kappa_{D1}$ is given by

$$n = \lambda(1, 1-\beta, \alpha)\left\{\sum_{l=0}^{m}\frac{[P_l(\kappa_{D1}) - P_l(\kappa_{D0})]^2}{P_l(\kappa_{D0})}\right\}^{-1}, \tag{7.3}$$

where $\lambda(1, 1-\beta, \alpha)$ is the tabulated non-centrality parameter and the $P_l(\kappa_{D0})$ and $P_l(\kappa_{D1})$ are the cell probabilities under H_0 and H_1 respectively. For the case of three raters, and using an expression for κ_{3D} suggested by Prentice (1988), this formula reduces to

$$n = \lambda(1, 1-\beta, \alpha)\left\{\frac{[\pi(1-\pi)(2-\pi)(\kappa_{D1} - \kappa_{D0})]^2}{(1-\pi)^3 + \kappa_{D0}\,\pi(1-\pi)(2-\pi)} + \frac{3[\pi(1-\pi)(\kappa_{D1} - \kappa_{D0})]^2}{(1-\kappa_{D0})\pi(1-\pi)} \right.$$
$$\left. + \frac{[\pi(1-\pi^2)(\kappa_{D1} - \kappa_{D0})]^2}{\pi^3 + \kappa_{D0}\,\pi(1-\pi^2)}\right\}^{-1} \tag{7.4}$$

For example, suppose we wish to ensure at $m = 3$, $\kappa_{D0} = 0.40$, $\kappa_{D1} = 0.60$ and $\pi = 0.5$, that H_0 is rejected at the 5% significance level (two-sided) with 80%

power. In this case, the required number of subjects may be obtained as $n = 87$. For a continuous outcome variable, we previously found that the required number of subjects at $\kappa_{C0} = 0.40$, $\kappa_{C1} = 0.60$ was 64. Thus the relative efficiency associated with the choice of a dichotomous rather than a continuous outcome variable is given in this case by 0.74. It is also interesting to note that the corresponding value of RE in the case of two raters can be shown (see Donner and Eliasziw, 1994) to be substantially smaller (0.67).

More generally, the ratio of the values of n obtained from (7.2) and (7.3) is a measure of the information loss associated with the choice of a dichotomous rather than a continuous trait in studies of interrater agreement. The complete results are given in Tables 2(a)–(d) for $m = 2, 3, 4, 5$, respectively, at selected levels of κ_{D0} and κ_{D1}.

Table 7.2(a) Effect on sample size requirements of selecting a dichotomous rather than a continuous outcome variable in the case of two raters (from Donner and Eliasziw, 1994). Test of $H_0 : \kappa = \kappa_0$ *vs.* $H_1 : \kappa = \kappa_1$ at $\alpha = 0.05$, $\beta = 0.20$

		Continuous outcome	Dichotomous outcome ($\kappa = \kappa_D$)			Relative efficiency (RE)		
κ_0	κ_1	($\kappa = \kappa_C$)	$\pi = 0.10$	$\pi = 0.30$	$\pi = 0.50$	$\pi = 0.10$	$\pi = 0.30$	$\pi = 0.50$
0.40	0.50	499	1617	762	660	0.31	0.65	0.76
	0.60	110	404	191	165	0.27	0.58	0.67
	0.70	42	180	85	74	0.23	0.49	0.57
	0.80	19	101	48	42	0.19	0.40	0.45
0.60	0.70	261	1340	593	503	0.19	0.44	0.52
	0.80	50	335	148	126	0.15	0.34	0.40
	0.90	15	149	66	56	0.10	0.23	0.27
0.80	0.90	58	779	336	283	0.07	0.17	0.20

Table 7.2(b) Effect on sample size requirements of selecting a dichotomous rather than a continuous outcome variable in the case of three raters. Test of $H_0 : \kappa = \kappa_0$ *vs.* $H_1 : \kappa = \kappa_1$ at $\alpha = 0.05$, $\beta = 0.20$

		Continuous	Dichotomous outcome			Relative efficiency (RE)		
κ_0	κ_1	outcome	$\pi = 0.10$	$\pi = 0.30$	$\pi = 0.50$	$\pi = 0.10$	$\pi = 0.30$	$\pi = 0.50$
0.40	0.50	283	1079	423	346	0.26	0.67	0.82
	0.60	64	270	106	87	0.24	0.60	0.74
	0.70	25	120	47	39	0.21	0.53	0.64
	0.80	12	68	27	22	0.18	0.44	0.56
0.60	0.70	167	900	358	294	0.19	0.47	0.57
	0.80	33	225	90	74	0.15	0.37	0.45
	0.90	10	100	40	33	0.10	0.25	0.30
0.80	0.90	41	522	215	178	0.08	0.19	0.23

Table 7.2(c) Effect on sample size requirements of selecting a dichotomous rather than a continuous outcome variable in the case of four raters. Test of $H_0 : \kappa = \kappa_0$ *vs.* $H_1 : \kappa = \kappa_1$ at $\alpha = 0.05$, $\beta = 0.20$

		Continuous	Dichotomous outcome			Relative efficiency (RE)		
κ_0	κ_1	outcome	$\pi = 0.10$	$\pi = 0.30$	$\pi = 0.50$	$\pi = 0.10$	$\pi = 0.30$	$\pi = 0.50$
0.40	0.50	215	872	323	256	0.25	0.67	0.84
	0.60	50	218	81	64	0.23	0.62	0.78
	0.70	20	97	36	29	0.21	0.56	0.69
	0.80	10	55	21	16	0.18	0.48	0.63
0.60	0.70	136	714	285	234	0.19	0.48	0.58
	0.80	27	179	72	59	0.15	0.38	0.46
	0.90	9	80	32	26	0.11	0.28	0.35
0.80	0.90	35	411	177	149	0.09	0.20	0.23

Table 7.2(d) Effect on sample size requirements of selecting a dichotomous rather than a continuous outcome variable in the case of five raters. Test of $H_0 : \kappa = \kappa_0$ *vs.* $H_1 : \kappa = \kappa_1$ at $\alpha = 0.05$, $\beta = 0.20$

		Continuous	Dichotomous outcome			Relative efficiency (RE)		
κ_0	κ_1	outcome	$\pi = 0.10$	$\pi = 0.30$	$\pi = 0.50$	$\pi = 0.10$	$\pi = 0.30$	$\pi = 0.50$
0.40	0.50	183	736	275	220	0.25	0.67	0.83
	0.60	43	184	69	55	0.23	0.62	0.78
	0.70	18	82	31	25	0.22	0.58	0.72
	0.80	9	46	18	14	0.20	0.50	0.64
0.60	0.70	121	598	250	210	0.20	0.48	0.58
	0.80	24	150	63	53	0.16	0.38	0.45
	0.90	8	67	28	24	0.12	0.29	0.33
0.80	0.90	32	344	158	137	0.09	0.20	0.23

A number of interesting trends can be noted in these results. Most notably, it is clear that the effect on RE of increasing the number of raters is very sensitive to the value of π (prevalence of the dichotomized trait). At $\pi = 0.5$ there is generally an improvement in *RE* as m increases from 2 to 5, an increase that frequently exceeds five percentage points. Much smaller improvements in the same direction are generally seen at $\pi = 0.3$. However at $\pi = 0.10$ the effect on RE of increasing m is either negligible or even leads to a decrease in *RE*. Moreover, the greatest impact on *RE* is always obtained when the number of raters increases from two to three, after which the results gradually stabilize.

It has been previously noted by several authors (e.g. Cohen, 1983, Kraemer, 1986; Donner and Eliasziw, 1994) that the effect of dichotomization on the power of significance tests for both product-moment and intraclass correlation coefficients can be profound. The results presented above, which essentially focus on the case of an intraclass correlation, confirm that these findings hold for any value of m, and that the loss of information increases sharply as π deviates from 0.50. This loss can be somewhat minimized by increasing m, but only if the prevalence of the dichotomized trait is close to 0.5. Otherwise the loss of information remains fairly stable or even increases with m.

These results can also be used to quantify the effect of dichotomization on the total number of raters required to preserve a prespecified level of statistical power. For example, suppose that approximately 50 subjects are expected to be available in a planned study of interrater agreement where it is of interest to test $H_0 : \kappa_0 = 0.6$ vs. $H_1 : \kappa_1 = 0.8$. Then for a continuous outcome measure, only two raters are required for this test to have 80% power with $\alpha = 0.05$ (two-sided) at $n = 50$ (Table 7.2(a)). However, if the outcome measure is instead a dichotomized trait having prevalence $\pi = 0.5$, then Table 7.2(d) shows that five raters are required to maintain this level of power with $n = 53$ subjects. Thus the cost of dichotomization in this case may be interpreted as the need to more than double the required number of raters.

It is also important to note that the results in Tables 7.2(a)–(d) apply only to the case of a truly dichotomous variable which categorizes members of a population into non-overlapping subpopulations on a nominal scale. A second type of dichotomous outcome variable arises when the inherently continuous variable Y_{ij} is categorized to either enhance interpretability or to correspond more closely to how the variable is used in practice. For example, patient blood pressure scores are frequently dichotomized into hypertensive and normotensive categories. In this case the process of dichotomization tends to attenuate the value of the interrater agreement coefficient κ_C. For example, suppose Y_{ij} is regarded as a success if $Y_{ij} > \mu + h\sigma$,where h is a selected multiplier. Then the prevalence of the resulting binary trait is given by $\pi = 1 - \Phi(h)$, where Φ denotes the cumulative normal distribution function, and the attenuated value κ_{CA} is given by (Kraemer, 1979)

$$\kappa_{CA} = \frac{1}{2\Pi\pi(1-\pi)} \int_0^{\kappa_C} \frac{1}{(1-y^2)^{1/2}} \exp\left(\frac{-h^2}{1+y}\right) \mathrm{dy}, \tag{7.5}$$

where $\Pi = 3.1415927$. It may also be shown that $0 < \kappa_{CA} \leq (2/\Pi)\arcsin \kappa_C < \kappa_C$, with equality only if $\pi = 0.5$, i.e. $h = 0$. The divergence between κ_{CA} and κ_C increases as π approaches 0 or 1. For example, if a normally distributed outcome variable with associated interrater agreement coefficient $\kappa_C = 0.8$ is dichotomized at the mean $(h = 0)$, the prevalence of the resulting binary trait is given by $\pi = 1 - \Phi(0) = 0.50$. The interrater agreement coefficient κ_{CA} associated with this trait may then be obtained from formula (7.5) as 0.59, considerably smaller than the value of κ_C. Application of formula (7.3) with $\kappa_{D1} = 0.59$, $\kappa_{D0} = 0.40$ and $\pi = 0.50$ is now seen to yield $n = 182$ as compared to a required sample size of only 19 subjects if Y_{ij} is measured on a continuous scale. Thus the loss of information resulting from this form of dichotomization can be very severe in practice. More generally, it follows that replacing κ_1 by κ_{CA} in Tables 7.2(a)–(d) has the effect of further lowering the corresponding values of *RE*, often substantially.

7.3 The Choice between a Polychotomous or Dichotomous Outcome Variable

The results in Section 7.2 quantify the loss of information that results when a dichotomous trait replaces a continuous trait in a study of interrater agreement. A similar problem arises when polychotomous outcome data are treated as dichotomous, either to enhance interpretability or to simplify the statistical analysis. For example, Westlund and Kurland (1953) provide data from two neurologists who classified 149 Winnipeg patients and 69 New Orleans patients on a four-point scale: (1) certain multiple sclerosis (MS), (2) probable MS, (3) possible MS or (4) doubtful, unlikely, or not MS. Barlow (1996) grouped categories (1)–(3) together and compared MS to non-MS.

Bartfay and Donner (2001) investigated the effects of such dichotomization on inference procedures for the kappa statistic in the case of two raters. In this section we extend their results to the case of multiple raters, focusing again on implications for required sample size.

Suppose that an outcome variable obtained on each of n individuals is classified into one of $U(> 2)$ mutually exclusive categories by each of m raters. The coefficient of interrater agreement κ may then be interpreted as the degree of within-subject correlation with respect to the outcome variable, assumed constant across pairs of raters. As shown by Altaye *et al.* (2001a), a model for such studies may be developed using a Dirichlet multinomial distribution. These authors developed a one degree of freedom goodness-of-fit procedure that may be used to provide sample size requirements for testing non-null hypotheses concerning κ. The results obtained, which may be regarded as an extension of formula (7.3) to the case of a multinomial outcome variable, can be used to estimate the number of subjects required in a study of interrater agreement involving a given number of raters and outcome categories.

The goodness-of-fit procedure for this problem is a direct extension of the procedure discussed in Section 7.2, for the case of a binary outcome variable. It will therefore not be discussed in detail here. However, the corresponding sample size requirements are again obtained by recognizing that the resulting test statistic has a limiting non-central chi-square distribution with one degree of freedom and non-centrality parameter that depends on cell probabilities given under $H_0 : \kappa = \kappa_0$ and $H_1 : \kappa = \kappa_1$, respectively.

We focus here on the case $U = 3$ and denote the expected probability of a rating falling into category U, $U = 1, 2, 3$, by π_U where $\sum_{U=1}^{3} \pi_U = 1$. Table 7.3, extracted from Altaye *et al.* (2001a), shows the number of subjects required in this case to test $H_0 : \kappa = \kappa_0$ vs. $H_1 : \kappa = \kappa_1$ at $\alpha = 0.05$ (two-sided) with 80% power.

Table 7.3 Number of subjects required to test $H_0 : \kappa = \kappa_0$ *vs.* $H_1 : \kappa = \kappa_1$ at $\alpha = 0.05$ (two-sided) with 80% power when the number of categories $u = 3$ (from Altaye *et al.*, 2001a)

				Number of raters					
π_1	π_2	κ_0	κ_1	2	3	4	5	6	7
0.1	0.1	0.4	0.6	205	113	83	68	59	53
		0.6	0.8	172	102	77	65	57	52
0.1	0.2	0.4	0.6	155	85	62	51	44	39
		0.6	0.8	128	75	57	48	42	38
0.1	0.3	0.4	0.6	135	73	53	43	37	33
		0.6	0.8	109	64	49	41	36	33
0.1	0.4	0.4	0.6	127	69	50	40	35	31
		0.6	0.8	102	60	46	38	34	30
0.2	0.2	0.4	0.6	126	69	50	41	36	32
		0.6	0.8	104	61	47	39	34	31
0.2	0.3	0.4	0.6	115	62	45	37	32	29
		0.6	0.8	94	55	42	35	31	28
0.2	0.4	0.4	0.6	111	61	44	36	31	28
		0.6	0.8	91	53	41	34	30	27
0.3	0.3	0.4	0.6	107	58	42	35	30	27
		0.6	0.8	87	52	39	33	29	26

It can be seen from Table 7.3 that the values of n and m required to achieve a given power depends greatly on the values of the π_U. As any of the π_U approach 0.1, the required value of n increases greatly. Moreover the greatest savings in the required number of subjects is observed when m increases from 2 to 3.

Now suppose we combine two of the three categories so as to create a dichotomous rather than a trichotomous outcome variable. For example, the trichotomous variable characterized by $\pi_1 = 0.2$, $\pi_2 = 0.3$, $\pi_3 = 0.5$ may be

Table 7.4 Number of subjects required to test $H_0 : \kappa = \kappa_0$ *vs.* $H_1 : \kappa = \kappa_1$ at $\alpha = 0.05$ (two-sided) with 80% power when three categories having probabilities π_1, π_2, $1 - \pi_1 - \pi_2$ are collapsed into two categories (probabilities $\pi = \pi_1 + \pi_2$, $1 - \pi$)

					Number of raters					
π_1	π_2	π	κ_0	κ_1	2	3	4	5	6	7
0.1	0.1	0.2	0.4	0.6	241	129	93	76	65	58
0.1	0.1	0.2	0.6	0.8	192	112	85	71	63	57
0.1	0.2	0.3	0.4	0.6	190	101	72	58	50	44
0.1	0.2	0.3	0.6	0.8	148	87	66	55	48	44
0.1	0.3	0.4	0.4	0.6	170	90	64	51	44	39
0.1	0.3	0.4	0.6	0.8	130	76	58	48	42	38
0.1	0.4	0.5	0.4	0.6	165	86	61	49	42	37
0.1	0.4	0.5	0.6	0.8	126	73	55	46	41	37

collapsed to create a dichotomous outcome variable with probabilities $\pi = 0.2$ and $1 - \pi = 0.8$, or, alternatively, with probabilities $\pi = 0.5$ and $1 - \pi = 0.5$. Table 7.4 shows the impact on the required sample size when two of the categories having probabilities π_1 and π_2 in Table 7.3 are combined into a single category having probability $\pi = \pi_1 + \pi_2$, with the values of κ_0 and κ_1 remaining unchanged.

A comparison of these results with those in Table 7.3 shows that collapsing a three-category outcome variable into two categories inevitably leads to a loss in statistical power. Although the percentage increase in the required number of subjects needed to maintain the desired levels of α and β tends to decline as the number of raters increases, it remains fairly substantial (>5%) even at $m = 7$.

7.4 Incorporation of Cost Considerations

Power considerations are clearly an important factor in choosing the number of subjects and raters that should be recruited for a study of inter-observer agreement. However, most studies are also subject to budget constraints, requiring that the investigator achieve specified power while minimizing the total study cost C. This cost will typically depend on C_1, the cost of recruiting a subject, C_2, the cost of recruiting a rater and C_3, the cost of obtaining a measurement on a subject that has already been recruited. This leads naturally to an overall linear cost model of the form

$$C = C_1 n + C_2 m + C_3 nm. \tag{7.6}$$

The values of n and m required to ensure that a test of $H_0 : \kappa = \kappa_0$ has specified statistical power while minimizing the overall cost C was essentially dealt with for the case of a quantitative outcome variable by Eliasziw and Donner (1987). They found that although the optimal n and m were highly stable in light of moderate changes in sampling costs, the cost per rater has a greater impact on the overall cost than the cost per subject for values of κ that are of practical interest, i.e. $\kappa > 0.2$. This would suggest that, in general, it is less expensive to minimize the cost per rater, rather than the cost per subject.

In this section we extend these results to the case of a dichotomous variable but from a different perspective, determining the optimal values of n and m by minimizing the variance of the kappa statistic. The optimal values will be derived under the condition that the total cost for sampling and measurement is fixed at C, with the variance minimized subject to the restriction on cost given by (7.6).

Note that the optimal values of n and m can be obtained by expressing either of these factors in terms of the other in (7.6), substituting this expression in the appropriate variance formula and then minimizing it. For this purpose we used a variance formula given by Lipsitz *et al.* (1994). Let R_i denote the sum of positive ratings for the ith subject, $i = 1, 2, \ldots, n$ and P the overall proportion of positive ratings. The estimate of the variance of $\hat{\kappa}_C$ is then given by

$$\widehat{\text{Var}}(\hat{\kappa}_C) = \sum_{i=1}^{n} U_{i,n}^2 \tag{7.7}$$

where

$$U_{i,n} = \frac{[\{R_i(R_i - 1) - P_{00}\, m(m-1)\} - \{2P + \kappa(1-2P)(m-1)\}(R_i - mp)]}{P(1-P)nm(m-1)}$$

and

$$P_{00} = P^2 + \kappa P(1-P)$$

The results of the minimization process (see Appendix for details) are presented in Table 7.5. They indicate that for a fixed value of C the effect of varying the allocation of costs among raters, subjects and measurements impacts the required number of subjects much more than the number of raters. For example at C = \$5000, and a cost ratio given by 10:1:1 at $\pi = 0.3$ and $\kappa = 0.6$, the optimal values of m and n are found to be 4 and 211 respectively. If we double the total cost while

Table 7.5 The optimal values of m and n obtained under the constraint of C = \$5000 with different cost allocation ratios for rater, subject and measurement at various values of π and κ

		Ratios									
		2:1:1		4:1:1		10:1:5		10:1:1		1:1:1	
π	κ	m	n	m	n	m	n	m	n	m	n
0.1	0.4	13	34	11	37	10	43	11	75	14	33
	0.6	5	88	4	88	4	104	4	176	5	88
	0.8	4	102	4	101	3	121	4	203	4	102
0.2	0.4	5	83	5	83	4	98	5	166	5	83
	0.6	4	98	4	97	4	116	4	194	4	98
	0.8	4	105	4	105	3	125	4	209	4	106
0.3	0.4	4	102	4	102	3	122	4	203	4	103
	0.6	4	107	4	106	3	127	4	211	4	107
	0.8	4	109	3	108	3	130	3	216	4	109
0.4	0.4	3	118	3	118	3	143	3	235	3	119
	0.6	3	117	3	117	3	141	3	233	3	117
	0.8	3	114	3	114	3	137	3	227	3	115
0.5	0.4	2	150	2	149	2	187	2	298	2	150
	0.6	2	149	2	148	2	185	2	295	2	149
	0.8	2	147	2	146	2	183	2	292	2	148

retaining the same cost ratio, the optimal value of m remains unchanged at 4, but the number of required subjects increases to 426. Furthermore, the optimal number of raters is fairly stable over varying values of κ for constant values of π. For example, except at $\pi = 0.1$ and $\kappa = 0.4$, the optimal value of m ranges from 5 to 2 as π increases from 0.1 to 0.5. This is consistent with results obtained by previous researchers, showing that as π approaches 0 the number of required subjects and raters increases. The optimal value of m also shows little sensitivity to variation in either the value of C or the cost allocation; thus the optimal value is almost invariably between 2 and 5. This is consistent with our earlier findings (Altaye *et al.*, 2001b), which showed the gain obtained from increasing the number of raters tends to diminish after the accrual of five raters.

7.5 Final Comments

As statistical methods for the analysis of data arising from studies of interrater agreement continue to develop, it becomes increasingly important to consider the design features that drive these analyses. These considerations include the need to select an appropriate scale for the outcome measure, as well as to recruit a suitable number of both subjects and raters.

In this chapter we have expanded on previous research that has examined these issues in a more limited context, usually restricted to two raters and a continuous or binary outcome variable. An overall conclusion from our more general results is that the dichotomization of either a continuous or polychotomous outcome variable can have severe effects on statistical power regardless of the number of raters recruited per subject. The expression of these results in terms of sample size requirements should allow investigators to more carefully appreciate the trade-offs incurred when selecting a scale of measurement.

A related question of interest is whether or not increasing the number of raters per subject can minimize the deleterious effect of dichotomization. With respect to dichotomizing a continuous outcome variable, the results in Section 7.2 show that increasing the number of raters m can in fact reduce this effect, but only if the prevalence of the resulting binary trait is not far from 0.5. Otherwise increasing m can cause the dichotomization of a continuous trait to actually lead to a further loss of information. On the other hand, the loss in power incurred from collapsing a polychotomous trait to form a binary outcome variable can always be reduced by increasing the number of raters. However, the information lost cannot be totally recovered even if as many as seven raters per subject are recruited. Although the results presented in Section 7.3 are restricted to the case in which three categories are collapsed into two, the loss in power is even greater when more than three categories are collapsed into a dichotomy.

Most studies of interrater agreement are subject to budgetary constraints, requiring that the investigators achieve specified statistical power while minimizing the total study cost C. This cost will usually depend upon several components, including the cost of recruiting a subject C_1, the cost of recruiting a rater C_2, and the cost C_3 obtaining a measurement on a subject that has already been recruited. In this case it may be of interest to determine the optimal values of n and m by maximizing precision holding the total cost fixed for various allocations of C_1, C_2 and C_3. Our results in Section 7.4, which focus on the case of a dichotomous outcome variable, show that the number of raters needed to minimize the variance of the kappa statistic is fairly insensitive to both C and the cost allocation ratio, usually lying within the range 2–5.

Appendix

We assume that the values of P and κ can be assessed at the design stage of the study. Terms that involve $\sum_{i=1}^{n} R_i$ and $\sum_{i=1}^{n} R_i^2$ can be expressed in terms of P, κ, n and m, where

$$\sum_{i=1}^{n} R_i = mn\,P$$

and

$$\sum_{i=1}^{n} R_i^2 = mn\,P + \{P^2 + \kappa P(1 - P)\}nm(m - 1)$$

However, terms that involve $\sum_{i=1}^{n} R_i^3$ and $\sum_{i=1}^{n} R_i^4$ cannot be easily obtained or expressed in terms of the other variables. The minimum possible value for $\sum_{i=1}^{n} R_i^3$ and $\sum_{i=1}^{n} R_i^4$ is zero (i.e. when all raters score a negative outcome) and the maximum value is nm^3 and nm^4 respectively (i.e. when all raters score a positive outcome). We therefore use the average of the minimum and maximum values to estimate these sums. We also conducted a sensitivity analysis using different values for $\sum_{i=1}^{n} R_i^3$and $\sum_{i=1}^{n} R_i^4$ ranging from $\sum_{i=1}^{n} R_i^2$ to the appropriate maximum value during minimization of $\mathrm{Var}(\hat{\kappa})$, noting that the optimal value of k changes very little. We also note that in this minimization process m and n were treated as continuous variables. However in the reported table both m and n are rounded based on the 0.5 rule (i.e. $m, n \geq 0.5$–1 and $m, n < 0.5$–0) and hence the total cost might be slightly higher or lower than its specified maximum.

Acknowledgement

Dr Donner's research was partially supported by a grant from the Natural Sciences and Research Council of Canada.

References

Altaye, M., Donner, A. and Eliasziw, M. (2001a). A general goodness-of-fit approach for inference procedures concerning the kappa statistic. *Statistics in Medicine*, **20**: 2479–2488.

Altaye, M., Donner, A. and Klar, N. (2001b). Inference procedures for assessing interobserver agreement among multiple raters. *Biometrics*, **57**: 584–588.

Bahadur, R. R. (1961). A representation of the joint distribution of responses to *n* dichotomous items. In: *Studies in Item Analysis and Prediction*. Solomon, H. (ed.). Stanford Mathematical Studies in the Social Sciences VI. Stanford, Calif.: Stanford University Press.

Banerjee, M., Capozzoli, M., McSweeney, L. and Sinha, D. (1999). Beyond kappa: a review of interrater agreement measures. *The Canadian Journal of Statistics*, **27**: 3–23.

Barlow, W. (1996). Measurement of interrater agreement with adjustments for covariates. *Biometrics*, **52**: 695–702.

Bartfay, E. and Donner, A. (2001). Statistical inferences for interobserver agreement studies with nominal outcome data. *The Statistician*, **50**: 135–146.

Cantor, A. B. (1996). Sample-size calculations for Cohen's kappa. *Psychological Methods*, **1**: 150–153.

Cohen, J. (1983). The cost of dichotomization. *Applied Psychological Measurement*, **7**: 249–253.

Donner, A. and Eliasziw, M. (1992). A goodness-of-fit approach to inference procedures for the kappa statistic: confidence interval construction, significance-testing and sample size estimation. *Statistics in Medicine*, **11**: 1511–1519.

Donner, A. and Eliasziw, M. (1994). Statistical implications of the choice between a dichotomous or continuous trait in studies of interobserver agreement. *Biometrics*, **50**: 550–555.

Eliasziw, M. and Donner, A. (1987). A cost-function approach to the design of reliability studies. *Statistics in Medicine*, **6**: 647–655.

Flack, V. F., Afifi, A. A. and Lachenbruch, P. A. (1988). Sample size determinations for the two rater kappa statistic. *Psychometrika*, **53**: 321–325.

Fleiss, J. L. *Statistical Methods for Rates and Proportions*, 2nd edn 1981. New York: John Wiley & Sons, Ltd.

Kraemer, H. C. (1979). Ramifications of a population model for κ as a coefficient of reliability. *Psychometrika*, **44**: 461–472.

Kraemer, H. C. (1986). A measure of 2×2 association with stable variance and approximately normal small sample distribution: planning cost-effective studies. *Biometrics*, **42**: 359–370.

Kraemer, H. C. (1992). Measurement of reliability for categorical data in medical research. *Statistical Methods in Medical Research*, **1**: 183–199.

Lipsitz, S. R., Laird, N. M. and Brennan, T. A. (1994). Simple moment estimates of the κ coefficient and its variance. *Applied Statistics*, **43**: 309–323.

Prentice, R. L. (1988). Correlated binary regression with covariates specific to each binary observation. *Biometrics*, **44**: 1033–1048.

Scheffé, H. *The Analysis of Variance*, 1959. New York: John Wiley & Sons, Ltd.

Shrout, P. E. (1998). Measurement reliability and agreement in psychiatry. *Statistical Methods in Medical Research*, **7**: 301–317.

Walter, S. D., Eliasziw, M. and Donner, A. (1998). Sample size and optimal designs for reliability studies. *Statistics in Medicine*, **17**: 101–110.

Westlund, K. B. and Kurland, L. T. (1953). Studies on multiple sclerosis in Winnipeg, Manitoba, and New Orleans, Louisiana. I. Prevalence, comparison between the patient groups in Winnipeg and New Orleans. *American Journal of Hygiene*, **57**: 380–396.

8

Restricted Optimal Design in the Measurement of Cerebral Blood Flow Using the Kety–Schmidt Technique

J.N.S. Matthews and P.W. James

School of Mathematics and Statistics, University of Newcastle upon Tyne, Newcastle upon Tyne NE1 7RU, UK

8.1 Introduction

The rate at which blood flows through the brain, the cerebral blood flow rate or simply the cerebral blood flow (CBF), is a quantity which is of wide interest. Its measurement is of potential value to physiologists and clinicians. The management of severe head injury can be informed by the value of CBF. However, it is not widely used in this context and this is, in part, due to the difficulty of measuring this quantity. There have been several developments in recent years in using imaging methods to determine CBF, but these methods all focus on measuring CBF in a particular region. For the purpose of clinical management of severe head injury there is interest in a measurement of global CBF. This can be provided by a method due to Kety and Schmidt (1945, 1948) which is now over 50 years old but which has been implemented in recent studies of CBF in children with severe head injury (Matthews *et al.*, 1995a, b). Further advantages of this method over many imaging

Applied Optimal Designs Edited by M.P.F. Berger and W.K. Wong
 ISBN: 0-470-85697-1 (HB)

methods are that it does not require the use of radioactive materials and can be carried out at the bedside.

The method, which will be described in detail in the next section, proceeds by serially taking blood samples from the patient, some samples being from the arterial circulation and some from the venous circulation. Most of this chapter is concerned with how to choose the nature and timings of these samples. Section 8.3 describes the statistical model used and Section 8.4 deals with locally optimal designs. Section 8.5 considers Bayesian issues and Section 8.6 describes optimal Bayesian designs. Neither type of design is practicable; designs that are more realistic are considered in Section 8.7.

8.2 The Kety–Schmidt Method

This method for measuring CBF involves the introduction of an inert marker gas into the inhalation mixture of the patient. The concentration of this gas in the patient's circulation will rise to an equilibrium level, determined by the proportion of the marker in the inhalation mixture. The marker can be one of a variety of gases but our experience is limited to the use of nitrous oxide (N_2O). The rate at which this equilibrium level is approached will be faster in the arterial circulation than in the venous circulation. If the venous circulation is sampled from a line inserted into the retrograde jugular bulb, then this will measure the concentration of the marker in the venous blood as it drains from the brain. An application of Fick's principle shows that the CBF can be obtained from the arterial and venous concentrations at time t, $C_a(t)$ and $C_v(t)$, respectively, as

$$\text{CBF} = 100\lambda \times \frac{C_v(t_{eq})}{\int_0^{t_{eq}} C_a(u) - C_v(u)\mathrm{d}u} \tag{8.1}$$

where λ is the blood : brain partition coefficient and the factor of 100 is required to give the conventional units of ml 100 g^{-1} min^{-1}. The upper limit t_{eq} is a time (in minutes) by which arterial, venous and cerebral tissue concentrations of N_2O have reached saturation. In this chapter we will ignore the factor of 100 (i.e. we measure CBF in ml g^{-1} min^{-1}) and as $\lambda \approx 1$ we can ignore the remaining multiplicative constant. An example of such curves is shown in Figure 8.1.

Matthews *et al.* (1999) used data obtained from a study of severe head injury in children (Matthews *et al.*, 1995a, b) to investigate suitable forms for $C_a(t)$ and $C_v(t)$. The most appropriate form, in terms of being in keeping with general physiological principles and providing a good fit to the data, was found to be

$$C_a(t) = A(1 - \exp(-k_a t))$$
$$C_v(t) = A(1 - \exp(-k_v t)),$$

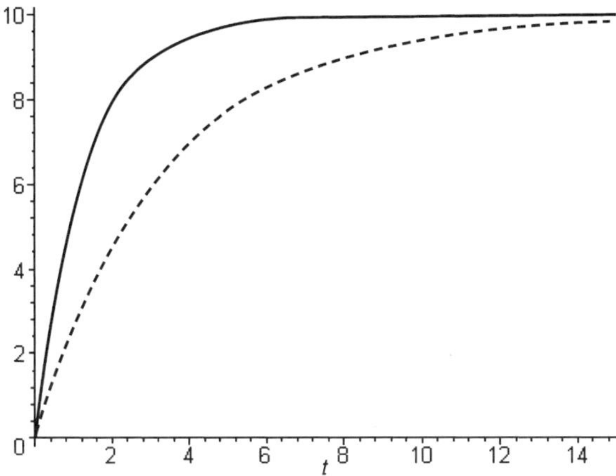

Figure 8.1 Example of the concentration time curves for arterial (solid line) and venous (dashed line) circulations, with $A = 10$, $k_a = 0.8\ \text{min}^{-1}$, $k_v = 0.3\ \text{min}^{-1}$

where $k_a > k_v$ and A are unknown positive parameters which need to be estimated from observations of the concentration of the marker.

This form for the concentration–time curves will be adopted in the present chapter. The CBF can be evaluated by substituting the above expressions into Equation (8.1). This requires a value for t_{eq}. Although it is unusual for observations to be made for more than 30 minutes, it makes no real difference if we set $t_{eq} = \infty$. If this is done the expression for CBF is

$$\text{CBF} = \frac{k_a k_v}{k_a - k_v}. \tag{8.2}$$

Note that this does not depend on A, which simply reflects the proportion of N_2O in the breathing mixture. As this value is an artefact of the measurement procedure it is appropriate that it does not feature in the expression for CBF.

In practice blood is sampled from the subject at a series of times t_i, some samples being taken from the arterial curve and the remainder from the venous curve. Practicalities of the method dictate that all observations need to be made within 30 minutes of introducing the marker gas into the inhalation mixture. At which times and from which curve the observations are best made will be examined in the following sections.

8.3 The Statistical Model and Optimality Criteria

In addition to the exponential forms for the concentration–time curves given above, Matthews *et al.* (1999) found that a model which assumes independent normal deviations from these curves, with common variance, was an adequate description of the data collected in the study of childhood head injuries. Thus we assume that the concentration, y_i observed at time t_i is

$$y_i = A(1 - \exp(-k_a t_i)) + \varepsilon_i,$$

if the observation is from the arterial line and

$$y_i = A(1 - \exp(-k_v t_i)) + \varepsilon_i,$$

if it is from the venous line. The residuals ε_i are independent and have a normal distribution with zero mean and variance σ^2. The model can be fitted by least squares, which for this model coincides with maximum likelihood. That is, we estimate A, k_a and k_v by the values which minimise

$$\sum_a [y_i - A(1 - \exp(-k_a t_i))]^2 + \sum_v [y_i - A(1 - \exp(-k_v t_i))]^2,$$

where $\sum_a$, $\sum_v$ denote summation over the observations from the arterial and venous curves respectively.

The expected information matrix for the parameters A, k_a, k_v under this model is

$$M(A, k_a, k_v) = N\sigma^{-2} \begin{pmatrix} S_{av} & & \\ AS_{2a} & A^2 S_{1a} & \\ AS_{2v} & 0 & A^2 S_{1v} \end{pmatrix}.$$

Here

$$S_{av} = \sum_a \eta_i p(t_i, k_a)^2 + \sum_v \eta_i p(t_i, k_v)^2$$

where $p(t, k) = 1 - \exp(-kt)$. Also

$$S_{1a} = \sum_a \eta_i t_i^2 (1 - p(t_i, k_a))^2, \qquad S_{2a} = \sum_a \eta_i t_i (1 - p(t_i, k_a)) p(t_i, k_a),$$

with analogous definitions for S_{1v} and S_{2v}. Here η_i is the number of observations made at time t_i divided by N, the total number of observations (counted over both the arterial and venous curves). Consequently each η_i is positive and $\sum_a \eta_i + \sum_v \eta_i = 1$. In the development of optimal designs which follows we will determine designs defined by the t_i and η_i. However, we will make no attempt to ensure all $N\eta_i$ are integers, i.e. we follow the usual practice of determining *continuous* designs. The extent to which it is possible in this application to make more than one observation at a given time is discussed in detail in Section 8.7. We would remark that designs which cannot be implemented in practice might be useful in the assessment of designs which can be used. At this stage we persist in deriving optimal continuous designs and leave consideration of the practicalities until later in the chapter.

As the parameter A is of no direct interest, it is useful to note that the dispersion matrix for the parameters k_a and k_v is the relevant sub-matrix of $M(A, k_a, k_v)^{-1}$, which we will write as $M(k_a, k_v)^{-1}$. It follows that

$$M(k_a, k_v) = N\sigma^{-2}\frac{A^2}{S_{av}}\begin{pmatrix} S_{av}S_{1a} - S_{2a}^2 & \\ -S_{2a}S_{2v} & S_{av}S_{1v} - S_{2v}^2 \end{pmatrix}.$$

We will investigate optimal designs with respect to two criteria. The first is D-optimality with respect to the parameters k_a and k_v, a type of design sometimes referred to as D_S-optimality. This is done because A is of no interest, so there is no virtue in considering full D-optimality of all the parameters in the model. We will seek designs which minimise the determinant of the dispersion matrix of k_a and k_v, which is equivalent to maximising $\det(M(k_a, k_v))$. Notice that because A enters the model linearly, the observations determining the D_S-optimal design will not depend on A. However the design will depend on k_a and k_v.

An approach which might, at first, appear more focused on the objective of the study is to find a design which minimises the variance of the estimate of CBF given in (8.2). However, data from the head injury studies cited above indicate that the distribution of the estimate of CBF obtained by substituting $\hat{k}_a, \hat{k}_v$, the estimates of k_a and k_v respectively, into (8.2) is rather skew. The value for CBF can also become rather large if k_v is only just smaller than k_a. Analysis of the data, and simulations from the model used for the observations, suggests that estimates of the reciprocal of the CBF have a distribution which is much closer to normal than is the case on the untransformed scale. In practice an interval estimate for CBF will be useful and it is preferable to obtain this by transforming an interval estimate established on a scale on which the sampling distribution is closer to normal.

A further advantage of working with $D = 1/\text{CBF}$ is that a simple reparameterisation of the models for C_a and C_v allows D to be written as a linear form in the parameters. If we write

$$C_a(t) = A(1 - \exp(-k_a t)) = A(1 - \exp(-t/c_a))$$
$$C_v(t) = A(1 - \exp(-k_v t)) = A(1 - \exp(-t/c_v)),$$

so $c_a = 1/k_a$ and $c_v = 1/k_v$, then $D = c_v - c_a$. The dispersion matrix of c_a, c_v is $M(c_a, c_v)^{-}$, where

$$M(c_a, c_v) = N\sigma^{-2}\frac{A^2k_a^2k_v^2}{S_{av}}\begin{pmatrix} r^2(S_{av}S_{1a} - S_{2a}^2) & \\ -S_{2a}S_{2v} & r^{-2}(S_{av}S_{1v} - S_{2v}^2) \end{pmatrix},$$

with $r = k_a/k_v$. The variance of $\hat{D}$, the estimate of D, can be written as $a^T M(c_a, c_v)^- a$ where B^- denotes a g-inverse of B and $a = (-1\,1)^T$. In most cases a simple inverse will suffice but it will be seen that there is a circumstance in which a g-inverse is needed.

8.4 Locally Optimal Designs

We will start by obtaining locally D$_S$-optimal designs and designs minimising the variance of $\hat{D}$. These can only be implemented if we assume values for some of the parameters or functions of parameters in the model and, as such, are unrealistic. However, in the next section we will adopt a Bayesian approach in which this problem is addressed by postulating a prior distribution for the unknown parameters. Greater insight into these designs can be obtained if the locally optimal designs are known, as these are the limiting case of their Bayesian counterparts when the prior has low dispersion.

8.4.1 D$_S$-optimal designs

The D$_S$-optimal design is found by maximising the determinant of $M(k_a, k_v)$, which is

$$S_{1a}S_{1v} - (S_{2a}^2 S_{1v} + S_{2v}^2 S_{1a})/S_{av}, \tag{8.3}$$

where the factor of $N^2\sigma^{-4}A^4$ has been omitted, as it has no bearing on the values of the observations which maximise (8.3). A combination of numerical and analytical investigations gives the locally D$_S$-optimal design as a three- or four-point design. The observations are:

$$\begin{array}{lll} \text{arterial curve}: & t_{a1} = 0.8420/k_a: & t_{a2} = \infty \\ \text{venous curve}: & t_{v1} = 0.8420/k_v: & t_{v2} = \infty \end{array}$$

The weights for both t_{a1} and t_{v1} are 0.3761 and the weights on the two observations taken at infinity can be any two non-negative numbers which add to 0.2478. A three-point design results if one of the weights on the infinite observations is zero. Here, and throughout this chapter, an observation taken 'at infinity' simply means that the observation is taken at the largest value which can be accommodated in practice, which in this application we take to be 30 minutes. The derivative plots which confirm that we have found optimal designs are shown in Figure 8.2.

The observations at the earlier time points are principally concerned with gathering information about the k parameters, while the observations at infinity are mainly concerned with collecting information about A. As it has previously been pointed

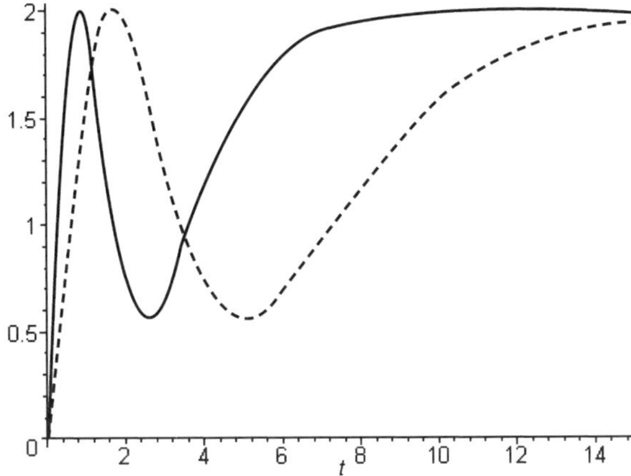

Figure 8.2 Derivative plot for locally D_S-optimal design with $k_a = 1$ min^{-1} and $k_v = 0.5$ min^{-1}. The solid and dashed lines plot possible observations at time t from the arterial and venous circulations respectively

out that A is of no direct interest, some explanation of this may be in order. While A is not of interest in itself, the equations estimating k_a and k_v from the observations will implicitly require information about A, so if the design provides poor information about A this could in turn lead to poor estimates for k_a and k_v. Therefore while A is of no direct interest in this chapter, the designs derived will need to obtain appropriate levels of information about this parameter in order provide optimal or good estimates of the parameter(s) which are of direct interest.

The D-optimal design for all the parameters (including A) has three points, two on the arterial curve, one of which is at infinity and the other just below $1/k_a$, and one observation at $1/k_v$ on the venous curve. Each point has weight of a third. The D_S-optimal design has given slightly less weight to observations at infinity, as might be expected when there is no direct interest in A.

8.4.2 Designs minimising var$(\hat{D})$

The variance of $\hat{D}$ can be written as

$$\frac{r^2(S_{av}S_{1a} - S_{2a}^2) + r^{-2}(S_{av}S_{1v} - S_{2v}^2) - 2S_{2a}S_{2v}}{S_{av}S_{1a}S_{1v} - (S_{2a}^2 S_{1v} + S_{2v}^2 S_{1a})}$$

provided that the denominator does not vanish, i.e. provided $M(c_a, c_v)$ is not singular. Analytical progress does not seem to be possible but numerical investigations can be made. However, a rather special circumstance arises since it turns out that the variance of $\hat{D}$ computed from a particular two-point design is smaller than for any design for which $M(c_a, c_v)$ is invertible.

A design in which $M(A, k_a, k_v)$ is singular cannot provide information on all parameters but there may still be the possibility of estimating some parameters or functions of them. Indeed even if the design is such that $M(c_a, c_v)$ is singular, $a^T(c_a, c_v)^T$ can still be estimated provided a is in the range space of $M(c_a, c_v)$. A design like this can be obtained analytically and a brief outline follows.

One point is at time x_a/k_a on the arterial curve and the other is at x_v/k_v with weights η_a and $\eta_v = 1 - \eta_a$ respectively. The quantities x_a and x_v can be thought of as scaled observation times. If for this design we write $T_{av} = S_{av}$, $S_{1z} = c_z^2 T_{1z}$ and $S_{2z} = c_z T_{2z}$, where z is either a or v, then T_{av}, T_{1z}, T_{2z} are all functions just of x_a, x_v, η_a and η_v. D can be estimated provided $a^T = (-1\,1)$ is in the range space of $M(c_a, c_v)$ and this is so if $h(x_v)/h(x_a) = r = k_a/k_v$, with $h(t) = t/(e^t - 1)$. If x_a and x_v satisfy this constraint then a suitable g-inverse can be found and $a^T M(c_a, c_v)^- a$ turns out to be

$$\frac{\eta_a p_a^2 + \eta_v p_v^2}{\eta_a \eta_v p_a^2 x_v^2} e^{-2x_a}, \tag{8.4}$$

where factors which do not determine the design have been ignored and where we write $p_z = 1 - \exp(-x_z)$, where z is either a or v. From this expression we obtain the optimal weights as $\eta_a = p_v/(p_a + p_v)$ and $\eta_v = p_a/(p_a + p_v)$. Substituting these into (8.4) we find that, apart from irrelevant factors, the minimum variance is

$$\left(\frac{\exp(x_v)}{x_v} + \frac{\exp(x_a)}{r x_a} \right)^2.$$

This quantity can be minimised numerically subject to the constraint $h(x_v)/h(x_a) = r$. The scaled observation times clearly depend on k_a and k_v only through the ratio r, although knowledge of each of k_a and k_v is necessary to obtain the actual observation times. A few scaled observation times and weights are shown in Table 8.1.

Table 8.1 Values determining the locally optimal design minimising var($\hat{D}$)

r	x_a	x_v	η_v	r	x_a	x_v	η_a
1.5	1.5021	0.8200	0.5814	3	2.5196	0.7760	0.6301
2	1.9266	0.7890	0.6102	4	2.9256	0.7720	0.6376

Although of some technical interest, these designs can only be implemented if we know the values of the parameters we seek to estimate. It is also unlikely that an experimenter would wish to carry out a design which could estimate only such limited aspects of the model. We will now turn to a Bayesian approach to try to avoid these difficulties.

8.5 Bayesian Designs and Prior Distributions

8.5.1 Bayesian criteria

In the absence of any knowledge about the parameters it will be very difficult to produce an efficient design. However, in most circumstances there will be prior knowledge about at least some of the parameters and if this can be encapsulated in a prior distribution then it can be incorporated into a Bayesian procedure to determine optimal designs. Formal decision theoretic methods can be employed but these can be technically difficult. A widely used approximation for Bayesian D-optimal designs is to find observations and weights which maximise $E_\pi(\log \det M)$, where the expectation is over the prior distribution. See Chaloner and Verdinelli (1995) for more discussion of this approximation and of the approach based on utilities which underlies it.

If we define a prior distribution $\pi(\sigma^2, A, k_a, k_v)$ for the unknown parameters, then approximate Bayesian D_S-optimal designs are found by minimising

$$-\int \log \det(M(k_a, k_v))\pi(\sigma^2, A, k_a, k_v)d\sigma^2 dA\, dk_a dk_v$$

as a function of t_i and η_i. Substituting the expression for $M(k_a, k_v)$ gives

$$4E_\pi \log \sigma - 4E_\pi \log A - 2\log N - \int \log\left(S_{1a}S_{1v} - \frac{(S_{2v}^2 S_{1a} + S_{2a}^2 S_{1v})}{S_{av}}\right)\pi(k_a, k_v)dk_a dk_v \quad (8.5)$$

where $\pi(k_a, k_v)$ is simply the (k_a, k_v) marginal of $\pi(\sigma^2, A, k_a, k_v)$. The first two expectations in (8.5) do not depend on t_i or η_i so have no effect on the determination of optimal designs nor on determinations of efficiencies for any of the designs considered in this chapter, and are ignored hereafter. The term 2 log N can also be ignored for the designs we consider in Section 8.6 but it will need to be kept in mind in Section 8.7.

It follows from (8.5) that for the purposes of determining Bayesian optimal designs, only the $\pi(k_a, k_v)$ marginal of the full prior needs to be specified.

In a similar way, optimal designs intended to minimise the logarithm of the variance of $\hat{D}$ can be found by minimising

$$\int \log\{a^T M(c_a, c_v)^- a\}\pi(\sigma^2, A, k_a, k_v)d\sigma^2 dA\, dk_a dk_v$$

which is found to be

$$
\begin{aligned}
&2\mathrm{E}_\pi \log \sigma - 2\mathrm{E}_\pi \log A - \log N \\
&\quad + \int \log(c_{\mathrm{a}}^4(S_{\mathrm{av}}S_{1\mathrm{v}} - S_{2\mathrm{v}}^2) + c_{\mathrm{v}}^4(S_{\mathrm{av}}S_{1\mathrm{a}} - S_{2\mathrm{a}}^2) - 2c_{\mathrm{a}}^2 c_{\mathrm{v}}^2 S_{2\mathrm{a}}S_{2\mathrm{v}})\pi(k_{\mathrm{a}}, k_{\mathrm{v}})\mathrm{d}k_{\mathrm{a}}\mathrm{d}k_{\mathrm{v}} \\
&\quad - \int \log(S_{\mathrm{av}}S_{1\mathrm{a}}S_{1\mathrm{v}} - (S_{2\mathrm{v}}^2 S_{1\mathrm{a}} + S_{2\mathrm{a}}^2 S_{1\mathrm{v}}))\pi(k_{\mathrm{a}}, k_{\mathrm{v}})\mathrm{d}k_{\mathrm{a}}\mathrm{d}k_{\mathrm{v}} \qquad (8.6)
\end{aligned}
$$

provided that $M(c_{\mathrm{a}}, c_{\mathrm{v}})$ is invertible. Designs based on singular $M(c_{\mathrm{a}}, c_{\mathrm{v}})$ do not arise because, for the priors we consider, D would only be estimable for sets of $(k_{\mathrm{a}}, k_{\mathrm{v}})$ which have zero probability.

8.5.2 Prior distribution

As we have just seen, it is necessary to consider only the marginal of the prior for k_{a} and k_{v}. In a genuine application the prior distribution should try to capture the prior knowledge concerning the parameters. Clearly the prior needs to respect aspects of the structure of the problem, so it must give zero probability to the regions $k_{\mathrm{a}} \leq 0$, $k_{\mathrm{v}} \leq 0$ and $k_{\mathrm{a}} \leq k_{\mathrm{v}}$. It may also be unwise to base a prior distribution too closely on estimates of the parameters obtained from previous studies. However, it may be useful to inspect values from previous studies. The estimates obtained by Matthews *et al.* (1995 b) for k_{a} and k_{v} from 109 determinations on 18 children are shown in Figure 8.3.

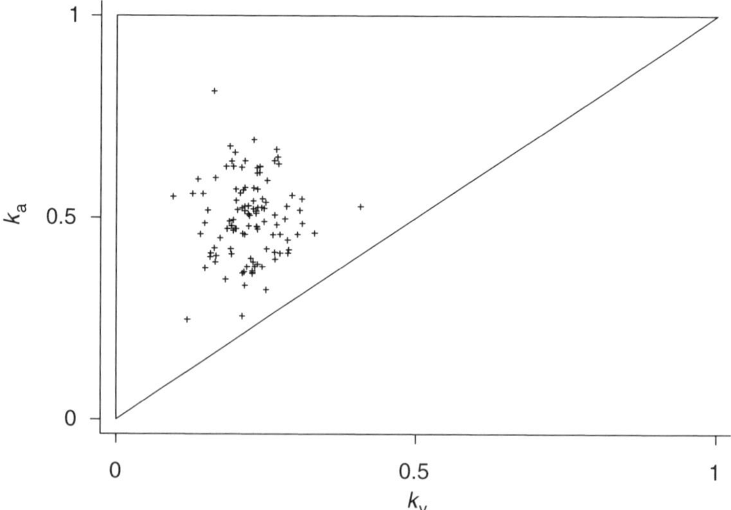

Figure 8.3 Plot of 109 estimates of k_{a}, k_{v} from 18 patients by Matthews, Matthews *et al.* (1995b)

The values in other studies may well differ somewhat from those in Figure 8.3. For example, adult patients may give slightly different values. Also the k parameters govern how quickly the inert marker is absorbed into the circulation and these are likely to be affected by the patient's respiratory function, which may differ for different groups of patients. Bearing all this in mind it is prudent to consider prior distributions which allow rather more variation than is apparent in Figure 8.3. In particular clinicians may not be prepared to use this technique on patients with poor respiratory function, thereby imposing a lower limit on the distribution of the k_a and k_v.

It is also useful to define the prior distribution in terms of parameters which can be varied to change the dispersion of the prior distribution. Theoretical results (Chaloner, 1993) show that Bayesian designs obtained using priors with limited dispersion differ little from locally optimal designs. Indeed, previous experience (Mehrabi and Matthews, 1998; Matthews and Allcock, 2004) shows that designs which have more support points than locally optimal designs only arise if the dispersion in the prior is substantially more than the limits given by these theoretical results. Consequently, we will consider a range of priors, some of which ascribe substantially more variation to k_a and k_v than is apparent in Figure 8.3. This is not meant to imply that such priors are realistic for practical use, it is simply a device to explore the range of designs this methodology can generate.

The initial idea for a prior is to consider a distribution that is uniform over the set

$$S = \{(k_a, k_v)|k_{max} > k_a > 0, k_{max} > k_v > 0 \quad \text{and} \quad k_a > k_v\}$$

and zero elsewhere. However, this has two deficiencies. First it allows values of either parameter to be arbitrarily close to zero, which represents arbitrarily slow absorption of the marker and this is plainly unrealistic. Therefore S is adjusted to

$$\{(k_a, k_v)|k_{max} > k_a > k_{min}, k_{max} > k_v > k_{min} \quad \text{and} \quad k_a > k_v\},$$

where in numerical work we take $k_{min} = 0.1\ \text{min}^{-1}$. However, even this version of S is not quite suitable as it allows k_v to be arbitrarily close to k_a and so can give rise to arbitrarily (and unrealistically) large CBFs. Therefore a final restriction we place on S is that no element of S gives rise to a CBF greater than CBF_{max}. Hence the priors we use are uniform on the following set

$$S = S(k_{min}, k_{max}, \text{CBF}_{max}), \{(k_a, k_v)|k_{max} > k_a > k_{min}, k_{max} > k_v > k_{min} \\ \text{and} \quad 0 < k_a k_v/(k_a - k_v) < \text{CBF}_{max}\},$$

where k_{max}, k_{min} and CBF_{max} are positive and can be adjusted by the investigator. In the following we always take $k_{min} = 0.1\ \text{min}^{-1}$ and allow all combinations of $k_{max} = 1$ or $8\ \text{min}^{-1}$ and $\text{CBF}_{max} = 1.5$ or $3\ \text{ml g}^{-1}\ \text{min}^{-1}$. Figure 8.4 illustrates S for $k_{max} = 1\ \text{min}^{-1}$ and both $\text{CBF}_{max} = 1.5$ and $3\ \text{ml g}^{-1}\ \text{min}^{-1}$. Values of CBF

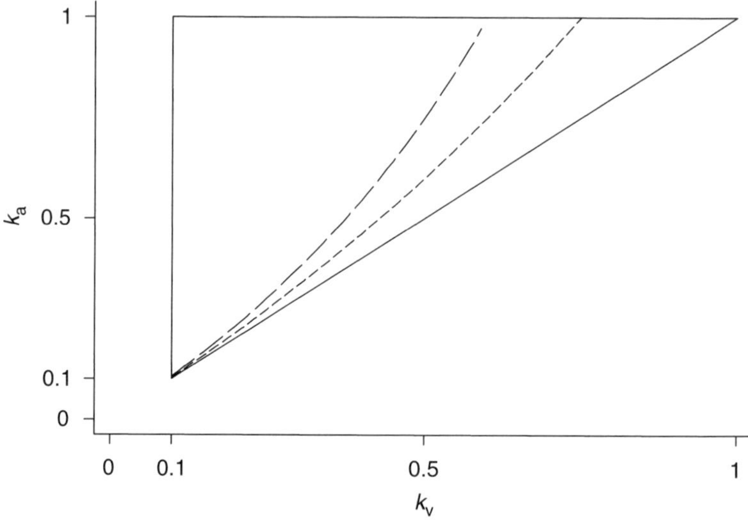

Figure 8.4 Prior is uniform over the region S: k_{min} =0.1, k_{max} =1 min^{-1} and CBF$_{max}$ = 1.5 (long dashes) and 3 (short dashes) ml g^{-1} min^{-1}: region S is within the triangle and to the left of the appropriate broken line

larger than 1.5 are quite unusual, and values larger than 3 are very unlikely to be encountered. Note that with this definition of S, k_v can never exceed $k_{max}\text{CBF}_{max}/(k_{max} + \text{CBF}_{max}) < k_{max}$, so the upper bound on k_v is never achieved.

8.6 Optimal Bayesian Designs

8.6.1 Numerical methods

Analytic optimisation of (8.5) and (8.6) is not possible so numerical methods have been employed. The expectation under the prior distribution is approximated by replacing the integral with the average over 400 points (k_a, k_v) randomly and independently drawn from the set S. The minimisation is performed by using routine E04UCF from the NAG Library (NAG, 1996), which minimises the objective function subject to each t_i being between 0 and 30 minutes and $\sum_a \eta_i + \sum_v \eta_i = 1$, with each $\eta_i \geq 0$. This program uses a sequential quadratic programming method (see, for example, Gill *et al.*, 1981, p. 237), with derivatives estimated by finite differences.

It is widely acknowledged that functions which define optimal designs often have many local minima, so it can be difficult to find a global minimum. To try to overcome this problem, number theoretic nets (Fang and Wang, 1994) with up to 5000 points were used to provide a large number of starting values for the NAG routine. Minimisations were performed for designs with m observations on each of the arterial and venous curves, with m being successively increased from $m = 2$, until the minimum of the objective function stabilised. Of course, this procedure can produce designs with fewer than m points on either or both of the curves because in numerical work it is convenient to allow the weights to be non-negative rather than strictly positive.

8.6.2 D$_S$-optimal designs

The D$_S$-optimal designs are shown in Table 8.2. The quantities quoted in Table 8.2 for the minimised value of the objective function are the values for the function

$$-\int \log\left(S_{1a}S_{1v} - \frac{(S_{2v}^2 S_{1a} + S_{2a}^2 S_{1v})}{S_{av}}\right)\pi(k_a, k_v)\mathrm{d}k_a\mathrm{d}k_v$$

at the optimal design.

Table 8.2 D$_S$-optimal designs for k_a, k_v for four distinct prior distributions

Parameters of prior distribution	Value of objective function		Arterial observations				Venous observations	
$k_{max} = 1$, $CBF_{max} = 1.5$	3.3626	Times	1.265		30		3.069	
		Weights	0.378		0.242		0.380	
$k_{max} = 1$, $CBF_{max} = 3$	3.6632	Times	1.242		30		2.644	
		Weights	0.378		0.241		0.381	
$k_{max} = 8$, $CBF_{max} = 1.5$	8.74075	Times	0.177	1.488	30		1.311	8.274
		Weights	0.368	0.066	0.131		0.355	0.080
$k_{max} = 8$, $CBF_{max} = 3$	9.5808	Times	0.175	1.653	5.410	30	0.836	6.003
		Weights	0.371	0.061	0.015	0.103	0.349	0.101

For the priors with $k_{max} = 1$, designs with three points are optimal. For these prior distibutions, the mean of (k_a, k_v) is (0.6809, 0.2787) ($CBF_{max} = 1.5$) and (0.6886, 0.3257) ($CBF_{max} = 3$). The observation times in Table 8.2 for the designs based on these priors are very similar to locally optimal designs with parameters set to the mean of the prior distribution.

For the priors with $k_{max} = 8$ the designs are now more dispersed, having three arterial and two venous points when $CBF_{max} = 1.5$, and four arterial and two venous points when $CBF_{max} = 3$. The priors with $k_{max} = 8$ give higher probability to larger values of k_a than do those with $k_{max} = 1$, and these will generate arterial concentration–time curves that rise to equilibrium levels much more quickly. In these circumstances it is clear that good designs need to take observations on the arterial curve much sooner than for the priors which do not allow such large values of k_a. In all the designs in Table 8.2, just over a third of the observations must be taken at the earliest prescribed time on the arterial curve. For the priors with $k_{max} = 8$, this time is substantially earlier than for priors with $k_{max} = 1$.

8.6.3 Optimal designs for var($\hat{D}$)

Designs which minimise the prior expectation of the logarithm of the variance of the estimator of D are shown in Table 8.3.

Table 8.3 Optimal designs for var($\hat{D}$) for four distinct prior distributions

Parameters of prior distribution	Value of objective function		Arterial observations				Venous observations	
$k_{max} = 1$ $CBF_{max} = 1.5$	5.9949	Times Weights	2.480 0.322		20.694 0.113		3.106 0.565	
$k_{max} = 1$ $CBF_{max} = 3$	5.7947	Times Weights	2.286 0.289		11.166 0.108		2.743 0.603	
$k_{max} = 8$ $CBF_{max} = 1.5$	4.6148	Times Weights	0.417 0.169	4.369 0.182		22.940 0.037	1.272 0.557	5.246 0.055
$k_{max} = 8$ $CBF_{max} = 3$	3.9383	Times Weights	0.338 0.112	0.793 0.085	16.76 0.110	25.63 0.009	0.873 0.659	6.275 0.024

The optimal designs for $k_{max} = 1$ each have two arterial points and one venous point, with the venous point having just over half of the total weight. Unlike the D_S-optimal designs, there is no longer a point at the upper end of the range of possible observations. This reflects the fact that a design which focuses exclusively on the estimation of D, rather than of k_a and k_v jointly, requires more limited information about A. For $k_{max} = 8$ the designs have two venous points, one earlier and one later than the single venous observation for the $k_{max} = 1$ designs. The later of these observations has very little weight although, as in the designs for $k_{max} = 1$, the venous points together attract more than half of the total weight. For $CBF_{max} = 1.5$, there are three arterial points: the earlier two are before and after the earlier arterial time for the $k_{max} = 1$ designs and each receives a weight of about 17 or 18%. The latest arterial observation receives little weight. When $CBF_{max} = 3$ the venous points are similar to the design for $CBF_{max} = 1.5$ but the arterial points are slightly different. Although Table 8.3 shows four arterial points, the last point receives very little weight and may not be reliable. The earliest point is similar to the design for $CBF_{max} = 1.5$ but the other points are rather earlier and later than the middle arterial point from the $CBF_{max} = 1.5$ design.

A general comment about these designs is that they were much more difficult to identify than those shown in Table 8.2. The D_S-optimal designs were identified with gradient-based numerical methods using number-theoretic nets to provide between 4000 and 5000 starting values. The designs in Table 8.2 were obtained from 96% of the starting values chosen. For the designs in Table 8.3 the different starting values gave rise to a much wider range of putative optima. Indeed, while the designs shown in Table 8.3 are the result of extensive numerical investigations, it would be prudent to entertain the possibility that designs with smaller minimum values exist. Despite the relative difficulty of finding designs focused specifically on estimating the reciprocal of CBF, it seems worthwhile. The efficiencies of the D_S-optimal designs evaluated using the criteria in (8.6) with respect to the designs in Table 8.3 are between 75 and 81%.

8.7 Practical Designs

8.7.1 Reservations about the optimal designs

The optimal designs obtained in Section 8.6 are more realistic than those in Section 8.3 in so far as they do not require precise values for the parameters to be known before the design can be implemented. However, there are other aspects of these designs which make them of questionable value to the practitioner. To see this it is necessary to reconsider exactly how observations are made in this application.

When an observation is taken a sample of a given volume of blood is drawn from either the venous or arterial line. The sample is placed in an evacuated tube for analysis immediately after the end of the period of data collection. For the study on children (Matthews *et al.*, 1995b) the volume withdrawn for each observation is 0.2 ml. As each determination of CBF requires N such samples and serial measurement of CBF is likely to be an important feature in the monitoring of the patients, it is important, particularly for children, to ensure that the total volume of blood withdrawn does not compromise the patient.

Implementing designs of the nature of those shown in Tables 8.2 and 8.3 in these circumstances could pose problems. The optimal designs require replication of observations at a limited number of times, rather than single observations at a series of different times. If the intention is to make a total of N observations then, in principle, an optimal design will require $N\eta_i$ observations to be taken at time t_i. The following problems could arise:

(a) The usual difficulties with implementing optimal designs apply. For example, $N\eta_i$ may not be an integer, or even close to one, for practical values of N. Also at the bedside it may be awkward to ensure the accuracy of the timing of the sampling that is implied by the t_i values shown in Tables 8.2 and 8.3. These difficulties attend the implementation of many optimal designs and trying to overcome them by use of simple approximations is certainly a possible approach.

(b) A further difficulty concerns the repeated observations at a given time. Suppose m observations are to be made at a time t. Two methods can be envisaged: taking m samples, each of 0.2 ml, as close together in time as possible, or withdrawing a single sample of $0.2m$ ml and then dividing this prior to analysing the N_2O concentration. Neither approach is practically attractive. Withdrawing a blood sample and putting it in a correctly labelled tube and then safely storing the tube at the bedside of a critically ill patient is more time consuming than might be imagined. As such sampling in this way would result in samples taken at a range of times, rather than the single time specified by the optimal design. Therefore this approach is probably infeasible. Withdrawing a single larger sample presents fewer problems at the bedside. This sample could then be divided after the data collection has finished and before being analysed to determine the N_2O concentration. However, this is also questionable practice for at least two reasons. First, it introduces an extra step (dividing the sample) into an already complicated procedure, thereby increasing the chance of errors. Second, it raises a difficulty with the statistical model used for our analysis. The errors are taken to be independent. Part of the error will be due to measurement errors in the gas analyser, and these are likely to be independent between different samples presented to the analyser, regardless of how close in time the samples have been obtained. However, there will be other components contributing to the error term in the model, such as how well the N_2O mixes with the breathing mixture, how well it is being absorbed and several other biological reasons for the concentration to depart from its mean value. While these effects are likely to fluctuate sufficiently rapidly that observations taken at least 30 s to 1 min apart might reasonably be treated as independent, this is very unlikely to be plausible for samples that originated in a single pooled sample. This strong degree of dependence could seriously undermine the value of the replication used in the optimal designs.

(c) A further problem, which applies to most applications of optimal design theory, is that the number of distinct points in the optimal designs is limited and prevents the data collected from providing evidence about the validity of the model. While this may not be important for processes which are well understood and for models which have the confidence of the users, the model used in this application is relatively new and it would be helpful if data to support the model could still be collected.

8.7.2 Discrete designs

To address these concerns optimal designs obtained by searching a more restricted class of designs were obtained. Unlike the optimal designs, these designs require the specification of the total number of observations, N, to be taken. The criteria in

(8.5) and (8.6) are then minimised with one observation permitted at each of a restricted set of times. Effectively this amounts to setting each $\eta_i = 1/N$. The times at which observations can be made start at $t = 1$ min and are separated by at least one minute. The first set of these designs assumes $N = 22$. The priors adopted above will be used. There is likely to be little value in many observations at the extreme end of the series, so the times are a regular sequence of one-minute intervals up to $t = 20$ min and two further observations, one at 25 min and one at 30 min, i.e the design space is

$$T = \{t | t = 1, 2, 3, \ldots, 19, 20, 25, 30\}.$$

A further analysis was performed using a reduced space

$$T' = \{t | t = 1, 2, 3, \ldots, 10, 25, 30\},$$

for which $N = 12$. This has been done because there may be circumstances when an investigator wishes to make fewer observations, perhaps to preserve blood volume in a particularly small patient.

The design problem is now simply a matter of deciding whether the sample at time $t \in T$ is to be taken from the arterial or venous curve, so the objective function can be evaluated at each of the $2^N - 2$ possible designs (the two designs which take observations solely on one of the curves are omitted). This is accomplished with the help of the algorithm by Chase (1970). There are 4,194,302 designs in T and it is feasible to perform this optimisation by complete evaluation on a PC in a few hours. The results of this analysis are shown in Tables 8.4 and 8.5.

Table 8.4 Practical designs: D_S-optimality criterion

Prior distribution	Objective function[a]	Efficiency[b]	Arterial times	Venous times
$N = 22$: times in T				
$k_{max} = 1, \text{CBF}_{max} = 1.5$	−1.39455	49	$t \leq 3, t \geq 12$	$4 \leq t \leq 11$
$k_{max} = 1, \text{CBF}_{max} = 3$	−0.89553	20	$t \leq 2, t \geq 10$	$3 \leq t \leq 9$
$k_{max} = 8, \text{CBF}_{max} = 1.5$	9.62036	0.1	$t = 1, t \geq 11$	$2 \leq t \leq 10$
$k_{max} = 8, \text{CBF}_{max} = 3$	11.38278	0.03	$t = 1, t \geq 10$	$2 \leq t \leq 9$
$N = 12$: times in T'				
$k_{max} = 1, \text{CBF}_{max} = 1.5$	−0.94408	72	$t \leq 2, t \geq 8$	$3 \leq t \leq 7$
$k_{max} = 1, \text{CBF}_{max} = 3$	−0.49606	67	$t \leq 2, t \geq 8$	$3 \leq t \leq 7$
$k_{max} = 8, \text{CBF}_{max} = 1.5$	9.89722	4	$t = 1, t \geq 8$	$2 \leq t \leq 6$
$k_{max} = 8, \text{CBF}_{max} = 3$	11.60434	3	$t = 1, t \geq 8$	$2 \leq t \leq 6$

[a]Value of (8.5) excluding terms in A and σ.
[b]Ratio of square root of geometric mean generalised variance under the prior as a percentage.

Table 8.5 Practical designs expected log var($\hat{D}$) criterion

Prior distribution	Objective function[a]	Efficiency[b]	Arterial times	Venous times
$N = 22$: times in T				
$k_{max} = 1, \text{CBF}_{max} = 1.5$	3.7096	45	$t \leq 2, t \geq 12$	$3 \leq t \leq 11$
$k_{max} = 1, \text{CBF}_{max} = 3$	3.6408	39	$t \leq 2, t \geq 12$	$3 \leq t \leq 11$
$k_{max} = 8, \text{CBF}_{max} = 1.5$	4.4338	5	$t = 1, t \geq 14$	$2 \leq t \leq 13$
$k_{max} = 8, \text{CBF}_{max} = 3$	4.4580	3	$t = 1, 12t \geq 14$	$2 \leq t \leq 11, t = 13$
$N = 12$: times in T'				
$k_{max} = 1, \text{CBF}_{max} = 1.5$	3.8996	68	$t \leq 2, t \geq 9$	$3 \leq t \leq 8$
$k_{max} = 1, \text{CBF}_{max} = 3$	3.8037	61	$t = 1, 3, t \geq 9$	$t = 2, 4 \leq t \leq 8$
$k_{max} = 8, \text{CBF}_{max} = 1.5$	4.5197	9	$t = 1, t \geq 9$	$2 \leq t \leq 8$
$k_{max} = 8, \text{CBF}_{max} = 3$	4.5206	5	$t = 1, 8, t \geq 10$	$2 \leq t \leq 7, t = 9$

[a]Value of (8.6) excluding terms in A and σ.
[b]Ratio of geometric mean variance under the prior as a percentage.

Several comments are in order:

(a) The designs have a similar form, with roughly equal numbers of observations being on each curve. A slightly lower proportion of observations are on the venous curve when assessed using the D_S-optimality criterion. Each design starts with one or two observations on the arterial curve, which are then generally followed by a block of venous observations, with the remaining arterial observations being taken after this block. This simple pattern is slightly disturbed for the most dispersed prior but even here the design with all venous observations in a single block, as in the optimal design for other priors, gives the second smallest value for the objective function (design not shown).

(b) Relative to the optimal continuous designs from previous sections, these designs are not particularly efficient. The efficiencies are much lower for the priors with $k_{max} = 8$ and are lower for the designs with $N = 22$ than $N = 12$. The latter observation is perhaps to be expected: the traditional optimal designs use only a few support points and such concentrations of observations are not possible for the practical designs. Designs which require 22 equally spaced observations cannot come close to emulating this, whereas designs with fewer points are in some sense more similar to the optimal designs. The former observation perhaps reflects the form of the criterion rather than that the designs are so poor as these figures suggest. It may be that parameter values in some region covered by the prior, for example regions with large values for k_a, could present problems for designs with the smallest observation at $t = 1$ (the optimal designs in the preceding section have the earliest arterial observation noticeably earlier than this), which give rise to very large variances for the estimate of D and hence dominate the efficiency calculation. In any case, the priors with $k_{\max} = 1$ are more realistic.

(c) While the designs with fewer points are more efficient, those with larger numbers of observations do give smaller variances: for $k_{\max} = 1$, the geometric mean variances of $\hat{D}$ are 17–20% larger for $N = 12$ than for $N = 22$. On the basis of the ratio of sample sizes, one might naively expect the variance for the former to be about $22/120 - 1 \approx 80\%$ larger. The much lower improvement as the sample size increases indicates the poorer designs which have to be used when the sample size increases.

The reduced design space T' has been chosen arbitrarily. The motivation was to take fewer samples, but there was no requirement to use this particular subset of T. Indeed, if T' had been obtained from T by deleting the even times less than or equal to 20, then the best design with respect to D_S-optimality (using the least dispersed prior) has an efficiency of only 8% as compared with 49% for T'. This shows that if there is a choice about which times are to be used, then how this is exercised can have an important effect on the efficiency of the chosen design. It may well be that an investigator needs to use fewer observations than are assumed when T is used (perhaps to limit the volume of blood taken from the patient). T' was chosen to illustrate the possible effects of reducing the number of observations, but this particular subset, while seemingly sensible, is admittedly arbitrary.

A full investigation of all possible designs based on observations from a subset of T with a particular size is not generally feasible. If we wish to take M observations from a set of N possible times, then the number of designs is

$$\binom{N}{M}(2^M - 2).$$

For $N = 22$ and $M = 12$ this is a number in excess of 2.6 billion and exhaustive searching is now impossible. It might be possible to develop more subtle approaches, perhaps using ideas based on exchange algorithms, but we will attempt a simple pragmatic approach which has no pretensions to optimality but which seems to be sensible. The procedure is as follows.

Designs which start with two observations and end with four observations on the arterial curve, with the middle six observations on the venous curve, are broadly in line with the designs in Tables 8.5 and 8.6. For each possible set of 12 times chosen from the 22 in T we will compute the objective functions (8.5) and (8.6) assuming that the first two and last four times are sampled from the arterial line with the others taken from the venous line. This gives a search of

$$\binom{22}{12}$$

Table 8.6 Practical designs with $N = 12$

Prior distribution ($k_{max} = 1$ in all cases)	Objective function[a]	Efficiency[a]	Type of design[b]	Times chosen
D_S-optimal criterion				
$CBF_{max} = 1.5$	−0.96689	73	5	$t \leq 8, t \geq 19$
$CBF_{max} = 1.5$	−0.93605	72	6	$t \leq 9, t \geq 20$
$CBF_{max} = 3$	−0.51704	67	5	$t \leq 8, t \geq 19$
$CBF_{max} = 3$	−0.48538	66	6	$t \leq 7, t \geq 18$
$\text{Var}(\hat{D})$ criterion				
$CBF_{max} = 1.5$	3.90372	67	5	$t \leq 11, t = 30$
$CBF_{max} = 1.5$	3.89013	68	6	$t \leq 12$
$CBF_{max} = 3$	3.80588	61	5	$t \leq 11, t = 30$
$CBF_{max} = 3$	3.80163	61	6	$t \leq 12$

[a]Defined as in Tables 8.4 and 8.5.
[b]Designs are designated by the number of observations on the venous curve: either five or six.

designs, which is entirely feasible. Note that T' is one of these subsets. This exercise will be restricted to the two priors with $k_{max} = 1$. The procedure may not give designs that have smaller D_S-optimality criteria than appear in Table 8.4, as those designs have only five points observed on the venous curve, so are not a subset of the designs in this search. Consequently the exercise was extended to consider 12-point designs starting with two and ending with five arterial observations, i.e. designs with five venous observations.

The results are shown in Table 8.6. The designs in Table 8.6 with five venous observations have slightly lower D_S-optimality criteria than the values shown in Table 8.4 for $N = 12$ designs with the same prior. The improvement has resulted from the relocation of arterial observations from T' at 9 and 10 minutes (Table 8.4) to 19 and 20 minutes (Table 8.6). The designs with six venous observations have slightly larger criteria than the corresponding designs in Table 8.4. However, in no case does the efficiency of a design in Table 8.6 differ by more than 1% from that in Table 8.4. This suggests that for this criterion at least, the arbitrary choice of T' was actually quite sensible.

For the $\text{var}(\hat{D})$ criterion the position is reversed, with designs having six venous observations doing slightly better than those on Table 8.5 and those with five venous points doing slightly worse. The improvements for the 'six-point' designs come from the relocation of the observations to the first 12 minutes. The improvement in efficiency is not noticeable for $CBF_{max} = 3$, but for $CBF_{max} = 1.5$ the efficiency of the new design if 68% as opposed to the 61% seen in Table 8.5.

8.8 Concluding Remarks

The specific forms of the optimal continuous designs and of practical designs which appear to perform well are shown in the tables. However, some general features of these designs can be discerned and these may be of value to the investigator:

- The earliest observations need to be taken on the arterial curve. If observations are taken too late then they will only provide information about A and not k_a. As k_a is larger than k_v observations gathering information about k_a must be taken before those on the venous curve.
- After initial observations on the arterial curve the next block of observations is on the venous curve, obtaining information about k_v.
- While A is not of direct interest the estimates of the k parameters are affected by its value, so a few observations are taken as late as possible in order to obtain this information. Provided the prior does not envisage very small values of k_v it is probably immaterial which curve is used for these observations. In designs with fixed numbers of observations on the two curves, it is sensible to use the arterial curve for this purpose because observations from the venous curve are likely to be informative about k_v over a wider time interval than is the case for informative observations about k_a taken from the arterial curve.
- Designs intended to minimise var$(\hat{D})$ need fewer (or no) late observations than those based on the D_S-optimality criterion.

Of course, changes in priors and criteria may lead to changes but these general features are likely to be discernible in most good designs for this problem.

The construction of the practical designs, in which effectively a minimum spacing was imposed between successive observations, raised an interesting feature. While good designs using more observations give better estimates than those with fewer, the improvement can be very modest. A rather unrigorous but helpful way to view this is to realise that the imposition of a minimum spacing forces the larger practical designs to be, in some sense, less like the optimal continuous designs in Tables 8.2 and 8.3 than do the smaller designs. This may well be a feature of optimal designs in other applications when practical constraints impose something akin to the minimum inter-observation interval we have used. A practical consequence in this application is that the investigator, mindful of the need not to take too much blood from a patient, may wish to use fewer observations, as the added benefit of more observations is too modest to justify the extra volume of blood being taken. If CBF determinations are to be taken serially during a patient's stay in hospital, this may make the limited benefit of using more observations for each determination even less justifiable.

The optimal continuous designs, especially those using the var$(\hat{D})$ criterion, were difficult to find. In this, as in many applications, the nature of optimal designs is such as to make their direct use impractical. However, the form of optimal designs, both in this and other applications, is such that we often cannot make direct use of them but they nonetheless provide very useful insights. We have used these designs in at least two ways. When the set of practical designs is too large to be searched exhaustively, the general form of the optimal designs can provide valuable guidance in selecting subsets which can feasibly be searched. Also optimal designs can be used to assess the efficiency of more practical designs. This latter use is especially helpful if it confirms that the practical alternatives are highly efficient. It should be acknowledged that it may be less helpful if the practical designs have low efficiencies while the optimal designs have a totally impractical form.

References

Chaloner, K. (1993). A note on optimal Bayesian design for nonlinear problems. *Journal of Statistical Planning and Inference*, **37**, 229–235.

Chaloner, K. and Verdinelli, I. (1995). Bayesian experimental design: a review. *Statistical Science*, **10**, 273–304.

Chase, P. J. (1970). Combinations of *M* out of *N* objects. *Communications of the Association of Computing Machinery*, **13**, 368.

Fang K.-T. and Wang Y. (1994). *Number-theoretic Methods in Statistics*. Chapman and Hall, London.

Gill, P. E., Murray, W. and Wright, M. H. (1981). *Practical Optimization*. Academic Press, London.

Kety, S. S. and Schmidt, C.F. (1945). The determination of cerebral blood flow in man by the use of nitrous oxide in low concentrations. *American Journal of Physiology*, **143**, 53–66.

Kety, S. S. and Schmidt, C. F. (1948). The nitrous oxide method for the determination of cerebral blood flow in man: theory, procedure and normal values. *Journal of Clinical Investigation*, **27**, 476–483.

Matthews, D. S. F., Bullock, R. E., Matthews, J. N. S., Aynsley-Green, A. and Eyre, J.A. (1995a). The temperature response to severe head injury and the effect on body energy expenditure and cerebral oxygen consumption. *Archives of Disease in Childhood*, **72**, 507–515.

Matthews, D. S. F., Matthews, J. N. S., Aynsley-Green, A., Bullock, R. E. and Eyre, J. A. (1995b). Changes in cerebral oxygen consumption are independent of changes in body oxygen consumption following severe head injury in childhood. *Journal of Neurology, Neurosurgery and Psychiatry*, **59**, 359–367.

Matthews, J. N. S. and Allcock, G. C. (2004). Optimal designs for Michaelis–Menten kinetic studies. *Statistics in Medicine*, **23**, 477–491.

Matthews, J. N. S., Matthews, D. S. F. and Eyre, J. A. (1999). Statistical methods for the estimation of cerebral blood flow using the Kety–Schmidt technique. *Clinical Science*, **97**, 485–492.

Mehrabi, Y. and Matthews, J. N. S. (1998). Implementable Bayesian designs for limiting dilution assays. *Biometrics*, **54**, 1398–1406.

NAG (Numerical Algorithms Group) (1996). *The NAG FORTRAN Library Mark 17*. The Numerical Algorithms Group Ltd, Oxford.

9

Optimal Experimental Design for Parameter Estimation and Contaminant Plume Characterization in Groundwater Modelling

James McPhee and William W-G. Yeh

Department of Civil and Environmental Engineering, UCLA, 5732B Boelter Hall, Los Angeles, CA 90066, USA

9.1 Introduction

Groundwater pollution has been a major subject of concern in the USA since the early 1980s, when the effects of thoughtless environmental policies concerning groundwater resources in previous decades began to be understood. Since then, researchers have directed a great deal of effort to increase knowledge about the fate and transport of contaminants in the subsurface. Most groundwater contamination cases are characterized by an initial spill of a pollutant, subsequent downward migration of the pure phase compound towards the water table, and the further spread of contamination due to dissolution of contaminants in the aqueous phase (Bedient *et al.*, 1999). The primary questions to be addressed on a remediation

Applied Optimal Designs Edited by M.P.F. Berger and W.K. Wong
 ISBN: 0-470-85697-1 (HB)

framework are: (1) where is the pollution's source, (2) what will happen if no action is taken, and (3) what are the most effective technological alternatives to address the clean-up of the contaminated water and soil.

To answer these questions we need to know as accurately as possible the dynamics and fate of pollutants in the subsurface. Simulation models solve groundwater flow and mass transport governing equations and henceforth predict the system's response to excitations. Groundwater systems' responses are governed by partial differential equations, and the embedded parameters usually are spatially dependent. Experimental design acquires special significance because appropriate estimation of the aquifer parameters depends upon the quantity and quality of available observations. At the same time, evaluation of an experimental design depends upon results of the simulation model for which parameters are to be estimated. Therefore, experimental design and parameter estimation are linked in a sequential process that generally terminates when the simulation model meets the reliability requirements in model application (Hsu and Yeh, 1989; Sun and Yeh, 1990). Considerations in designing a monitoring network relate to: (1) the scale of the problem to be addressed, (2) objectives of the design, (3) complexity of field conditions and uncertainty associated with hydrology and geology of the site where the design is to be implemented, and (4) applicability of the solution proposed (Loaiciga *et al.*, 1992).

This chapter focuses on the application of optimal experimental design for parameter estimation and contaminant plume characterization, and it is organized as follows. Section 9.2 reviews the physics that governs the groundwater flow and contaminant transport in saturated porous media, and discusses the problem of parameter estimation and how it is closely related to experimental design. Section 9.3 surveys relevant mathematical formulations for the problem of experimental design, highlighting some popular optimality criteria. Section 9.4 presents algorithms that may be suitable for solving the mixed-integer programming problem inherent to optimal experimental design. Section 9.5 analyses in detail two examples, a numerical example and a real-world case, of optimal experimental design. Section 9.6 provides a summary.

9.2 Groundwater Flow and Mass Transport in Porous Media: Modelling Issues

9.2.1 Governing equations

Groundwater movement is produced by energy gradients that may be natural or induced by human action, such as pumping from or recharge to the aquifer. Combining Darcy's law – which describes the average water velocity in a cross-sectional area of a porous medium – with the equation of mass conservation, the governing equation for groundwater flow is

$$S_s(\mathbf{x})\frac{\partial h(\mathbf{x},t)}{\partial t} = \nabla \cdot [K(\mathbf{x})\nabla h(\mathbf{x},t)] - q_s(\mathbf{x},t) \quad \mathbf{x} \in \Omega \tag{9.1}$$

subject to the initial condition

$$h(\mathbf{x},0) = h_0(\mathbf{x}) \quad \mathbf{x} \in \Omega \tag{9.2}$$

and the generalized boundary condition

$$K(\mathbf{x}) \cdot \nabla h(\mathbf{x},t) \cdot \mathbf{n} = \alpha(H - h(\mathbf{x},t)) + Q \quad x \in \Gamma \tag{9.3}$$

where $\mathbf{x}$ is the position vector; $h(\mathbf{x}, t)$ is hydraulic head, as a function of position and time (L); $K(\mathbf{x})$ is hydraulic conductivity (L/T); Ω is the flow domain of the aquifer; Γ is the domain's boundary; $S_s(\mathbf{x})$ is the specific storage $(L)^{-1}$; $q_s(\mathbf{x}, t)$ is an internal sink/source term (L^3/T); t is time; $\mathbf{n}$ is unit vector normal to Γ pointing out of the aquifer; H is a prescribed boundary head (L); Q is a prescribed boundary flux (L/T); and α is a parameter controlling the type of boundary condition. In Equations (9.1)–(9.3) and hereinafter, '$\nabla\cdot$' and '∇' are the divergence and gradient operators, respectively.

The solution of Equations (9.1)–(9.3) requires knowledge about the 'system parameters', K, S_s, q_s, H, Q and α over the entire flow domain and boundary. In general, these parameters are continuous functions of the spatial variables $\mathbf{x}$. The parameters H, Q and α describe the boundary conditions and are supposed to be known when building the conceptual model. Likewise, q_s represents a sink or source term that may be known with some precision, thanks to field data. The problem of parameter identification has dealt mostly with the estimation of parameters K and S_s. Experience suggests that state variable $h(\mathbf{x}, t)$ generally is not sensitive to values of S_s, so in general parameter identification reduces to estimating the spatial distribution of hydraulic conductivity over the flow domain. In practice, Equations (9.1)–(9.3) are solved by either the finite difference or the finite element method, while the continuous hydraulic conductivity field is 'parameterized' by either the zonation or an interpolation method. The process by which an infinite parameter dimension is reduced to a finite dimensional form is called parameterization (Yeh, 1986), and the corresponding set of discretized parameters are referred to as 'model parameters' (Carrera and Neuman, 1986).

Contaminant transport in porous media relates to a number of different processes. At a macroscopic level, mass released into the porous matrix is carried away via advective transport by the mean velocity field. Along its way, Brownian movement and mechanical dispersion introduce perturbations in the mean flow path, thus favouring spread of the contaminant plume (Corapcioglu and Choi, 1996). Water interacts with soil grains and adsorption/desorption processes occur during this interaction.

The governing equation for transport of a conservative (no sorption, no reaction) compound in groundwater results from the combination of the conservation-of-mass equation and Fick's law:

$$\frac{\partial(\theta C(\mathbf{x},t))}{\partial t} = \nabla \cdot [\theta \mathbf{D} \nabla C(\mathbf{x},t)] - \nabla[\theta \mathbf{v}(\mathbf{x}) C(\mathbf{x},t)] + q_s C_s \qquad (9.4)$$

where $C(\mathbf{x}, t)$ is the compound's concentration at location $\mathbf{x}$, time t (M/L^3); θ is soil porosity (−); $\mathbf{D}$ is the hydrodynamic dispersion coefficient tensor (L^2/T); $\mathbf{v}(\mathbf{x})$ is the velocity vector (L/T); q_s is a source flux (L^3/T); and C_s is the concentration at the source (M/L^3).

The first term on the right-hand side of Equation (9.4) accounts for the solute's tendency to spread out from the path it would follow according to the advective transport induced by the flow field. The spreading is called hydrodynamic dispersion, and occurs because of mechanical mixing due to the tortuous path that groundwater follows, and because of molecular diffusion. The second term on the right-hand side of Equation (9.4) represents advective transport that groundwater flow induces over the dissolved contaminants. The third term represents a source that adds contaminant mass at specified locations.

To solve Equation (9.4), $\mathbf{v}$, θ, q_s, C_s and $\mathbf{D}$ have to be known over the entire model domain. This may represent a burdensome task and may not be attainable at all for most cases. However, the velocity field can be estimated using a groundwater flow model, although the field will be estimated at discrete locations according to the flow model discretization associated with a numerical scheme. Source terms can be determined with some degree of certainty from field information. Porosity and dispersivity terms are continuous physical properties that are parameterized into a finite dimension form and incorporated into the simulation model.

9.2.2 Parameter estimation

As stated in the previous sections, the solution of the governing equations for groundwater flow and mass transport in the subsurface requires knowledge about model parameters and initial and boundary conditions. Unfortunately, these parameters are seldom directly observable in the field, and when they are, usually only at certain specific locations. Therefore, traditionally the determination of aquifer parameters, or model calibration, relies on comparing model output with state variable measurements such as groundwater head or concentration, which are observable. Model calibration thus consists of finding a set of parameter values that produces the best model output fit to observed data. Calibration can be performed by trial and error or systematically, using an automated algorithm, in what is called the 'Inverse Problem'. For a complete review on the Inverse Problem and its application to groundwater modelling, refer to Yeh (1986).

In principle, in addition to the parameter values, the Inverse Problem should also determine the parameter structure. If the unknown parameters are parameterized using a zonation scheme, then the number and shape of zones represent parameter structure. If interpolation is used, then the number and location of basis points play this role. In the zonation case one would like to define many zones, so as to improve the model fit to the observed data and thus reduce the modelling error (residuals). However, since calibration is performed against a limited number of observations, increasing parameter dimension (number of zones) leads to greater parameter uncertainty. Parameter uncertainty error can be quantified by a norm of the covariance matrix of the estimated parameters, which is defined by

$$\mathrm{Cov}(\hat{P}) = E\left\{(\bar{P} - \hat{P})(\bar{P} - \hat{P})^T\right\} \tag{9.5}$$

where $\hat{P}$ and $\bar{P}$ are the estimated and true parameters, respectively, E is the mathematical expectation; and T is the transpose of a vector when used as a superscript. The covariance matrix of the estimated parameters can be approximated, according to Yeh and Yoon (1976, 1981), as

$$\mathrm{Cov}_h^m(\hat{P}_m) \approx \left(\frac{1}{\sigma_h^2} J_h^T J_h\right)^{-1} \tag{9.6}$$

where m is the parameter dimension, i.e. the number of zones in the zonation scheme; σ_h^2 is the variance of the head observations; and J_h is the Jacobian matrix of state variable (h) with respect to each of the parameters. The elements of the Jacobian matrix are

$$(J_h)_{ij} = \frac{\partial h_j}{\partial p_i} \tag{9.7}$$

where h_j is the state variable at the jth observation location and p_i is the ith parameter. The elements of the Jacobian matrix are the sensitivity coefficients. The dimension of the covariance matrix is $(m \times m)$. The governing equation for groundwater flow helps in evaluating the sensitivity coefficients by either the influence coefficient, the sensitivity equation, or the variational method (Becker and Yeh, 1972; Yeh, 1986, 1992). Note that the Jacobian matrix is predicated on prior parameter estimates. If both the flow and mass transport models are used in calibration using head and concentration observations, Wagner and Gorelick (1987) propose the following equation for the covariance matrix of the estimated parameters:

$$\mathrm{Cov}_{h,c}^m(\hat{P}_m) \approx \left(\frac{1}{\sigma_h^2} J_h^T J_h + \frac{1}{\sigma_c^2} J_c^T J_c\right)^{-1} \tag{9.8}$$

where J_c is the Jacobian matrix of estimated concentration values with respect to changes in the model parameters. In Equation (9.8), σ_c^2 represents the variance of the concentration observations.

Tsai *et al.* (2002) further extend the approximation of the parameter covariance matrix to incorporate prior information about the model parameters,

$$\mathrm{Cov}_{h,c,T}^{n}(\hat{P}_n) \approx G_m \left(\frac{1}{\sigma_h^2} J_h^T J_h + \frac{1}{\sigma_c^2} J_c^T J_c + \frac{1}{\sigma_p^2} G_m^T G_m \right)^{-1} G_m^T \tag{9.9}$$

where G_m is a structure matrix as defined by Yeh and Sun (1984), and σ_p^2 is the variance of model parameter measurements based on prior information. Additionally, Equation (9.9) further transforms the uncertainty to each computational node by the structure matrix. Hence, the resulting covariance matrix has a dimension of $(n \times n)$, where n is the number of computational nodes.

From Equations (9.6)–(9.9) it is easy to appreciate the influence of field data quality and quantity in parameter uncertainty. Since J_h and J_c are evaluated at those locations and times for which observations are available, any experimental design should aim at sampling at those locations and times where state variables are most sensitive to the estimated parameters (Knopman and Voss, 1987). Such a design is said to provide the maximum amount of information about the unknown parameters.

9.3 Problem Formulation

9.3.1 Experimental design for parameter estimation

Experimental design for parameter estimation deals with the problem of defining experimental conditions that increase the reliability of a simulation model. This procedure consists of exciting the aquifer by pumping and/or recharge so as to include the resulting information into a parameter estimation method (Inverse Problem). Plausible ways of measuring the quality of an inference made upon a model are either measuring the variance of the parameter estimates, or measuring the variance of the estimated response or model predictions (Wagner, 1999). The variance of model predictions can be approximated by using a first-order expansion

$$\mathrm{Cov}(y) \approx [J_y][\mathrm{Cov}(\hat{P})][J_y]^T \tag{9.10}$$

where y is the vector of predicted state variables and J_y is the prediction Jacobian, the sensitivity of predicted heads and concentrations to changes in model parameters (Wagner, 1995).

To illustrate the concept, we only consider head observations. However, the methodology applies to concentration observations as well. A norm is generally used to measure the magnitude of the covariance matrix, defined by Equations (9.6), (9.8) and (9.9). It is intuitively obvious that the experimental design objective should be aimed at minimizing the norm of the covariance matrix; that is, to make matrix $\left(J_h^T J_h\right)^{-1}$ as small as possible. Since the variance term (σ_h^2) is a constant it does not affect optimization and is therefore dropped from the formulation. Steinberg and Hunter (1984) list the following measures of this 'smallness', which are widely used as criteria for experimental design in linear statistical models:

- ***A*-optimality**: A design is said to be A-optimal if it minimizes $\mathrm{tr}\left(J_h^T J_h\right)^{-1}$, where tr denotes trace.
- ***D*-optimality**: A design is said to be D-optimal if it minimizes $\det\left(J_h^T J_h\right)^{-1}$, where det denotes determinant.
- ***E*-optimality**: A design is said to be E-optimal if it minimizes the maximal eigenvalue of $\left(J_h^T J_h\right)^{-1}$.
- ***G*-optimality**: A design is said to be G-optimal if it minimizes max $\sigma(\mathbf{x})$, where $\sigma(\mathbf{x})$ is the variance of the estimated response, and the maximum is taken over all possible vectors $\mathbf{x}$ of predictor variables.
- **I_λ-optimality**: A design is said to be I_λ-optimal if it minimizes $\int \sigma(\mathbf{x})\lambda(\mathrm{d}\mathbf{x})$, where λ is a probability measure on the space of predictor variables. This criterion, which is also called average integrated variance, belongs to a more general class of L-optimality criteria (Fedorov, 1972).

Among these, the most widely used design criteria are the A-optimality, D-optimality and E-optimality (Sciortino *et al.*, 2002). Using different norms leads to slightly different conclusions regarding the optimal design. The D-optimality criterion minimizes the volume of a hyper-ellipsoid in the parameter space, with no consideration of the relationship between the ellipsoid's axes lengths, which are in turn proportional to the square root of the covariance matrix eigenvalues. This may lead to less robust designs where the volume of the confidence region is small, but where one or more of the parameters might be poorly estimated. For an extended discussion on this topic refer to Sciortino *et al.* (2002).

It is apparent that if resources to accurately measure model parameters or to observe system responses at all locations and for all times were unlimited, the problem of model reliability would be trivial. Such measurements are costly, though, and therefore the task of experimental design is to characterize explicitly the trade-off between sampling costs and the reduction in uncertainty provided by the new measurement(s). Some researchers have formulated this problem as the minimization of a norm of the covariance matrix, subject to a cost constraint (Cleveland and Yeh, 1990; McKinney and Loucks, 1992; Wagner, 1995; Sciortino *et al.*, 2002). Others have reversed this formulation and posed problems where the cost is minimized while the reliability of the model is kept above a required limit (Hsu and Yeh, 1989; Nishikawa and Yeh, 1989; Cleveland and Yeh, 1991;

Tucciarelli and Pinder, 1991). McCarthy and Yeh (1990) adopted a different concept, the δ-identifiability of Yeh and Sun (1984), to design a field data collection strategy that ensures required model accuracy.

9.3.2 Monitoring network design for plume characterization

The particular objective of a characterization effort greatly influences the mathematical approach to groundwater quality monitoring network design. For instance, a monitoring network can be included in a landfill project in order to detect any potential contamination to the aquifer. On the other hand, if knowledge already exists about a contamination episode, a monitoring network will be required to characterize the plume and evaluate the performance of a remediation alternative.

In their comprehensive literature review, Loaiciga *et al.* (1992) classify monitoring network design objectives into the following groups: (1) ambient monitoring, (2) detection monitoring, (3) compliance monitoring, and (4) research monitoring. Ambient monitoring seeks to understand the characteristics of regional groundwater quality variations over time. Detection monitoring aims at identifying the presence of targeted contaminants as soon as their concentration exceeds prespecified levels. Compliance monitoring pertains to groundwater quality monitoring requirements at a disposal facility after detecting the presence of chemical compounds. Research monitoring concerns detailed spatial and temporal groundwater quality sampling design to meet specific goals. This chapter will concentrate on reviewing the problem formulation related to detection and compliance monitoring.

Detection monitoring

Detection monitoring networks are designed to identify groundwater contamination as soon as it is released at a hazardous waste site. For those sites, a compliance boundary around the site is usually defined, such that targeted compound concentrations may not exceed the threshold values at the compliance boundary. A detection monitoring network has to be designed with the following three objectives in mind (Angulo and Tang, 1999):

- Maximization of the probability of detecting contaminants (network reliability)
- Minimization of plume volume when detected
- Minimization of the total cost of the system (well location and operation, remediation measures if required).

These three objectives conflict with each other. To maximize detection probability, a large number of wells should be installed, thus implying large construction and operation costs. An alternative to this is to locate the monitoring wells further apart from the source of contamination, for as the plume advances its volume increases and it becomes easier to detect. In this case, however, by the time the plume is detected the aquifer volume that needs remediation is bigger, hence increasing the remediation costs. To minimize remediation costs it is better to sample close to the source. But sampling close to the source implies that smaller volumes of contamination have to be detected. Consequently, the monitoring network must be denser, thus increasing construction costs.

The reliability of a particular design, defined as its capacity to detect contamination in groundwater early, must be estimated *before* actual contamination occurs. Therefore simulation models are used to predict the behaviour of a contaminant plume once it is released. As stated previously, groundwater and mass transport simulation models require the specification of model parameter values that are almost never known with certainty. In the event of contaminants reaching the ground, the uncertainty associated with model parameters makes it impossible to exactly predict *where* and *when* pollution will appear at a particular location. To overcome this difficulty, a stochastic framework is used to define probabilities of success or failure of the monitoring design.

The stochastic approach relies on assuming that the model parameters are random variables with certain underlying statistical characteristics (Rubin, 1991). Through running a simulation model several times for different realizations of the model parameters, one can generate many probable contaminant plumes. This procedure is called Monte Carlo simulation. A particular monitoring design may be evaluated in terms of the number of times it is able to detect a contaminant plume, with respect to the generated plumes.

To deal with the multi-objective nature of detection monitoring design, researchers have proposed diverse mathematical formulations to be solved by optimization algorithms. A *minimum cost approach* (Angulo and Tang, 1999) seeks to find the optimal trade-off between monitoring and remediation by considering three types of costs: (a) construction and monitoring (sampling plus analysis) costs; (b) remediation costs when a plume is detected at early stages, i.e. before it reaches the compliance boundary; and (c) remediation costs when the plume fails to be detected by the monitoring network. Typically remediation costs in case (c) will be higher, for by the time the plume reaches the compliance boundary its size has increased. Another possible approach is to maximize the probability of detecting a contaminant plume. Datta and Dhiman (1996) present a *chance-constrained* formulation to attain this goal, where the objective function is to minimize the positive difference between simulated concentrations and a prescribed standard at those locations, out of the set of potential locations, where monitoring wells are not installed. Meyer *et al.* (1994), on the other hand, propose an explicit *multi-objective optimization* problem with the following objectives: (1) minimization of monitoring network size (number of wells), (2) maximization of detection probability, and (3) minimization of plume area (two-dimensional model) at detection time.

A different approach for monitoring network relies on *distance-based* algorithms. These essentially are based on locating monitoring wells at those sites where their impact is estimated to be higher. The impact of a monitoring site can be defined in terms of many attributes, among which are the prospect of plume detection and exposure hazard criteria. Hudak and Loaiciga (1993) define the value of a monitoring well as an inverse function of the distance from the contamination source and the distance to the nearest contamination envelope. A contamination envelope is the outer limit of the region that may be contaminated due to a pollution episode, considering the uncertainty associated with mass transport in the subsurface. To prevent the algorithm from clustering monitoring wells only around potential pollution sources, Hudak *et al.* (1995) combine a *covering-weight* approach with the aforementioned distance-based weighting method. In the covering model, wells 'cover' the corresponding node and adjacent neighbouring nodes. A well that covers a node contributes some fraction of that well's weight to the objective function, and the fractions are denoted w_{sik} for the well's own node (primary sampling domain) and w_{tik} for neighbouring nodes (secondary sampling domain). For details on the calculation of nodal fractions see Hudak *et al.* (1995). The optimization model is posed as a linear integer-programming model that maximizes the sum of covered weights.

Compliance monitoring

After detecting groundwater pollution, action must be taken to mitigate the impacts of contaminants on the environment and public health. Once a mitigation or remediation programme is designed and put into action, one would like to be sure that the clean-up objectives are accomplished. Remediation methods include containment methods for source control, hydraulic controls, bioremediation, soil vapour extraction systems, etc. (Bedient *et al.*, 1999; GWRTAC, 1999). No matter which remediation technique is selected, adequate monitoring of the contaminant plume is required. Compliance monitoring must take into account not only the existence of a plume, but also its approximate location, shape, size and dynamics.

The fact that plume dynamics are important makes the problem of network design considerably more complicated than detection monitoring. If the plume is moving, intense monitoring in clustered locations may be inefficient. Sampling frequency and sampling period become important issues, since the interval between samples should be large enough to minimize sampling and sample analysis costs. On the other hand, care must be taken to sample frequently enough so as to keep close track of the plume's movement and concentration evolution. When extending the network design problem to a space–time domain, the combinatorial nature of the problem makes it computationally burdensome. For example, to select 6 sampling locations out of 100 potential sites, for 10 time periods, a complete enumeration algorithm should evaluate roughly 12 billion combinations (Montas *et al.*, 2000). Therefore innovative approaches, directed search and heuristics that take into account plume dynamics must be considered.

James and Gorelick (1994) adopt the concept of expected value of sample information (EVSI), which constitutes the foundation of the data-worth analysis. For a single sample taken at location b_0, the EVSI equals the difference between the remediation cost given prior information and the expected remediation cost given the information added by the new sample. The methodology selects the best locations for new sampling by estimating the ESVI at every location where contamination could occur. The best location has the maximum ESVI. The procedure as described is repeated several times, each iteration using the previously determined remediation cost and contamination envelope as the prior estimates for the updating procedure. It constitutes a Bayesian approach that allows determining of the optimal number of samples. It is important to note, nevertheless, that many times the combined information of two or more samples is greater than the stepwise summation of the information provided by each sample individually. Therefore, global optimum is not guaranteed.

Montas *et al.* (2000) present a methodology that aims at incorporating explicitly the time dimension in the objective function of a directed search optimization algorithm for monitoring network design and sampling scheduling. The overall objective of the design process is to determine an optimal set of monitoring well locations, M, and a well activity schedule,$M^k \subset M$, such that contaminant plumes present in the vicinity of these wells are characterized with maximum accuracy while minimizing the cost. The cost constraint is replaced by minimizing the total number of wells and constraining the maximum number of active wells for a given time period. Plume characterization is dealt with by minimizing the maximum plume characterization error, $e(M)$, defined as the maximum difference between the true plume's zeroth-, first- and second-order spatial moments, and their estimates based on concentration values sampled at discrete locations. The 'true' plume is represented by the ensemble average of Monte Carlo plume realizations.

To alleviate the computational expense inherent to the combinatorial problem, Montas *et al.* (2000) divide the optimization procedure into two steps that combine a partial enumeration aimed at reducing the size of the search space and a full enumeration over the resulting reduced search space. The preliminary network, M_1, is selected out of a discrete set of network densities (wells/m^2) and geometric permutations by selecting the network with minimum density that keeps the characterization error below e_1 for the entire simulation period. The final network, M_2, is determined by selecting wells from M_1 such that $e(M_2)$ is minimized and at any time the number of active wells is less than N_{aw}. This is achieved by a full enumeration of N_{aw} well combinations from M_1.

Numerical examples presented by Montas *et al.* (2000) show that characterization error decreases *approximately exponentially* with the number of active wells. It is clear that numerical results are a function of the model parameters such as the statistical properties of the hydraulic conductivity field. Characterization error increases when using a single plume rather than the ensemble average of multiple realizations because the averaging procedure smoothes the shape of the plume. Although using an average plume instead of many single plume realizations results in increased uncertainty, Montas *et al.* (2000) prefer this method in view of the savings in computational cost.

9.4 Solution Algorithms

As we have seen in the previous sections, the problem of experimental design for parameter estimation requires decisions to be made with regard to the location of pumping/recharge wells that will excite the system, the pumping/recharge rate that will be applied, and the duration of the experiment. Simulation models typically treat space and time in a discrete fashion, so the location and duration variables are integers. On the other hand, pumping rates are continuous. The combination of integer and continuous variables makes the problem computationally intensive. A very helpful assumption is to consider the pumping variable as belonging to a finite set of discrete rates, therefore dramatically reducing the search space.

The discrete, or integer, programming problems resulting from these assumptions may have a finite number of feasible solutions. Typically this number is very large, and a complete enumeration is prohibitive, making these problems very difficult to solve. Bertsimas and Tsitsiklis (1997) propose three categories of algorithms to solve integer-programming problems:

- **Exact algorithms** that are guaranteed to find an optimal solution, but may take a very large number of iterations. These include cutting plane, branch-and-bound and branch-and-cut, and dynamic programming methods.
- **Approximation algorithms** that provide in bounded time a suboptimal solution together with a bound on the degree of suboptimality.
- **Heuristic algorithms** that provide a suboptimal solution, but without a guarantee on its quality. Although the running time is not guaranteed to be not excessively large, evidence suggests that some of these algorithms find a good solution fast. Among these, local search methods and evolutionary algorithms may be highlighted.

When nonlinearity is present, some of the aforementioned algorithms rely on linearization and iteration, and convergence is not necessarily guaranteed. Among the examples available in the literature, it is common to find combinations of algorithms that seek to use a heuristic approach to reduce the search space of the optimization problem, and then apply exact algorithms or explicit enumeration schemes to the reduced feasible sets. Cleveland and Yeh (1991) use dynamic programming (DP) to estimate the optimal location and scheduling of sampling in a groundwater tracer test. Sciortino *et al.* (2002) use a genetic algorithm (GA) coupled with an analytical model for dense non-aqueous phase liquids (DNAPLs) dissolution in groundwater to identify with maximum reliability the location and extent of a rectangular DNAPL pool. Wagner (1995) compares a branch-and-bound algorithm and a GA to solve the problem of finding a cost-optimal sampling strategy that minimizes prediction uncertainty while keeping sampling costs at a prescribed level. His results suggest that near-optimal strategies can be found with the GA in a fraction of the time required to obtain a global optimal solution with the branch-and-bound algorithm.

Note that the experimental design for parameter estimation depends on information about the parameter values that themselves are subject to estimation. To overcome this predicament, sequential sampling procedures have been proposed in the literature where initial parameter values estimated from prior information are utilized to produce a first 'generation' of experimental design, and data collected accordingly in the field. With the acquired data, parameter values are updated and the procedure repeated until the prescribed experimental goal is satisfied (Hsu and Yeh, 1989; Sciortino *et al.* 2002).

9.5 Case Studies

9.5.1 Experimental design for parameter estimation

Hsu and Yeh (1989) present an iterative solution procedure for experimental design and parameter estimation. The procedure, as shown in Figure 9.1, begins by setting an experimental goal. In this case the goal consists of parameter estimation and model prediction subject to a specified degree of precision which is directly related to the required reliabilities of the estimated parameters and the predicted system response. The procedure requires prior information of model parameters, which can be obtained from hydrogeological data. Based upon the prior information, optimization of experimental design is carried out and data collected in the field accordingly. Using the collected data, model parameters are estimated by an inverse algorithm, which involves the estimation of the parameter structure as well as the parameter values. The calibrated model is then used to verify the experimental goal, and a second iteration begins if the goal is not met. The procedure is consistent with the concept of sequential sampling.

Decision variables are the number and locations of pumping and observation wells and the pumping rate used to excite the aquifer. The wells are assumed to be members of a finite set of discrete locations, and the pumping rate belongs to a set of discrete pumping rates. Conceptually, the problem is expressed as

$$\min C\left(np, no, lp_j, lo_i \mid i \in I_{no}, j \in I_{np}\right) \tag{9.11}$$

subject to

$$s(i,k) \le Sal(i,k), \quad i \in I_{NC},\ k \in I_{NT} \tag{9.12}$$

$$\|\operatorname{cov}(\hat{P})\| \le Ral \tag{9.13}$$

$$lp_j \in Pal \quad j \in I_{np} \tag{9.14}$$

$$lo_i \in Oal \quad i \in I_{no} \tag{9.15}$$

$$Q(lp_j) \in Qal \quad j \in I_{np} \tag{9.16}$$

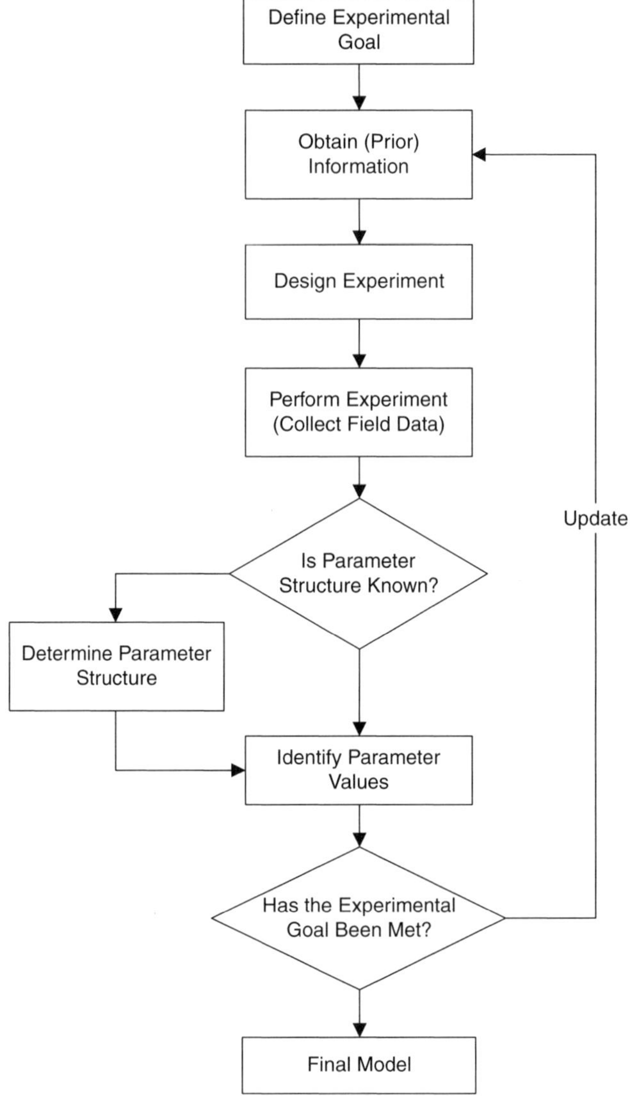

Figure 9.1 Iterative solution for experimental design and parameter estimation (modified from Hsu and Yeh, 1989)

where C is the cost; np, no are the number of pumping and observation wells, respectively; lp_j and lo_i are the location of pumping well j and observation well i, respectively; $\| \operatorname{cov}(\hat{P}) \|$ is a norm of the covariance matrix of $\hat{P}$, Ral is the required reliability of the estimated parameters; Pal is the set of possible sites of pumping wells; Oal is the set of possible sites of observation wells; Q is the pumping rate; Qal is the set of allowable pumping rates; $s(i, k)$ is the drawdown at location i at the end of time step *k*; and *Sal(i,k)* is the maximum allowable drawdown at location *i* at the end of time step *k*.

Hsu and Yeh (1989) use a Cramer–Rao bound on the covariance matrix, in order to replace it with Fisher's information matrix. Assuming that the estimated parameters are uncorrelated, the authors define $(Irq)_n = \sigma^2/(Ral)_n$, the required information for identifying parameter p_n at a level of reliability $(Ral)_n$. Observation errors are assumed to be Gaussian and uncorrelated random variables, with zero mean and constant variance σ^2. Therefore, the reliability term $(Ral)_n$ in Equation (9.13) can be replaced by the information requirement, $(Irq)_n$.

The required information term, $(Irq)_n$, can be computed if a maximum allowable variance *Val(j,l)* for the predicted response at node *j* and time period *l* is given. The adoption of a particular value of *Val(j,l)* is a management decision and reflects the accuracy water managers require from the simulation model results. By taking a first-order approximation of the predicted system response, $s^p(j,l)$, about the true-parameter system response,

$$(Irq)_n = \max_{\substack{j \in I_{NC'} \\ l \in I_{NT'}}} \left[\frac{\left[\frac{\partial s^p(j,l)}{\partial p_n}\right]^2}{\frac{Val(j,l)}{n\sigma^2} - \frac{1}{n}} \right] \qquad n \in I_{NZ} \tag{9.17}$$

where $I_{NC'}$ and $I_{NT'}$ are the sets of nodes and times, respectively, at which the system response is to be predicted. I_{NZ} is the set of model parameters with *n* elements. In Equation (9.17), $\partial s^p(j,l)/\partial p_n$ represents the sensitivity of model output (in this case, hydraulic heads) to changes in model parameters.

Hsu and Yeh (1989) propose a heuristic approach that finds an optimal solution (something not always guaranteed by search methods) and reduces computational cost. Figure 9.2 depicts the algorithm as proposed by Hsu and Yeh (1989). Loop 1 chooses the total maximum discharge (*Q*) for the system, while loop 2 reflects the fact that a feasible design may not require the maximum discharge. Loop 3 evaluates the possible number and combinations of pumping wells corresponding to the current discharge level. For all these combinations installation and operation costs are the same, since the methodology assumes uniform costs. Within loop 3, drawdowns are computed using the unit response method (Maddock, 1972; Gorelick, 1982) and feasibility of constraint (9.12) is verified. If this constraint is feasible, loops 4 and 5 determine the number and combinations of observation wells and the information term is computed for each combination by superposition using unit sensitivities (for details of this process, see Hsu and Yeh 1989). If the information constraint as in Equation (9.13) is satisfied, then the current solution with its pumping rates and number and combinations of pumping wells and observation wells is a feasible solution. The results and associated cost are printed. Otherwise, the searching procedure continues.

The methodology proposed by Hsu and Yeh (1989) makes use of the indices INDEX0, INDEX1 AND INDEX2. INDEX0 indicates the maximum discharge that

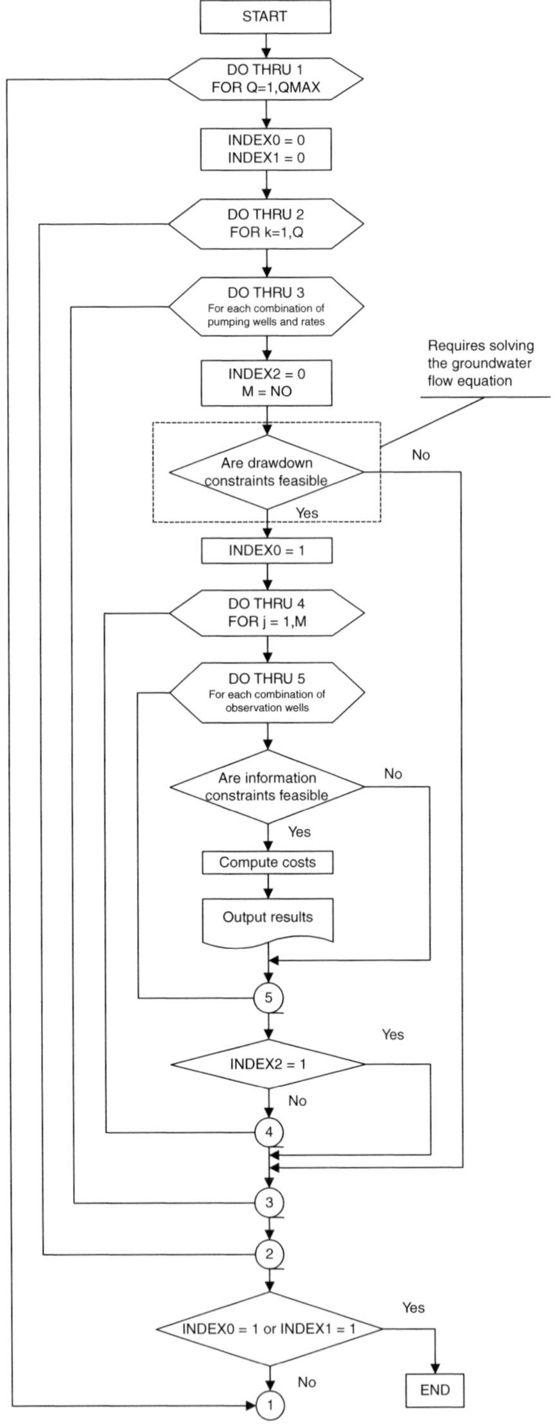

Figure 9.2 Flow chart of proposed heuristic algorithm (modified from Hsu and Yeh, 1989)

can be pumped from the system subject to the drawdown constraints. However, if the problem is feasible then it is not necessary to find the maximum rate. Therefore, INDEX1 indicates that if there is a feasible solution for the current choice of pumping rate, the algorithm should stop at loop 1. INDEX2 reduces computational work. For a given pumping rate Q, if there is a feasible solution with M observation wells, then the number of observation wells to be checked for the next combination is less than or equal to M.

Hsu and Yeh (1989) present the following numerical example to demonstrate their methodology. Figure 9.3 shows a hypothetical two-dimensional, confined aquifer with five transmissivity zones. It has a surface area of approximately 50 km^2.

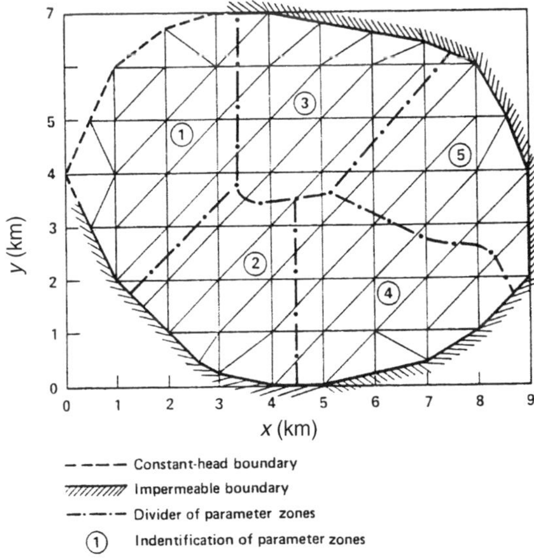

Figure 9.3 Aquifer system and finite element discretization (Source: Hsu and Yeh, 1989)

Prior information about the corresponding transmissivities indicates values of 500, 1000, 1500, 2000 and 2500 m^2/day, respectively. To compute the amount of information required for designing a pumping test, the user must specify a future operation of the aquifer a priori. In their example, the authors include three pumping wells, the characteristics of which are shown in Table 9.1. This configuration constitutes the 'production system' as depicted in Figure 9.4.

In their example, the authors assume that pumping and observation wells can be installed at any of the 10 potential sites, which are shown in Figure 9.5. Each pumping well is allowed to pump at any of the five prescribed pumping rates: 2500, 5000, 7500, 10 000, and 12 500 m^3/day. Likewise, a unit installation cost of 1.0 unit per well is chosen for all observation wells. The unit operational cost for each

Table 9.1 Production system and required information (Source: Hsu and Yeh, 1989)

	Pumping wells				
Location	(2, 3)		(4, 5)		(7, 3)
Rate (m^3/day)	25 000		25 000		12 500
	Parameter 1	Parameter 2	Parameter 3	Parameter 4	Parameter 5
Information	8.64×10^{-4}	4.71×10^{-5}	5.99×10^{-7}	2.66×10^{-6}	5.83×10^{-7}

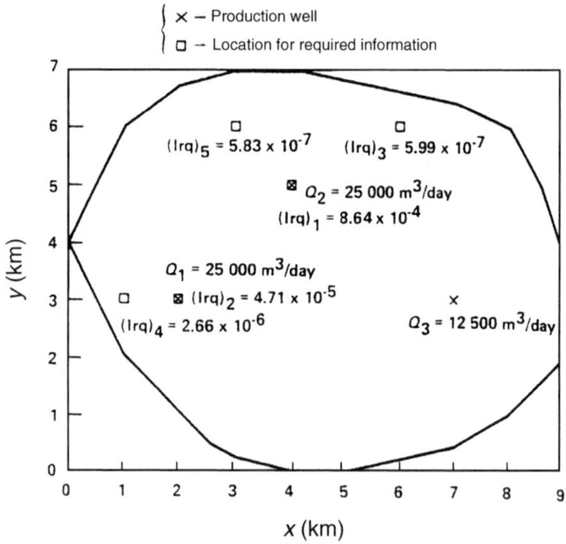

Figure 9.4 Layout of production system and the required information (Source: Hsu and Yeh, 1989)

pumping well is 0.1 unit per unit pumping rate. The unit installation cost for pumping wells is set as 20.0 units per unit pumping rate. Additionally, an 8.0 m limit on allowable drawdown is set for the entire model domain.

By applying the heuristic methodology to the hypothetical problem, Hsu and Yeh (1989) find that four solutions have the same cost of 126.6 units. Each of these solutions include two pumping wells, at locations (2, 5) and (3, 2), and six observation wells as shown in Figure 9.6. Despite their similar configuration, each solution provides different information about the unknown parameters. A summary of the results is presented in Table 9.2. From the initial four alternatives, solution 1 is optimal because it maximizes the sum of total information.

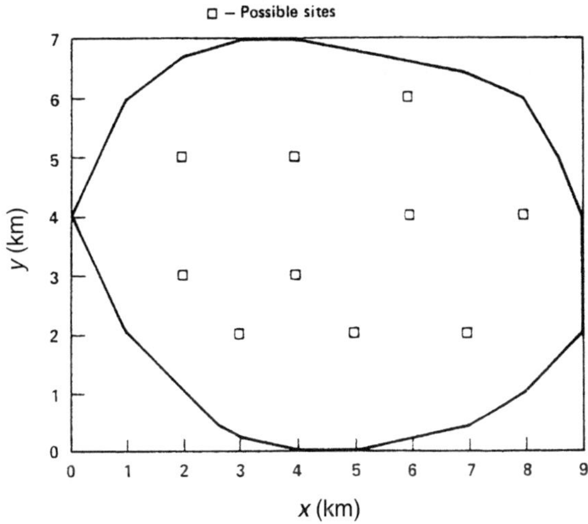

Figure 9.5 Possible sites for pumping and observation wells (Source: Hsu and Yeh, 1989)

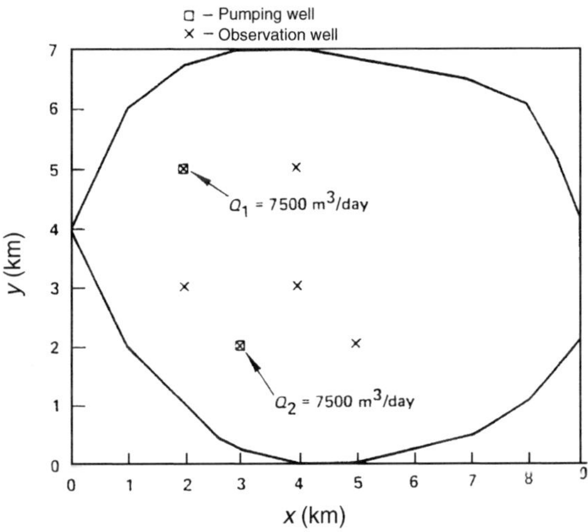

Figure 9.6 Layout of the optimal design (Source: Hsu and Yeh, 1989)

Table 9.2 Solutions with minimum cost (Source: Hsu and Yeh, 1989)

	Observation well						
Solution	1	2	3	4	5	6	Sum of information, 1.0×10^{-3}
1	(2, 5)	(2, 3)	(4, 5)	(3, 2)	(4, 3)	(5, 2)	1.298*
2	(2, 5)	(2, 3)	(4, 5)	(3, 2)	(4, 3)	(8, 4)	1.292
3	(2, 5)	(2, 3)	(4, 5)	(3, 2)	(4, 3)	(7, 2)	1.293
4	(2, 5)	(2, 3)	(3, 2)	(4, 3)	(8, 4)	(7, 2)	1.270

Note: * = optimal solution.

9.5.2 Experimental design for contaminant plume detection

Storck *et al.* (1997) present a methodology for optimally designing monitoring networks to detect initial contamination in three-dimensional heterogeneous aquifers. Their work is an extension of Meyer *et al.* (1994), which presents a multi-objective optimization approach to the problem of monitoring network design. The optimization problem considers explicitly three objectives: (1) minimization of monitoring network size (number of wells), (2) maximization of detection probability, and (3) minimization of plume area (two-dimensional model) at detection time. Storck *et al.* (1997) incorporate the effect of parameter uncertainty by using a Monte Carlo approach that generates a large number of equally likely plumes resulting from leakage from a landfill. The two sources of uncertainty are the exact location of the contaminant leak and the spatial distribution of hydraulic conductivity. Each Monte Carlo simulation consists of (1) random generation of leak location, (2) generation of a realization of a random saturated hydraulic conductivity field, (3) solution of the steady state groundwater flow model to obtain the velocity field, and (4) transport simulation of the resultant contaminant plume until it reaches the compliance boundary. Each time concentration of the contaminant exceeds a limiting value at a potential well location j, the area of the plume, $A_{i,j}$, is calculated for that time and assigned to that specific well location. The optimization problem of minimizing the plume area at detection time while maximizing the reliability of the system is posed as

$$\min R = \sum_{i=1}^{I} \sum_{j \in N_i} A_{i,j} z_{i,j} + B \sum_{i=1}^{I} w_i \tag{9.18}$$

subject to

$$w_i + \sum_{j \in N_i} z_{i,j} = 1 \quad i = 1, \ldots, I \tag{9.19}$$

$$x_j \geq z_{i,j} \quad i = 1, \ldots, I; j \in N_i \tag{9.20}$$

$$Ww_i + \sum_{j \in N_i} x_j \leq W \quad i = 1, \ldots, I \tag{9.21}$$

$$\sum_{j=1}^{J} x_j = W \tag{9.22}$$

$$x_j, w_i \in \{0, 1\} \quad j = 1, \ldots, J; i = 1, \ldots, I$$

$$z_{i,j} \in \{0, 1\} \quad i = 1, \ldots I; j \in N_i \tag{9.23}$$

where R is the sum of (1) the plume areas at the time of initial detection for all plumes that are detected and (2) a weighting factor times the number of undetected plumes; I is the number of plume realizations; J is the number of potential well locations; $x_j = 1$ if a well is located at j, 0 otherwise; $z_{i,j} = 1$ if the plume i is initially detected at location j, 0 otherwise; $w_i = 1$ if plume i is undetected, 0 otherwise; $A_{i,j}$ is the area of plume i at the time of initial detectability at location j; B is the weight assigned to an undetected plume; W is the number of wells in the network; and N_i is the set of well locations at which plume i is detectable.

Equation (9.18) is effectively a weighted objective function that incorporates two objectives: minimizing the volume of detected plumes and minimizing the number of undetected plumes. The third objective, minimizing the number of monitoring wells, is included as a constraint in Equation (9.22). By varying the values of B and W, the three-way trade-off between the objectives is quantified and the Pareto set of efficient solutions is constructed. The parameter B can be interpreted as the expected area of an undetected plume. As B is increased, the optimal network will tend to a solution in which more plumes are detected. By varying the value of W in constraint (9.22), a family of area/detection non-inferior sets is obtained. The other constraints work as follows. Constraint (9.19) ensures that each plume will be either detected at some monitoring location, or will leave the model domain undetected. Equation (9.20) prevents a plume from being detected at a location in which no well is located. When a plume is detectable by the network, but the value of B is smaller than the area of the plume when it is initially detected, the optimization program sets $w_i = 1$ for this plume. Equation (9.21) prevents this by requiring that a plume must be detected (i.e. $w_i = 0$) if there is at least one well in the network in which the plume is detectable. The integer-programming problem formulated in Equations (9.18)–(9.23) was solved using a simulated annealing code.

Storck *et al.* (1997) apply the monitoring network design methodology to a sanitary landfill located in Will County, Illinois, USA. Figure 9.7 depicts the layout of the landfill site, showing the 18 previously existing wells and the boundary of the groundwater model used to simulate the system. The model domain is discretized and is 60 cells long, 40 cells wide and 30 cells deep. The network of potential monitoring sites consists of six rows of nine equally spaced wells. The multi-objective linear integer-programming formulation presented in Equations (9.18)–(9.23) is applied to the system. Figure 9.8 compares the expected performance of the existing wells to that obtained by optimally locating networks of various sizes from the set of potential wells. It can be seen that the existing P and G series wells are clearly suboptimal with respect to the design objectives. The lower curve in Figure 9.8 shows that the 18 P and G series wells can be replaced by a network of 5 wells located optimally, achieving similar results in terms of detection probability, and considerably reducing the average contaminated volume at time of detection. Likewise, 10 optimally located wells achieve a reliability of 92% and 17 optimally placed wells can detect all of the 600 generated plumes. The trade-off curves shown in Figure 9.8 can help designers and decision-makers in selecting a monitoring scheme that will balance the conflicting objectives involved in the problem.

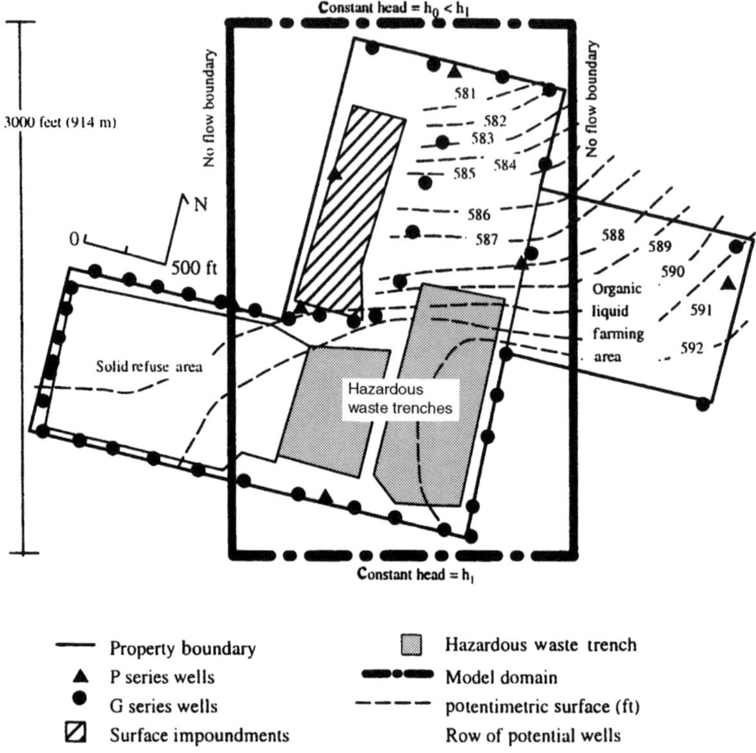

Figure 9.7 Map of ESL site showing area included in groundwater model (Source: Storck *et al.*, 1997)

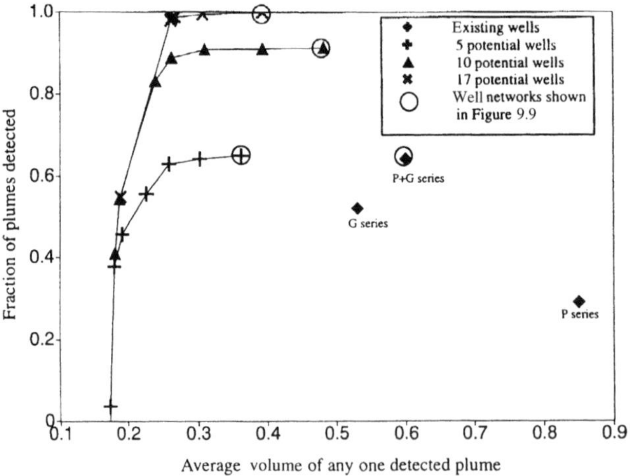

Figure 9.8 Pareto-optimal trade-off curves in objective space for well networks containing 5, 10 and 17 wells. Performance of existing networks is also shown for comparison (Source: Storck *et al.*, 1997)

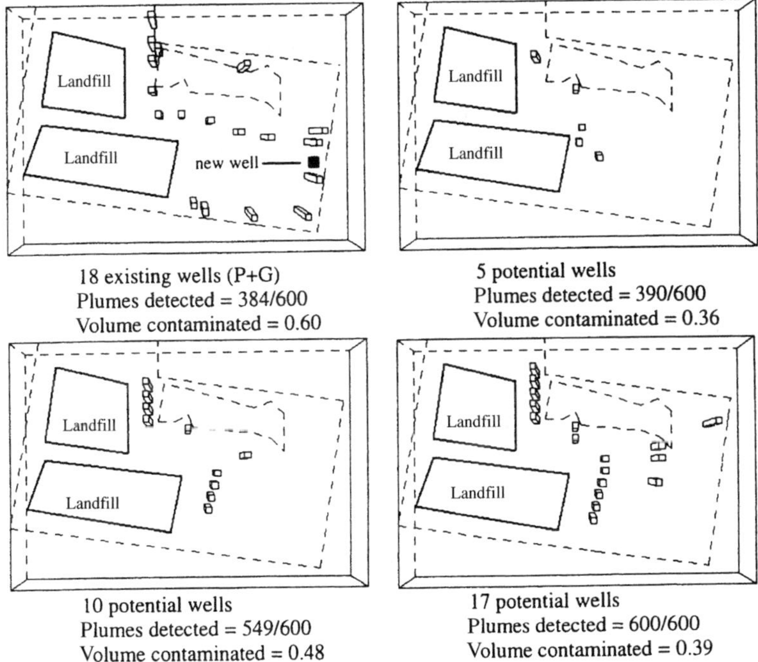

Figure 9.9 Circled networks from Figure 9.8 shown in decision space. Existing well network image also shows location of the one potential well which most increases detection percentage (Source: Storck *et al.*, 1997)

Figure 9.9 shows four points from Figure 9.8, plotted in the decision space. Storck *et al.* (1997) point out that no direct comparison can be made between the simulated networks and the existing wells because of the many simplifications adopted within the modelling process.

9.6 Summary and Conclusions

This chapter reviews the application of experimental design in two areas of groundwater modelling: parameter estimation and contaminant plume characterization. Experimental design for parameter estimation aims at selecting the samples that minimize the variance in either the parameter space or the prediction space. Since model parameters are distributed, accurate site characterization and adequate parameterization are paramount for improving the reliability of model predictions. Model predictions, in turn, are used to evaluate the performance of remediation designs before actually implementing them. All the approaches reviewed rely on estimating the parameter covariance matrix, which is generally unknown before samples are taken. Therefore, prior information about the unknown parameters and Bayesian approaches for updating parameter estimates gain relevance in improving design reliability.

Monitoring network design for plume characterization is a key issue in dealing with the protection of groundwater systems against contamination. Since it is impossible to directly observe a plume of dissolved compounds in groundwater, characterization relies on samples taken at limited locations and at discrete times. The design problem concerns basically the determination of sampling locations and sampling times that provide maximum reliability and accuracy in plume detection and characterization, while keeping construction and sampling costs within reasonable limits.

This chapter reviewed a few examples of monitoring network design, classified in two major groups: detection monitoring and compliance monitoring. Detection monitoring deals with the problem of early identification of groundwater pollution, nearby hazardous sites like landfills, combustible reservoirs, etc. Detection design usually implies a trade-off between system reliability and remediation costs. Locating monitoring wells close to the source best suits the latter objective. On the other hand, the system reliability increases when wells are located further away from the source to take advantage of plume spreading. Compliance monitoring aims at characterizing plume dynamics for evaluating the performance of remediation plans. In this case the objective is to minimize plume characterization error. Design complexity increases since the monitoring configurations are time-dependent, i.e. the time dimension has to be considered in the optimization program.

The complexity of monitoring network design for groundwater quality assessment stems from two main sources, site characterization and computational expense. Accurate site characterization is required since simulation models are a cornerstone of the design process. In most cases the monitoring network must be designed before actual contamination occurs, or in cases when pollutants' distribution in the subsurface is unknown. Therefore, simulation models are used to predict a contaminant plume's behaviour. Inherent to the simulation process is the uncertainty that arises from attempting to represent the real world in a simplified fashion, and the uncertainty related to source location and other assumptions.

Monitoring network reliability is linked with the fact that hydrogeologic conditions and source characteristics cannot be known with certainty, and therefore the plume that has to be monitored is not completely described. Uncertainty has been dealt with in statistical fashion, usually employing Monte Carlo simulations and mass transport numerical models to generate multiple equally likely realizations of dissolved plumes. The design process is then carried out considering the ensemble of these possible plumes. This approach relies on the assumptions that the hydraulic conductivity field follows a known spatial distribution with known statistical parameters. It is easy to see that uncertainty also stems from this assumption. Future work should aim at incorporating other measures of model uncertainty into the design process; for example, by considering model structure error as a key element affecting the reliability of the optimal network configurations.

Computer-aided experimental and monitoring network design is computationally intensive because of its combinatorial nature. The optimal sampling configuration and sampling schedule come from a vast discrete space (in the simulation model) of location combinations and times. Although no general-purpose software exists for solving the experimental design problem in groundwater modelling, most published techniques involve variations of the following procedure:

1. Select a design criterion, e.g. *D*-optimality.
2. Among the possible candidate designs, choose the set that minimizes the design criterion. The algorithms used to search for the optimal design include dynamic programming, the branch-and-bound algorithm, simulated annealing, genetic algorithms and directed search. Commercial optimization software with integer-programming solving capabilities include GAMS (www.gams.com) and LINGO (www.lindo.com). Genetic algorithms and simulated annealing can be found in open source form in the World Wide Web.
3. Within the search algorithm, solve the groundwater flow and transport governing equation using either a finite difference or finite element approach to generate the Jacobian matrix. By solving the governing equation over the entire spatial and temporal domain, the sampling value of the candidate design is evaluated according to the design criterion. Commonly used software for solving the groundwater flow and transport equations are MODFLOW (McDonald and Harbaugh, 1988), MT3D (Zheng and Wang, 1999) and FEMWATER (Lin *et al.*, 1996), among others.

Acknowledgements

This material is based on work supported in part by SAHRA (Sustainability of semi-Arid Hydrology and Riparian Areas) under the STC Programme of the National Science Foundation, Agreement No. EAR-9876800.

References

Angulo, M. and Tang, W. H. (1999). Optimal ground-water detection monitoring system design under uncertainty, *J. Geotech. Geoenviron. Eng.*, **125**(6), 510–517.

Becker, L. and Yeh, W. W-G. (1972). Identification of parameters in unsteady open-channel flows, *Water Resour. Res.*, **8**(4), 956–965.

Bedient, P. B. *et al.* (1999). *Ground Water Contamination: Transport and Remediation*, 2nd edn, Prentice Hall PTR.

Bertsimas, D. and Tsitsiklis, J. N. (1997). *Introduction to Linear Optimization*, Athena Scientific.

Carrera, J. and Neuman S. P. (1986). Estimation of aquifer parameters under transient and steady state conditions: 1. Maximum likelihood method incorporating prior information, *Water Resour. Res.*, **22**(2), 199–210.

Cleveland, T. G. and Yeh, W. W-G. (1990). Sampling network design for transport parameter identification, *J. of Water Resour. Plng and Mgmt, ASCE*, **116**(6), 764–782.

Cleveland, T. G. and Yeh, W. W-G. (1991). Optimal configuration and scheduling of groundwater tracer test, *J. of Water Resour. Plng and Mgmt, ASCE*, **117**(1), 37–51.

Corapcioglu, M. Y. and Choi, H. (1996). Modeling colloid transport in unsaturated porous media and validation with laboratory column data, *Water Resour. Res.*, **32**(12), 3437–3449.

Datta, B. and Dhiman, S. (1996). Chance-constrained optimal monitoring network design for pollutants in ground water, *J. of Water Resour. Plng and Mgmt, ASCE*, **122**(3), 180–188.

Fedorov, V. V. (1972). *Theory of Optimal Experiments*, translated and edited by Studeen, W. J. and Klimko, E. M. Academic, San Diego, Calif.

Gorelick, S. M. (1982). A model for managing resources of groundwater pollution, *Water Resour. Res.*, **18**(4), 773–781.

GWRTAC(Groundwater Remediation Technologies Analysis Center), *Technology Overview Report: In Situ Bioremediation*, 1999. (http://www.gwrtac.org)

Hsu, N-S. and Yeh, W. W-G. (1989). Optimum experimental design for parameter identification in groundwater hydrology, *Water Resour. Res.*, **25**(5), 1025–1040.

Hudak, P. F. and Loaiciga, H. A. (1993). An optimization method for monitoring network design in multilayered groundwater flow systems, *Water Resour. Res.*, **29**(8), 2835–2845.

Hudak, P. F., Loaiciga, H. A. and Marino, M. A. (1995). Regional-scale groundwater quality monitoring via integer programming, *J. Hydrol.*, **164**, 153–170.

James, B. R. and Gorelick, S. M. (1994). When enough is enough: the worth of monitoring data in aquifer remediation design, *Water Resour. Res.*, **30**(12), 3499–3513.

Knopman, D. S. and Voss, C. I. (1987). Behavior of sensitivities in the one-dimensional advection-dispersion equation: implications for parameter estimation and sampling design, *Water Resour. Res.*, **23**(2), 253–272.

Loaiciga, H. A., Charbeneau, R. J., Everett, L. G., Fogg, G. E., Hobbs, B. F. and Rouhani, S. (1992). Review of ground-water quality monitoring network design, *J. Hydraul. Eng.*, **118**(1), 11–37.

Lin, H. C., Richards, D. R., Yeh, G. T., Cheng, J. R., Chang, H. P. and Jones, N. L. (1996). *FEMWATER: A Three-Dimensional Finite Element Computer Model for Simulating Density Dependent Flow and Transport*, US Army Engineer Waterways Experiment Station Technical Report, 129 pp.

McCarthy, J. M. and Yeh, W. W-G. (1990). Optimal pumping test design for parameter estimation and prediction in groundwater hydrology, *Water Resour. Res.*, **26**(4), 779–791.

McDonald, M. G. and Harbaugh, A. W. (1988). A modular three-dimensional finite-difference ground-water flow model. In *U.S. Geological Survey Techniques of Water-Resources Investigations*, Book 6, Ch. A1, 586 pp.

McKinney, D. C. and Loucks, D. P. (1992). Network design for predicting groundwater contamination, *Water Resour. Res.*, **28**(1), 133–147.

Maddock III, T. (1972). Algebraic technological function from a simulation model, *Water Resour. Res.*, **8**(1), 129–134.

Meyer, P., Eheart, J. W. and Valocchi, A. J. (1994). Monitoring network design to provide initial detection of groundwater contamination, *Water Resour. Res.*, **30**(9), 2647–2659.

Montas, H. J., Mohtar, R. H., Hassan, A. E. and Alkhal, F. A. (2000). Heuristic space-time design of monitoring wells for contaminant plume characterization in stochastic flow fields, *J. Contam. Hydrol.*, **43**, 271–301.

Nishikawa, T. and Yeh, W. W-G. (July 1989). Optimal pumping test design for the parameter identification of groundwater systems, *Water Resour. Res.*, **25**(7), 1737–1747.

Rubin, Y. (1991). Transport in heterogeneous porous media: prediction and uncertainty, *Water Resour. Res.*, **27**(7) 1723–1738.

Sciortino, A., Harmon, T.C. and Yeh, W. W-G. (2002). Experimental design and model parameter estimation for locating a dissolving dense nonaqueous phase liquid pool in groundwater, *Water Resour. Res.*, **38**(5), art. no. 1057.

Steinberg, D. M. and Hunter, W. G. (1984). Experimental design: review and comment, *Technometrics*, **26**(2), 71–97.

Storck, P., Eheart, J. W. and Valocchi, A. J. (1997). A method for the optimal location of monitoring wells for detection of groundwater contamination in three-dimensional heterogeneous aquifers, *Water Resour. Res.*, **33**(9), 2081–2088.

Sun, N-Z. and Yeh, W. W-G. (1990). Coupled inverse problems in groundwater modeling 2. Identifiability and experimental design. *Water Resour. Res.*, **26**(10), 2527–2540.

Tsai, F. T-C., Sun, N-Z. and Yeh, W. W-G. (2002). A combinatorial optimization scheme for parameter structure identification in ground-water modeling. Groundwater (accepted for publication),

Tucciarelli, T. and Pinder, G. (1991). Optimal data acquisition strategy for the development of a transport model for groundwater remediation, *Water Resour. Res.*, **27**(4), 577–588,

Wagner, B. J. (1995). Sampling design methods for groundwater modeling under uncertainty, *Water Resour. Res.*, **31**(10), 2581–2591.

Wagner, B. J. (1999). Evaluating data worth for ground-water management under uncertainty, *J. of Water Resour. Plng. and Mgmt, ASCE*, **125**(5), 281–288.

Wagner, B. J. and Gorelick, S. M. (1987). Optimal groundwater quality management under parameter uncertainty, *Water Resour. Res.*, **23**(7), 1162–1174.

Yeh, W. W-G. (1986). Review of parameter identification procedures in groundwater hydrology: the inverse problem, *Water Resour. Res.*, **22**(1), 95–108.

Yeh, W. W-G. (1992). Systems analysis in ground-water planning and management, *J. of Water Resour. Plng and Mgmt, ASCE*, **118**(3), 224–237.

Yeh, W. W-G. and Sun, N-Z. (1984). An extended identifiability in aquifer parameter identification and optimal pumping best design, *Water Resour. Res.*, **20**(12), 1837–1847.

Yeh, W.W-G. and Yoon, Y. S. (1976). A systematic optimization procedure for the identification of inhomogeneous aquifer parameters, in *Advances in Groundwater Hydrology*, edited by Z. A. Saleem, pp. 72–82. American Water Resources Association, Minneapolis, Minn.

Yeh, W. W-G. and Yoon, Y. S. (1981). Aquifer parameter identification with optimum dimension in parameterization, *Water Resour. Res.*, **17**(3), 664–672.

Zheng, Chunmiao, and Wang, P. (1999). Patrick MT3DMS, *A Modular Three-dimensional Multi-species Transport Model for Simulation of Advection, Dispersion and Chemical Reactions of Contaminants in Groundwater Systems; Documentation and User's Guide*, US Army Engineer Research and Development Center Contract Report SERDP-99-1, Vicksburg, Miss., 202 pp.

10

The Optimal Design of Blocked Experiments in Industry

Peter Goos, Lieven Tack and Martina Vandebroek

Department of Applied Economics, Katholieke Universiteit Leuven, Naamsestraat 69, B-3000 Leuven, Belgium

10.1 Introduction

Quite often, the response under investigation during an experiment is not only influenced by the experimental factors but also by extraneous sources of variation. For example, in the chemical industry, the yield of a process not only depends on the factors temperature or pressure, but also on the batch of raw material used. Similarly, the outcome of experimental runs can be affected by substantial day-to-day variation. Such extraneous sources of variation are referred to as blocking factors. Other examples of blocking factors include weeks, suppliers, plots of land, sets of twins, or pairs of eyes. From these examples, it is clear that the problem of designing experiments in the presence of blocking variables is common to many fields of study.

Although, in principle, the assumption of fixed block effects is valid only if the blocks were chosen by design, block effects were usually assumed to be fixed in the statistical literature. However, in cases where the blocks can be considered a random sample from a population, the block effects should be assumed random.

Applied Optimal Designs Edited by M.P.F. Berger and W.K. Wong
 ISBN: 0-470-85697-1 (HB)

The effect of these different assumptions on the analysis of response surface experiments was investigated by Khuri (1992, 1994), Ganju (2000), Gilmour and Trinca (2000) and Goos and Vandebroek (2004). Their effect on the design of an experiment was investigated by Goos and Vandebroek (2001).

In this chapter, we will give an overview of the statistical literature on the design of blocked response surface experiments. Firstly, attention will be restricted to blocking experiments for factorial, fractional factorial and standard response surface designs. Both theoretical and computational results from the optimal design literature will be discussed. Secondly, the methodology will be extended to allow for results influenced by unwanted time trend effects. Cost considerations will also be taken into account during the design phase. A pastry dough mixing experiment will be used throughout the chapter to illustrate the design options available to the experimenter.

10.2 The Pastry Dough Mixing Experiment

The pastry dough mixing experiment, which was carried out at the University of Reading in 1995 and described by Trinca and Gilmour (2000) and Gilmour and Trinca (2000), is a nice example of an experimental situation in which the optimal design approach described in this chapter can be applied. In the experiment, the effects of the factors moisture content, screw speed of the mixer and flow rate on the colour of a pastry are investigated. Several response variables were measured for each experimental run. The response variables referred to the light reflectance in several bands of the spectrum. For the experiment, seven days were available, and four runs could be performed per day. A full quadratic model in the three explanatory factors was to be estimated. Therefore, three levels were used for each factor.

Table 10.1 Factor levels used in the pastry dough mixing experiment

Moisture content (%)	Screw speed (rpm)	Flow rate (kg/h)
76	300	30.0
79	350	37.5
82	400	45.0

These levels are displayed in Table 10.1. Typically, the factor levels are rescaled to lie between -1 and $+1$. The coded level x can be obtained from the original measurement u by applying

$$x = \frac{u - u_0}{\Delta}$$

where u_0 is the midpoint of the interval $[u_{\min}, u_{\max}]$ and Δ is half the difference between $u_{\max}$ and $u_{\min}$. The coded levels for the pastry dough experiment can thus be computed in the following fashion:

$$x_1 = \frac{\text{Moisture content} - 79}{3},$$
$$x_2 = \frac{\text{Screw speed} - 350}{50},$$

and

$$x_3 = \frac{\text{Flow rate} - 37.5}{7.5}.$$

In this type of food experiment, there is often an important day-to-day variation due to the use of biological materials such as wheat, yeast and water. It is well known that the quality of these materials can vary considerably from day to day, from batch to batch or from supplier to supplier. This variation has to be taken into account when analysing the experimental results, but also when designing the experiment. In this chapter, we will show how this can be done. The pastry dough mixing experiment will be revisited in Section 10.6 and 10.9. In Section 10.6, we will describe how Trinca and Gilmour designed the original experiment and we compute a $\mathcal{D}$-optimal design for the experiment. In Section 10.9, it is investigated how the presence of a time trend affects the design.

10.3 The Problem

In order to design a blocked experiment, a number of questions need to be solved. Firstly, levels for all experimental factors have to be determined for each experimental run. In the literature on experimental design, a combination of factor levels is referred to as a treatment or a design point. Secondly, these design points have to be assigned to the blocks of the experiment. In the absence of trend effects, the blocks of the experiment are randomized, as well as the runs within each block. However, if trends effects are suspected, the design problem becomes even more complicated because the experimenter then also has to decide on the sequence in which the different design points will be run.

In this chapter, we start by considering the problem in which no time trend effects and costs are considered. In that case, the design problem can be graphically displayed. This is done in Figure 10.1 for an experiment with nine observations and three blocks of size three. Note that, as in all other figures with square and cubic design regions in this chapter, coded factor levels are used to represent the design points. The design on top of the figure displays the design points to be run in the

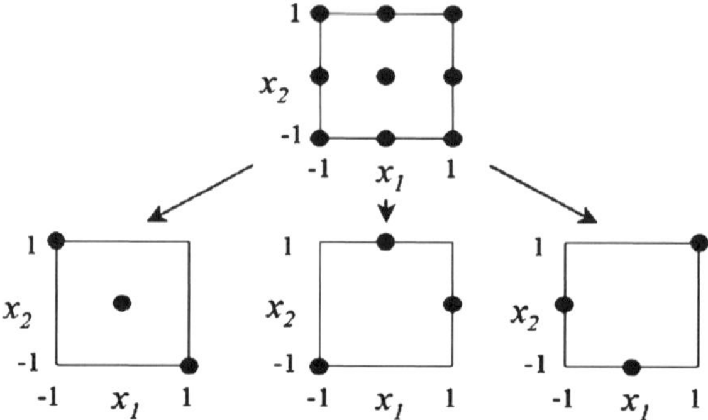

Figure 10.1 Graphical representation of the problem of designing a blocked response surface experiment in the absence of time trend effects (Reproduced from 'The optimal design of blocked and split-plot experiments', 2002, by permission of Springer-Verlag GmbH)

experiment. It can also be interpreted as the projection of the blocked experiment when the block effects are ignored. The nine points are then assigned to the three blocks of the experiment. In this chapter, we will discuss how the design points of a blocked experiment can be chosen and how they can be assigned to the different blocks of the experiment.

In the literature on blocked experiments, several solutions have been proposed for these two questions. One solution is to use standard response surface designs, as, for instance, factorial, fractional factorial and central composite designs, and assign the runs to the blocks such that the factor effects of interest can be estimated independently of the block effects. Another possible solution, advocated by Trinca and Gilmour (2000), is to determine the design points and to assign them to the blocks in two separate steps. A common feature of these approaches is that the blocked nature of the experiment is not taken into account when determining the design points of the experiment. This is in contrast with the optimal design approach adopted here.

In practice, the number of runs, the number of blocks and the block sizes are usually dictated by the experimental situation. The total number of runs available is usually restricted by time and/or cost constraints, while the block sizes are determined by, for instance, the size of a furnace, the number of runs that can be performed on one day, or the size of a batch of material. The number of blocks can be given by the number of suppliers or by the number of days available for the experiment, for example. We will denote the total number of observations in an experiment by n, the number of blocks by b, and the block sizes by k_i $(i = 1, 2, \ldots, b)$ and we assume that these parameters are given. We also assume that the factor effects do not vary from block to block, i.e. we assume that the block-by-treatment interactions are negligible. Wu and Hamada (2000) point out that the validity of this assumption has been confirmed in many empirical studies. This issue also receives attention by Gilmour and Trinca (2000). In the next section, we focus on the case where the block effects are assumed to be fixed. Next, the case of random effects is addressed.

10.4 Fixed Block Effects Model

When the blocks of the experiment are chosen by design, the block effects are usually assumed to be fixed. Typical examples of fixed blocking factors are gender or supplier. In this section, we first describe the statistical model corresponding to a blocked response surface experiment. Next, we discuss the use of standard designs in blocked experiments and introduce the concept of optimal design. Finally, we discuss some theoretical and computational results for the fixed block effects model.

10.4.1 Model and estimation

When the block effects are assumed to be fixed, the model corresponding to a blocked experiment is usually represented as

$$y_{ij} = \mathbf{f}'(\mathbf{x}_{ij})\boldsymbol{\beta} + \gamma_i + \varepsilon_{ij}, \quad i = 1, 2, \ldots, b; j = 1, 2, \ldots, k_i, \tag{10.1}$$

where y_{ij} is the response of the jth observation within the ith block, $\mathbf{f}(\mathbf{x})$ is the polynomial expansion of the settings of the experimental factors $\mathbf{x}$, $\boldsymbol{\beta}$ is the p-dimensional vector of polynomial effects of the experimental factors, γ_i is the ith block effect and ε_{ij} is the random error. It is assumed that the random errors are independent and that they have zero mean and variance σ_ε^2.

In matrix notation, the model can be written as

$$\mathbf{y} = \mathbf{X}\boldsymbol{\beta} + \mathbf{Z}\boldsymbol{\gamma} + \boldsymbol{\varepsilon}, \tag{10.2}$$

where $\mathbf{y}$ is the n-dimensional vector of responses, $\mathbf{X}$ is the $n \times p$ design matrix, $\mathbf{Z}$ is a $n \times b$ matrix of zeros and ones assigning the runs to the blocks, $\boldsymbol{\gamma}$ is the b-dimensional vector of block effects, and $\boldsymbol{\varepsilon}$ is the n-dimensional vector of random errors. In order to avoid multicollinearity, this model cannot contain an intercept because all rows of $\mathbf{Z}$ sum to one. As a result, the design matrix $\mathbf{X}$ does not contain a column of ones if the block effects are treated as fixed. The block effect γ_i $(i = 1, 2, \ldots, b)$ then serves as the intercept of the ith block.

The fixed model parameters $\boldsymbol{\beta}$ and $\boldsymbol{\gamma}$ can be estimated using ordinary least squares:

$$\begin{bmatrix} \hat{\boldsymbol{\beta}}^* \\ \hat{\boldsymbol{\gamma}}^* \end{bmatrix} = \begin{bmatrix} \mathbf{X}'\mathbf{X} & \mathbf{X}'\mathbf{Z} \\ \mathbf{Z}'\mathbf{X} & \mathbf{Z}'\mathbf{Z} \end{bmatrix}^{-1} \begin{bmatrix} \mathbf{X}' \\ \mathbf{Z}' \end{bmatrix} \mathbf{y}. \tag{10.3}$$

The variance-covariance matrix of this estimator is

$$\text{var} \begin{bmatrix} \hat{\boldsymbol{\beta}}^* \\ \hat{\boldsymbol{\gamma}}^* \end{bmatrix} = \sigma_\varepsilon^2 \begin{bmatrix} \mathbf{X}'\mathbf{X} & \mathbf{X}'\mathbf{Z} \\ \mathbf{Z}'\mathbf{X} & \mathbf{Z}'\mathbf{Z} \end{bmatrix}^{-1}. \tag{10.4}$$

The inverse of this matrix is the information matrix on the unknown parameters $\boldsymbol{\beta}$ and γ. The way of analysing a blocked experiment described here is referred to as fixed block effects analysis or as intra-block analysis. The latter name is used because the present estimation method only takes into account treatment comparisons within one block and ignores information contained in the differences of the block totals. It is important to point out that the estimated factor effects and their variances depend on the way the design $\mathbf{X}$ is partitioned into blocks.

10.4.2 The use of standard designs

The design of blocked experiments receives attention in most textbooks on experimental design. However, the attention is typically focused on blocking factorial or fractional factorial designs, and on the orthogonal blocking of standard response surface designs. In this section, these topics will be briefly reviewed.

Blocking factorial and fractional factorial designs

In screening experiments, the purpose of which is to detect important factors influencing the response, only two levels are used for each factor under investigation. The most popular designs for this situation are undoubtedly the two-level factorial and fractional factorial designs, commonly denoted by 2^m and 2^{m-f} designs respectively. The topic of blocking 2^m factorial designs is included in most books on response surface methodology (e.g. Khuri and Cornell, 1987; Myers and Montgomery, 2002), where arrangements of 2^m designs in $b = 2^q$ blocks are listed in table format. Sun *et al.* (1997) and Sitter *et al.* (1997) extended these results to 2^{m-f} fractional factorial designs. A selection of their results can be found in Wu and Hamada (2000). The arrangement of experimental runs in blocks is such that as many potentially important factorial effects, e.g. the main effects and the two-factor interactions, can be estimated as possible. This objective is formally expressed by the so-called minimum-aberration criterion introduced by Fries and Hunter (1980) to distinguish between 2^{m-f} designs with equal resolution.

When the experimental factors influence the response in a nonlinear fashion, the use of 3^m and 3^{m-f} designs can be considered. The arrangement of these designs in $b = 3^q$ blocks is discussed in Cheng and Wu (2002). A condensed version of their results is given in Wu and Hamada (2000). The arrangement of mixed-level factorial designs in blocks is discussed by Mukerjee (2002).

Although the results described in this section are very useful, they are only valid for a limited number of experimental situations. For example, the results on 2^m and 2^{m-f} designs cannot be applied when the number of runs available, the number of blocks available or the block sizes are not a power of two or when the block sizes are unequal.

Orthogonal blocking

When the experimental factors are expected to have a quadratic effect on the response of interest, standard response surface designs like the central composite designs or the Box–Behnken designs can be considered. The topic of blocking standard response surface designs is described in Khuri and Cornell (1987) and Myers and Montgomery (2002) who stress the importance of orthogonal blocking. In an orthogonally blocked experiment, the block effects do not affect the usual estimates of $\boldsymbol{\beta}$. Khuri (1992) showed that an experiment is orthogonally blocked if the average run in every block of the experiment is the same and equal to the average run in the entire experiment. Mathematically, the condition for orthogonal blocking is

$$\frac{1}{k_i}\mathbf{X}'_i\mathbf{1}_{k_i} = \frac{1}{n}\mathbf{X}'\mathbf{1}_n, \tag{10.5}$$

where $\mathbf{X}_i$ is the part of $\mathbf{X}$ corresponding to the ith block, and $\mathbf{1}_{k_i}$ and $\mathbf{1}_n$ represent k_i- and n-dimensional column vectors of ones. Examples of orthogonally blocked designs are the blocked 2^m and 2^{m-f} designs proposed by Sun *et al.* (1997) and Sitter *et al.* (1997), and the orthogonally blocked central composite designs described by Box and Hunter (1957). The blocked design displayed in Figure 10.1 is orthogonally blocked if the interest is in estimating a main effects model in x_1 and x_2. This is because the average value of x_1 and x_2 is equal to zero in the three blocks. When the interest is in estimating a model in x_1, x_2, x_1^2 and x_2^2, the design is still orthogonally blocked because the average values of x_1^2 and x_2^2 equal 2/3 in every block. If, however, a model containing the interaction x_1x_2 has to be estimated, the design is no longer orthogonally blocked. This is because the average value of x_1x_2 equals −2/3 in the first block and 1/3 in the second and third block.

The popularity of orthogonally blocked experiments is due to the fact that the estimation of the factor effects does not depend on the estimation of the block effects. In addition, the intra-block estimator $\hat{\boldsymbol{\beta}}^*$ is equivalent to the estimator $\hat{\boldsymbol{\beta}}_{\text{OLS}}$ obtained by ignoring the block effects. Ignoring the block effects leads to the following model:

$$\mathbf{y} = \mathbf{1}_n\beta_0 + \mathbf{X}\boldsymbol{\beta} + \mathbf{v}, \tag{10.6}$$

where β_0 is an overall intercept and $\mathbf{v}$ is the vector of random errors. An estimator for $\boldsymbol{\beta}$ is therefore obtained from the ordinary least squares estimator

$$\begin{bmatrix} \hat{\beta}_{0,\text{OLS}} \\ \hat{\boldsymbol{\beta}}_{\text{OLS}} \end{bmatrix} = \begin{bmatrix} \mathbf{1}'_n\mathbf{1}_n & \mathbf{1}'_n\mathbf{X} \\ \mathbf{X}'\mathbf{1}_n & \mathbf{X}'\mathbf{X} \end{bmatrix}^{-1} \begin{bmatrix} \mathbf{1}'_n \\ \mathbf{X}' \end{bmatrix} \mathbf{y} = (\mathbf{W}'\mathbf{W})^{-1}\mathbf{W}'\mathbf{y}, \tag{10.7}$$

where $\mathbf{W} = [\mathbf{1}_n\,\mathbf{X}]$. For the design in Figure 10.1, the estimates of the coefficients of x_1, x_2, x_1^2 and x_2^2 are independent of the estimation method used, whereas the estimate of the interaction effect depends on the estimator used.

10.4.3 Optimal design

The optimal design of tailor-made experiments for estimating model (10.2) has received attention from Atkinson and Donev (1989) and Cook and Nachtsheim (1989), who developed two different algorithms to search for designs that minimize the determinant of the variance-covariance matrix (10.4). The designs obtained in this way minimize the volume of the confidence ellipsoid about $\boldsymbol{\beta}$ and γ and are called $\mathcal{D}$-optimal.

In many cases, the interest will be only in estimating the factor effects $\boldsymbol{\beta}$. Instead of minimizing the determinant of (10.4), the determinant of $\text{var}(\hat{\boldsymbol{\beta}}^*)$ should then be minimized. From Harville's (1997) Theorem 8.5.11, we have that

$$\text{var}(\hat{\boldsymbol{\beta}}^*) = \sigma_\varepsilon^2\{\mathbf{X}'\mathbf{X} - \mathbf{X}'\mathbf{Z}(\mathbf{Z}'\mathbf{Z})^{-1}\mathbf{Z}'\mathbf{X}\}^{-1}, \tag{10.8}$$

which, after some matrix algebra, can be rewritten as

$$\text{var}(\hat{\boldsymbol{\beta}}^*) = \sigma_\varepsilon^2\left[\mathbf{X}'\mathbf{X}' - \sum_{i=1}^{b}\frac{1}{k_1}(\mathbf{X}'_i\mathbf{1}_{k_i})(\mathbf{X}'_i\mathbf{1}_{k_i})'\right]^{-1}. \tag{10.9}$$

From Harville's (1997) Theorem 13.3.8, we also have that

$$\begin{vmatrix}\mathbf{X}'\mathbf{X} & \mathbf{X}'\mathbf{Z}\\ \mathbf{Z}'\mathbf{X} & \mathbf{Z}'\mathbf{Z}\end{vmatrix} = |\mathbf{Z}'\mathbf{Z}|\,|\mathbf{X}'\mathbf{X} - \mathbf{X}'\mathbf{Z}(\mathbf{Z}'\mathbf{Z})^{-1}\mathbf{Z}'\mathbf{X}|. \tag{10.10}$$

Since the block sizes are dictated by the experimental situation, $\mathbf{Z}'\mathbf{Z} = \text{diag}(k_1, k_2, \ldots, k_b)$ and thus $|\mathbf{Z}'\mathbf{Z}| = \Pi_{i=i}^{b} k_i$ are constant for every feasible design. Minimizing the determinant of (10.8) or (10.9) is therefore equivalent to minimizing the determinant of (10.4). In other words, the $\mathcal{D}$-optimal design does not change when the interest is in estimating the factor effects $\boldsymbol{\beta}$ only.

Besides the $\mathcal{D}$-optimality criterion, many other criteria have been proposed in the literature. For example, an $\mathcal{A}$–optimal design minimizes the trace of the variance-covariance matrix (10.4). Unfortunately, designs that are optimal with respect to one criterion are in many cases not optimal with respect to other criteria. Many design construction algorithms proposed in the literature, however, use the $\mathcal{D}$-optimality criterion because $\mathcal{D}$-optimal designs usually perform well with respect to other criteria, and because it is computationally simpler than other criteria.

10.4.4 Some theoretical results

A limited number of theoretical results on the optimal design of blocked experiments under the assumption of fixed block effects can be found in the literature.

One result involves product designs and another is concerned with orthogonally blocked designs.

Product designs

Kurotschka (1981) shows that $\mathcal{D}$- or $\mathcal{A}$-optimal continuous[1] designs for the fixed block effects model can be represented by a product of two design measures – hence the name product designs. In some specific cases, this result can be used to construct exact $\mathcal{D}$- or $\mathcal{A}$-optimal designs with b blocks of size k. Consider, for example, an experiment with blocks of size 3 carried out to estimate the impact of the initial potassium/carbon (K/C) ratio, denoted by x, on the desorption of carbon monoxide (CO), denoted by y, in the context of coal gasification. The model of interest in the study was quadratic in nature:

$$y_{ij} = \beta_0 + \beta_1 x_{ij} + \beta_2 x_{ij}^2 + \gamma_i + \varepsilon_{ij}, \quad i = 1, 2, \ldots, b; j = 1, 2, 3. \qquad (10.11)$$

It is well known that the $\mathcal{D}$-optimal continuous design for estimating the quadratic model

$$y_i = \beta_0 + \beta_1 x_i + \beta_2 x_i^2 + \varepsilon_i \qquad (10.12)$$

has weight 1/3 at three design points: $x = -1$, $x = 0$ and $x = 1$. A $\mathcal{D}$-optimal design with b blocks of size 3 is then given by b identical blocks containing one run at -1, 0 and 1. An interesting property of these designs is that they are orthogonal. As a matter of fact, it can be verified that the average values of x and x^2 in each block are equal to 0 and 2/3, respectively.

The applicability of Kurotschka's result is limited for two technical reasons. Firstly, the block size k has to be greater than or equal to $p + 1$. In practice, however, the block size is often quite small. Secondly, it is required that, for all weights w_i in the continuous design for the model without block effects, kw_i should be an integer number. This is only possible in cases where all w_i are rational numbers, for example in polynomial regression on one variable or in mixture experiments.

Optimality of orthogonal blocking

Goos and Vandebroek (2001) have been able to show that, when the interest is in the factor effects $\boldsymbol{\beta}$ only, orthogonal blocking is an optimal strategy to assign the runs from a given set of design points to the b blocks of an experiment. This result is valid for a range of optimality criteria, including the $\mathcal{D}$- and $\mathcal{A}$-optimality criteria. When the interest is in estimating both the factor effects $\boldsymbol{\beta}$ and the block effects $\boldsymbol{\gamma}$,

[1]In the literature on the optimal design of experiments, much effort has been spent on continuous designs, which are represented using measures and assume that the number of observations is infinitely large. Even though the number of runs in real-life experiments is finite, some of the theoretical results on continuous designs are helpful in practical instances.

orthogonal blocking is a $\mathcal{D}$-optimal strategy for assigning the runs of a given design to the blocks.

Although this is an interesting result, it is of limited practical use because it does not help experimenters in choosing the design points. Moreover, the number of blocks b and the block sizes k_i dictated by the experimental situation seldom allow orthogonal blocking.

10.4.5 Computational results

In order to keep the exposition simple, suppose that nine runs are available to estimate the effect of two quantitative factors x_1 and x_2 on a response, and that the nine runs have to be assigned to three blocks of size three. The model of interest is given by

$$y_{ij} = \beta_1 x_{1ij} + \beta_2 x_{2ij} + \beta_3 x_{1ij} x_{2ij} + \beta_4 x_{1ij}^2 + \beta_5 x_{2ij}^2 + \gamma_i + \varepsilon_{ij}, \quad i,j = 1,2,3. \tag{10.13}$$

Using the points of the 3^2 factorial design as the set of candidates, the blocking algorithm of Atkinson and Donev (1989) produced the design in Figure 10.2 as a

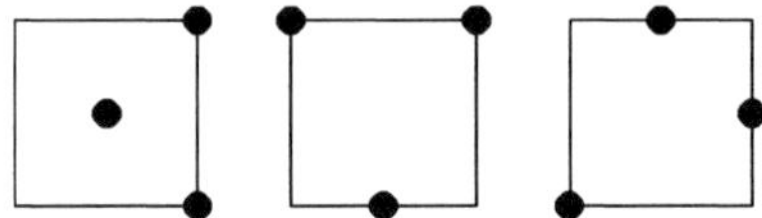

Figure 10.2 $\mathcal{D}$-optimal design with three blocks of size three for estimating the full quadratic model (10.13) (Reproduced from 'The optimal design of blocked and split-plot experiments', 2002, by permission of Springer-Verlag GmbH)

solution to this design problem. It can easily be verified that the projection of this design is different from the 3^2 factorial design. Instead, it only has eight distinct design points, one of which is duplicated. This result is important because, in contrast with the projection of the design from Figure 10.1, the 3^2 design is the $\mathcal{D}$-optimal nine-point design for estimating a full quadratic model in two factors in the absence of blocking:

$$y_i = \beta_0 + \beta_1 x_{1i} + \beta_2 x_{2i} + \beta_3 x_{1i} x_{2i} + \beta_4 x_{1i}^2 + \beta_5 x_{2i}^2 + \varepsilon_i. \tag{10.14}$$

An immediate consequence of this result is that a two-stage approach to designing blocked experiments does not lead to an optimal design for a blocked experiment. Trinca and Gilmour (2000) propose a two-stage approach to design experiments with small blocks. In the first stage, they suggest selecting a good projection design independent of the blocking structure of the experiment. This can be done, for instance, by computing a $\mathcal{D}$-optimal design for model (10.14). The design points obtained in this manner can then be assigned to the blocks such that the resulting blocked experiment is as orthogonal as possible. Applying this approach to the nine-point experiment with three blocks of size three, we would obtain the design points from the 3^2 design in the first stage and, consequently, the resulting block design would not be optimal. For this small example, the loss of efficiency incurred by using the two-stage approach would be small. As we will show in Section 10.6, the loss can be considerably larger for more complicated design problems.

10.5 Random Block Effects Model

In many practical design problems, the blocks of the experiment are randomly drawn from a population of blocks. For example, the batches of raw material used in experiments are typically randomly chosen from the batches available from the warehouse. In this section, we first introduce the random block effects model and describe a number of theoretical results. Next, we describe an algorithm to compute $\mathcal{D}$-optimal designs in the presence of random block effects and discuss some computational results.

10.5.1 Model and estimation

The most common notation for the random block effects model is

$$\mathbf{y} = \beta_0 \mathbf{1}_n + \mathbf{X}\boldsymbol{\beta} + \mathbf{Z}\boldsymbol{\gamma} + \boldsymbol{\varepsilon}, \tag{10.15}$$

where $\mathbf{X}$, $\mathbf{Z}$ and $\boldsymbol{\varepsilon}$ are defined as in the fixed block effects model, β_0 is the intercept, and $\boldsymbol{\gamma}$ is assumed to have zero mean and variance-covariance matrix $\sigma_\gamma^2 \mathrm{I}_n$, where I_n is the n-dimensional unity matrix. Note that, unlike in the fixed block effects model (10.2), an intercept can be estimated in the random effects model (10.15). Also, it is assumed that $\boldsymbol{\gamma}$ and $\boldsymbol{\varepsilon}$ are independent of each other. Under these assumptions, the variance-covariance matrix of the responses $\mathbf{y}$ is

$$\mathbf{V} = \sigma_\varepsilon^2 \mathbf{I}_n + \sigma_\gamma^2 \mathbf{Z}\mathbf{Z}', \tag{10.16}$$

and, if the entries of $\mathbf{y}$ are grouped per block,

$$\mathrm{V} = \mathrm{diag}(\mathbf{V}_1, \mathbf{V}_2, \ldots, \mathbf{V}_b), \tag{10.17}$$

where

$$\mathbf{V}_i = \sigma_\varepsilon^2 (\mathbf{I}_{k_i} + \eta \mathbf{1}_{k_i} \mathbf{1}'_{k_i}) \tag{10.18}$$

and $\eta = \sigma_\gamma^2/\sigma_\varepsilon^2$. The block-diagonal structure of $\mathbf{V}$ implies that the random block effects model assumes observations within the same block to be correlated, whereas observations from different blocks are assumed to be uncorrelated. The variance ratio η serves as a measure for the extent to which observations within one block are correlated. It is therefore referred to as the degree of correlation.

The unknown fixed model parameters can be estimated using the generalized least squares estimation equation

$$\begin{bmatrix} \hat{\beta}_{0,\text{GLS}} \\ \hat{\boldsymbol{\beta}}_{\text{GLS}} \end{bmatrix} = \left(\begin{bmatrix} \mathbf{1}'_n \\ \mathbf{X}' \end{bmatrix} \mathbf{V}^{-1} [\mathbf{1}_n \ \mathbf{X}] \right)^{-1} \begin{bmatrix} \mathbf{1}'_n \\ \mathbf{X}' \end{bmatrix} \mathbf{V}^{-1}\mathbf{y} = (\mathbf{W}'\mathbf{V}^{-1}\mathbf{W})^{-1}\mathbf{W}'\mathbf{V}^{-1}\,\mathbf{y}, \tag{10.19}$$

where, as previously, $\mathbf{W} = [\mathbf{1}_n \, \mathbf{X}]$. This estimator is also called the combined intra- and interblock estimator because it recovers the information contained within the block totals. Its variance-covariance matrix is

$$\text{var} \begin{bmatrix} \hat{\beta}_{0,\text{GLS}} \\ \hat{\boldsymbol{\beta}}_{\text{GLS}} \end{bmatrix} = \left(\begin{bmatrix} \mathbf{1}'_n \\ \mathbf{X}' \end{bmatrix} \mathbf{V}^{-1} [\mathbf{1}_n \, \mathbf{X}] \right)^{-1} = (\mathbf{W}'\mathbf{V}^{-1}\mathbf{W})^{-1}. \tag{10.20}$$

Of course, the variance components σ_γ^2 and σ_ε^2 unknown in practice and can be estimated by using restricted maximum likelihood (REML). A detailed discussion of the use of REML to analyse blocked experiments is given in Gilmour and Trinca (2000).

In a number of special cases, knowledge or estimation of σ_γ^2 and σ_ε^2 (or η) are not needed to estimate β_0 and/or $\boldsymbol{\beta}$. When the orthogonality condition (10.5) is satisfied, $\hat{\boldsymbol{\beta}}_{\text{GLS}}$ does not depend on η and is equal to $\hat{\boldsymbol{\beta}}_{\text{OLS}}$ and $\hat{\boldsymbol{\beta}}^*$. If, in addition, the block size is homogeneous, i.e. if $k_1 = k_2 = \cdots = k_b$, the intercept β_0 can also be estimated independently of η and $\hat{\beta}_{0,\text{GLS}} = \hat{\beta}_{0,\text{OLS}}$. Proofs of these results can be found in Khuri (1992) and Goos and Vandebroek (2004). It should be stressed, however, that even in an orthogonally blocked experiment, knowledge or estimation of σ_γ^2 and σ_ε^2 are needed to compute standard errors and to perform statistical tests.

A $\mathcal{D}$-optimal design for estimating the fixed parameters β_0 and $\boldsymbol{\beta}$ in the random block effects model minimizes the determinant of the variance-covariance matrix (10.20). An $\mathcal{A}$-optimal design minimizes the trace of (10.20). A $\mathcal{V}$-optimal design minimizes the average prediction variance $\mathbf{f}'(\mathbf{x})(\mathbf{W}'\mathbf{V}^{-1}\mathbf{W})^{-1}\,\mathbf{f}(\mathbf{x})$ over the design region. In general, the optimal designs depend on the relative magnitude η of the variance components σ_γ^2 and σ_ε^2 through $\mathbf{V}$.

10.5.2 Theoretical results

As for the fixed block effects model, a number of theoretical results have been described in the literature for the random block effects model. Firstly, it has been shown that $\mathcal{D}$-optimal designs for the random block effects model are also $\mathcal{D}$-optimal for the fixed block effects model provided η is large. Secondly, a number of cases have been identified in which the optimal designs do not depend on the degree of correlation η.

$\mathcal{D}$-optimal designs for large η

Goos and Vandebroek (2001) show that $\mathcal{D}$-optimal designs for the random block effects model are also $\mathcal{D}$-optimal for the fixed block effects model provided η is large. In order to show this, they start by rewriting the variance-covariance matrix $(\mathbf{W}'\mathbf{V}^{-1}\mathbf{W})^{-1}$ of $\hat{\beta}_0$ and $\hat{\boldsymbol{\beta}}$ as

$$(\mathbf{W}'\mathbf{V}^{-1}\mathbf{W})^{-1} = \sigma_\varepsilon^2 \left\{ \mathbf{W}'\mathbf{W} - \sum_{i=1}^{b} \frac{\eta}{1 + k_i\eta} (\mathbf{W}'_i \mathbf{1}_{k_i})(\mathbf{W}'_i \mathbf{1}_{k_i})' \right\}^{-1}, \quad (10.21)$$

where $\mathbf{W}_i$ is that part of $\mathbf{W}$ corresponding to the ith block. When the model contains no intercept and η becomes infinitely large, then $\mathbf{W}$ reduces to $\mathbf{X}$ and $\eta/(1 + k_i\eta)$ goes to $1/k_i$ so that we obtain Equation (10.9). As a result, minimizing the determinant of $(\mathbf{W}'\mathbf{V}^{-1}\mathbf{W})^{-1}$ when $\eta \to \infty$ is equivalent to finding a $\mathcal{D}$-optimal design for the fixed block effects model (10.2) in the absence of an intercept. For a formal proof of the equivalence when the random block effects model contains an intercept, we refer the interested reader to Goos and Vandebroek (2001).

Optimal designs that do not depend on η

Although optimal designs for the random block effects model in general depend on the degree of correlation η, some instances exist in which this is not the case. From a practical point of view, this is important because η is not known prior to the experiment.

First-order orthogonally blocked designs. Goos and Vandebroek (2001) show that orthogonally blocking a first-order design $\mathbf{X}$, which is $\mathcal{D}$-optimal for estimating a model without block effects, yields a $\mathcal{D}$-optimal design for the random block effects model. The resulting design is also optimal for the fixed block effects model. An important consequence of this result is that the classical blocked 2^m and 2^{m-f} designs, as described in Section 10.4.2, are $\mathcal{D}$-optimal. Unfortunately, this result cannot be used when the block sizes are odd numbers. This is because $\mathcal{D}$-optimal designs for first-order models have factor levels -1 and $+1$ only, and these designs cannot be blocked orthogonally when the block sizes are odd numbers.

Product designs. For the random block effects model, Atkins and Cheng (1999) obtain a theoretical result similar to that of Kurotschka (1981) for the fixed block effects model. For an example and a discussion of the relevance of this result, we refer the reader back to Section 10.4.4.

Minimum support designs. Cheng (1995) provides a method to construct the best possible minimum support designs for the random block effects model with respect to the $\mathcal{D}$-optimality criterion. A minimum support design is a design in which the number of distinct design points is equal to the number of fixed model parameters $p + 1$. Cheng shows that combining a $\mathcal{D}$-optimal $(p + 1)$-point design for model (10.6) and a $\mathcal{D}$-optimal design with b blocks of size k for comparing $p + 1$ treatments, for instance a balanced incomplete block design (BIBD), yields a $\mathcal{D}$-optimal minimum support design for the random block effects model. For an introduction to BIBDs, we refer the reader to Cox (1958). An important property of the minimum support designs constructed in this way is that their optimality does not depend on the degree of correlation η.

In general, minimum support designs are substantially less efficient than overall optimal designs. In mixture experiments, however, minimum support designs perform reasonably well. Consider a modified version of a constrained mixture experiment described by Atkinson and Donev (1992). In the experiment, the impact of three factors,

x_1 copper sulphate ($CuSO_4$)
x_2 sodium thiosulphate ($Na_2S_2O_3$)
x_3 glyoxal ($(CHO)_2$),

on the electric resistivity of a modified acrylonitrile powder was investigated. The following constraints were imposed on the levels of these factors:

$$\begin{array}{ccccc} 0.2 & \leq & x_1 & \leq & 0.8, \\ 0.2 & \leq & x_2 & \leq & 0.8, \\ 0.0 & \leq & x_3 & \leq & 0.6, \end{array}$$

and the model of interest was given by the second-order Scheffé polynomial

$$y_{ij} = \sum_{s=1}^{3} \beta_s x_{sij} + \sum_{s=1}^{2} \sum_{t=s+1}^{3} \beta_{st} x_{sij} x_{tij} + \gamma_i + \varepsilon_{ij}, \quad i = 1, 2 \ldots, 10; j = 1, 2, 3. \tag{10.22}$$

Suppose that 10 blocks of size three are available. The design region for this problem is graphically displayed in Figure 10.3. Each point in the triangle corresponds to a mixture of the three components x_1, x_2 and x_3. For instance, the point labelled 1 corresponds to a mixture containing 80% of the first component, 20% of the second component and 0% of the third component. It is a well-known result that the six-point second-order simplex lattice design, also displayed in the figure, is a $\mathcal{D}$-optimal design for the Scheffé model in the absence of blocking. The six-point simplex lattice design is a minimum support design since the Scheffé model has exactly six unknown fixed parameters. As it is a minimum support design, the six-point simplex lattice design can be combined with a BIBD with 10 blocks of

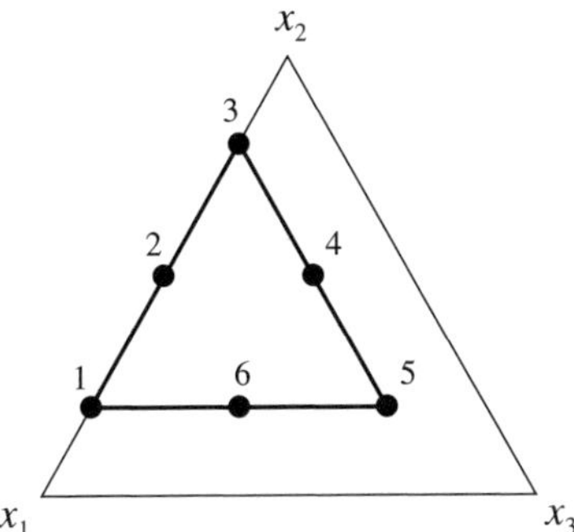

Figure 10.3 $\mathcal{D}$-optimal design for constrained mixture experiment (Reproduced from 'The optimal design of blocked and split-plot experiments', 2002, by permission of Springer-Verlag GmbH)

size three for comparing six treatments. The resulting design is $\mathcal{D}$-optimal for the random block effects model (10.22) in the class of minimum support designs. It is displayed in Table 10.2.

Table 10.2 $\mathcal{D}$-optimal minimum support design for the mixture experiment

Block	Design points			Block	Design points		
1	1	2	5	6	2	3	4
2	1	2	6	7	2	3	5
3	1	3	4	8	2	4	6
4	1	3	6	9	3	5	6
5	1	4	5	10	4	5	6

(Reproduced from 'The optimal design of blocked and split-plot experiments', 2002, by permission of Springer-Verlag GmbH)

Saturated designs. A saturated design is a design in which the total number of runs n is equal to the number of unknown fixed parameters $p+1$. It is easy to see that the $\mathcal{D}$-optimal saturated design for the random block effects model does not depend on the degree of correlation η. The optimal design points are those from a $\mathcal{D}$-optimal saturated design for model (10.6). The design matrix $\mathbf{W}$ is a $(p+1)\times(p+1)$ square matrix and its $\mathcal{D}$-criterion value becomes

$$|\mathbf{W}'\mathbf{V}^{-1}\mathbf{W}|^{-1} = |\mathbf{V}|\,|\mathbf{W}'\mathbf{W}|^{-1}.$$

Since we assume that the block sizes are dictated by the experimental situation, $|\mathbf{V}|$ is a constant for all design options. Hence, the $\mathcal{D}$-optimal saturated design for the random block effects model can be found by minimizing $|\mathbf{W}'\mathbf{W}|^{-1}$. This is nothing else than determining a $\mathcal{D}$-optimal design for a model without block effects. Of course, the practical value of this result is quite small since saturated designs are used as little as possible.

10.5.3 Computational results

We have applied the algorithm of Goos and Vandebroek (2001) to the experiment with nine runs and three blocks of size three for estimating the full quadratic model (10.13) in two variables including an intercept. As already pointed out, the optimal design in the presence of random block effects depends on the degree of correlation η. Using several values of η between 0 and $+\infty$ produced only two different designs. If $\eta \leq 3.8790$, the blocked design from Figure 10.1 is the $\mathcal{D}$-optimal design. It is important to note that the projection of this design is the 3^2 design, which is the $\mathcal{D}$-optimal nine-point design for model (10.14), i.e. the model without block effects. As a result, the optimal blocked experiment for small values of η is obtained from an efficient design for the model without block effects. For larger values of η, the design from Figure 10.2 is $\mathcal{D}$-optimal. This illustrates the theoretical results from Section 10.5.2.1.

The fact that the $\mathcal{D}$-optimal designs from Figures 10.1 and 10.2 are optimal for wide ranges of η is extremely important from a practical point of view because it implies that precise knowledge of η is not needed to compute the $\mathcal{D}$-optimal design for a particular design problem. For small degrees of correlation, the design from Figure 10.1 turns out to be 3% more efficient than that from Figure 10.2. For η-value around 3.8790, both designs are nearly equivalent, and for larger values of η, the design from Figure 10.2 is slightly better. The relative efficiencies reported here are computed as

$$\left(\frac{|\mathbf{W}\mathbf{V}^{-1}\mathbf{W}|}{|\widetilde{\mathbf{W}}\mathbf{V}^{-1}\widetilde{\mathbf{W}}|}\right)^{\frac{1}{p+1}}, \qquad (10.23)$$

where $\mathbf{W}$ and $\widetilde{\mathbf{W}}$ represent two competing designs. A relative efficiency of, for instance, 50% signifies that the design $\mathbf{W}$ has to be duplicated in order to obtain the same amount of information about $\boldsymbol{\beta}$ as by running design $\widetilde{\mathbf{W}}$.

10.6 The Pastry Dough Mixing Experiment Revisited

The set of treatments chosen by Trinca and Gilmour (2000) and Gilmour and Trinca (2000) is a central composite design with the two-level factorial portion, i.e. the corner-points of the cubic design region, duplicated and with six centre points. These treatments were assigned to the blocks according to Figure 10.4(a). It is easy to verify that the average levels of x_1, x_2 and x_3 are 0 such that the design is orthogonally blocked with respect to the main effect terms. This is not the case for the interactions and the quadratic effects. The projection of this design obtained by ignoring the blocks is symmetric. It is displayed in Figure 10.5(a). The $\mathcal{D}$-optimal design for the pastry dough experiment is given in Figure 10.4(b). This design turns out to be optimal for every value of η between 1 and 10, which are the η-values obtained by Gilmour and Trinca (2000). It only contains two centre points, and

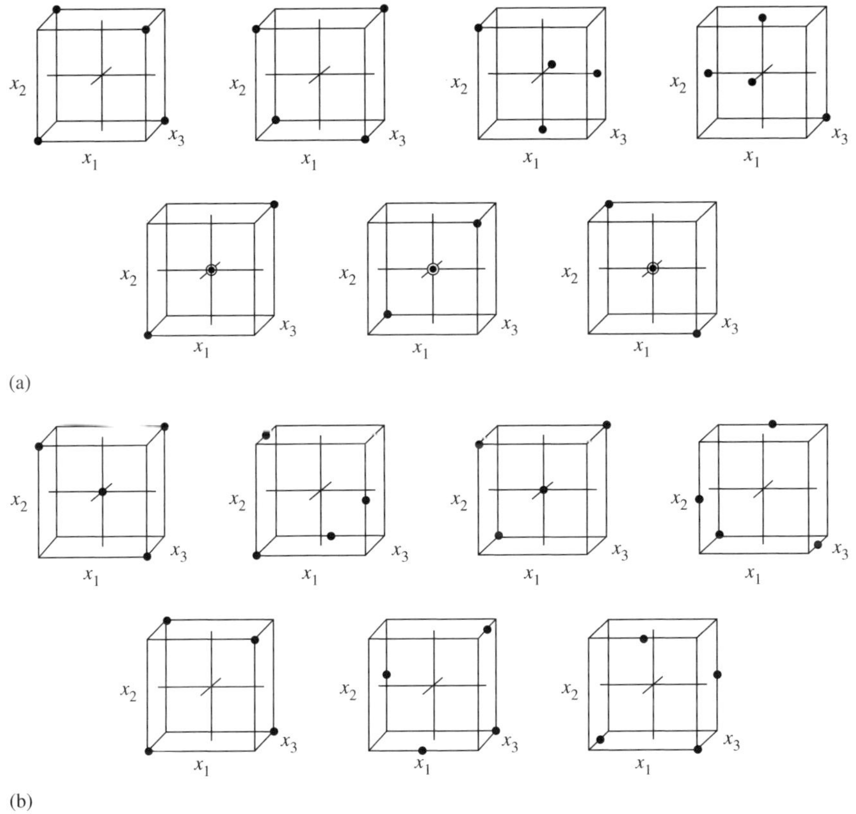

(a)

(b)

Figure 10.4 Design options for the pastry dough mixing experiment: (a) design used by Gilmour and Trinca (2000); (b) $\mathcal{D}$-optimal design (• is a design point, ◉ is a design point replicated twice (Reproduced from 'The optimal design of blocked and split-plot experiments', 2002, by permission of Springer-Verlag GmbH)

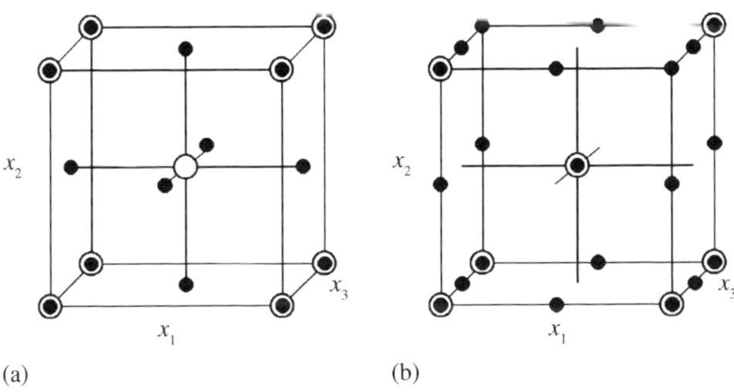

Figure 10.5 Projections of the designs from (a) Figure 10.4(a) and (b) Figure 10.4(b) obtained by ignoring the blocks (• is a design point, ◉ is a design point replicated twice, and ○ is a design point replicated six times) (Reproduced from 'The optimal design of blocked and split-plot experiments', 2002, by permission of Springer-Verlag GmbH)

midpoints of the edges of the cube instead of midpoints of its faces. This can be verified by examining its projection in Figure 10.5(b). The projection is nearly symmetric and, although it is not orthogonal with respect to the main effect terms, the blocked design is almost orthogonal with respect to all model terms. This can be verified by computing the mean efficiency factor (see John and Williams (1995) and Trinca and Gilmour (2000)), which is a measure for the orthogonality of a blocked design. The $\mathcal{D}$-optimal design has an efficiency factor of 96.10%, whereas the modified central composite design yields an efficiency factor of only 91.90%.

It is no surprise that the $\mathcal{D}$-optimal design outperforms the modified central composite design in terms of $\mathcal{D}$-efficiency. However, it is also much better in terms of $\mathcal{A}$-efficiency and $\mathcal{V}$-efficiency. In Figure 10.6, the relative $\mathcal{D}$-, $\mathcal{A}$- and

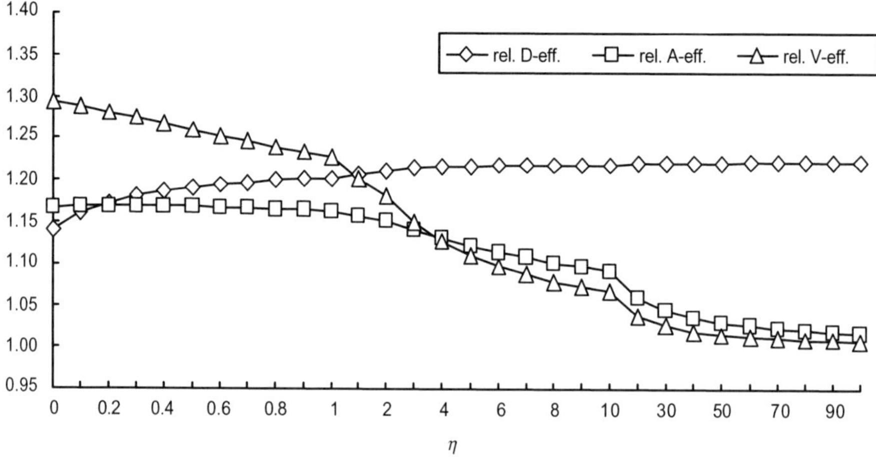

Figure 10.6 Relative $\mathcal{D}$-, $\mathcal{A}$- and $\mathcal{V}$-efficiencies of the $\mathcal{D}$-optimal design from Figure 10.4(b) with respect to the modified central composite design from Figure 10.4(a)

$\mathcal{V}$-efficiencies of the $\mathcal{D}$-optimal design from Figure 10.4(b) with respect to the modified central composite design from Figure 10.4(a) are displayed. The relative $\mathcal{D}$-efficiency increases with the degree of correlation η and is well above 115% unless η is close to zero. The relative $\mathcal{A}$- and $\mathcal{V}$-efficiencies decrease with η. They are substantially greater than 100% for small degrees of correlation and tend to 100% when η approaches 100.

The pastry dough mixing experiment is an excellent illustration of the usefulness of the optimal design approach to construct blocked industrial experiments. The $\mathcal{D}$-optimality criterion produces a nearly orthogonal design which performs excellently with respect to several design criteria. As a result, this example provides evidence that it is in many cases not a good idea to modify standard response surface designs in practical situations.

10.7 Time Trends and Cost Considerations

In this section, the optimal design of blocked experiments in the presence of unwanted time trend effects will be discussed first. Such time trends can be due to, for example, the ageing of material and analyst fatigue. Next, cost considerations will be taken into account to compute cost-efficient designs. Finally, the pastry dough mixing experiment will be revisited.

10.7.1 Time trend effects

One of the basic principles of good experimental design is randomization. The best way to do this is to randomly assign the experimental units to the different treatment combinations and to randomize the run order. Randomization leads to unbiased parameter estimates. If, however, over the course of sequential application of the treatments to experimental units, a systematic effect or trend influences the observations, then it may be more appropriate to use a systematic ordering of the experimental runs. For example, when a batch of material is created at the beginning of an experiment and treatments are to be applied to experimental units formed from the material over time, there could be an unwanted effect due to ageing of the material which influences the responses obtained. Other examples include laboratory experiments in which the responses are affected by instrument drift or analyst fatigue, trends due to poisoning of a catalyst, steady build-up of deposits in a test engine, heating and lighting elements in a laboratory, etc. Anderson and Bancroft (1952) describe an experiment to evaluate the effect of three variables on the amount of vitamin B_2 in turnip greens. A close examination of the data set reveals that the observations are affected by a linear time trend, maybe due to vitamin decay through time. Another example of trend effects comes from Joiner and Campbell (1976) who mention an experiment in which the amount of nitrogen in steel rods is recorded. Looking at the data as a function of time shows that the measured nitrogen contents decrease dramatically as time proceeds.

An experimenter who is aware of the nature of the time trend should construct a run order in which the estimates of the factor effects are little disturbed by the presence of the time trend. Often, the time dependence is represented by a polynomial. The objective is then to construct a run order such that the estimates of the important factor effects are orthogonal or nearly orthogonal to the postulated polynomial trend. If the estimator of a factorial effect is the same as when the time trend of qth order is not present, that effect is said to be q-trend-free or orthogonal to the time trend.

The literature on the construction of trend-resistant designs without blocking factors is quite extensive. Cox (1951) was the first to study the construction of designs for the estimation of treatment effects in the presence of polynomial trends. Based on complete enumeration, Draper and Stoneman (1968), Dickinson (1974) and Joiner and Campbell (1976) introduced the dual problem of constructing trend-robust run orders and minimum cost run orders. Cheng (1985), Coster and Cheng (1988) and Coster (1993) formulate the Generalized Foldover Scheme for generating run orders for two- and three-level factorial designs which minimize, or nearly minimize, the number of factor-level changes and for which all main effects are orthogonal to a polynomial time trend. A nice review of constructing trend-free run orders can be found in Cheng (1990) and Jacroux (1990). Based on Daniel and Wilcoxon (1966), Cheng and Jacroux (1988) derive a method for the construction of run orders that yield trend-free main effects and two-factor interactions by redesignating the factors to higher-order interactions in the standard order.

The case of trend-free designs in the presence of block effects goes back to Bradley and Yeh (1980). They consider binary block designs[2] to compare several treatments in b blocks of size k, assume a common polynomial trend in each block and define an experiment to be q-trend-free if and only if every treatment contrast has the same best linear unbiased estimator as under the model without the trend of order q. They show that this condition boils down to the requirement that each trend component is orthogonal to the treatment allocations throughout the experiment. As $k - 1$ is the highest possible degree to which a block design with block size k can be orthogonal, a design that is orthogonal to a $(k - 1)$th order time trend will simply be referred to as a trend-free design. Jacroux *et al.* (1995, 1997) and Jacroux (1998) tackle the problem of identifying and constructing block designs for comparing several treatments in the presence of possibly different linear trends within the blocks. However, it is known that for certain combinations of the design parameters trend-free block designs do not exist. Yeh *et al.* (1985) introduce the concept of nearly trend-free block designs for treatment comparisons. A nearly trend-free design is a design with an overall treatment arrangement as orthogonal as possible to all specified trend components. Lin and Dean (1991) show that for factorial experiments, certain designs which are not completely trend-free are nevertheless trend-free for estimating a subset of the main effects and interaction effects.

10.7.2 Cost considerations

As a second extension, cost considerations will be taken into account. Although many experiments are properly designed to possess good statistical properties, they may not be fit for use because of economic reasons. It is therefore important to incorporate cost considerations in optimal design theory. However, the literature on that topic is very scarce.

Generally speaking, two cost approaches are found in the literature. Firstly, a few authors deal with the costs associated with particular factor-level combinations.

[2]A binary block design is a balanced incomplete block design in which each treatment occurs at most once in each block. A balanced incomplete block design is a design for the comparison of v treatments in b blocks of size $k < v$ in which each treatment occurs the same number of times r and the number of times two different treatments occur together in a block is the same for all pairs of treatments.

Tack and Vandebroek (2001) refer to these costs as measurement costs. Examples of measurement costs include the equipment cost, the cost of material, the cost of personnel and the cost of spending time during the measurement. Secondly, it is usually expensive to alter some of the factor levels such as oven temperature or line set-up from one observation to the next. Tack and Vandebroek (2001) call the cost to change a factor level a transition cost. It is obvious that the number of factor-level changes has to be as small as possible in order to minimize the total transition cost.

Many authors assume that the total measurement cost of an experimental design only depends on the total number of observations and that this cost is independent of the factor-level combinations. This means that for fixed design size, the total measurement cost is also fixed. In order to better reflect practical situations, a few authors treat cases in which different measurement costs have to be assigned to different factor-level combinations. Kiefer (1959) suggests a complete enumeration of appropriate designs and selects the design that has the lowest total cost. Lawler and Bell (1966) and Neuhardt and Bradley (1971) avoid the unmanageable enumeration task for moderately sized problems by partial enumeration. However, partial enumeration suffers from the fact that the solution found may be far from the exact optimum. Other approaches come from Yen (1985) who uses integer programming and Pignatiello (1985) who provides a procedure for finding cost-efficient fractional factorial designs which permit the estimation of specified main and interaction effects.

An interesting approach to the minimization of the total transition cost can be found in Ju (1992) and Anbari and Lucas (1994). They take into account transition costs when designing split-plot experiments and show that these designs, besides their advantage in terms of efficiency, are much cheaper to perform than completely randomized experiments.

10.7.3 The trade-off between trend resistance and cost-efficiency

Designs that are optimally balanced for time trends can often not be used in practice because of cost restrictions. However, cost-efficient run orders are usually poorly balanced for time trend effects. As a result, a trade-off will have to be found between cost-efficiency and balance for time trends. Draper and Stoneman (1968) use the 2^3 factorial design, the standard run order of which is displayed in Table 10.3, to illustrate this.

Suppose that there is a linear time trend which is simulated by setting the observations $y_1, y_2, \ldots, y_8$ equal to $1, 2, \ldots, 8$ respectively. The effect of the time trend on the estimation of the factor effects can be obtained by taking the inner product of a contrast column of Table 10.3 with the simulated observation column. This inner product is referred to as the time count. Small time counts ensure a good balance for time trends whereas large time counts reflect a poor trend resistance. For the three main effects, the time counts for the standard order of the 2^3 factorial design amount to 4, 8 and 16. Clearly, the estimation of the effect of x_3 will be

Table 10.3 Standard run order of the full 2^3-factorial design

Run	x_1	x_2	x_3
1	−1	−1	−1
2	+1	−1	−1
3	−1	+1	−1
4	+1	+1	−1
5	−1	−1	+1
6	+1	−1	+1
7	−1	+1	+1
8	+1	+1	+1

influenced by the time trend much more than the effects of x_1 and x_2. It can also be verified that running the 2^3 factorial in the standard order requires 7, 3 and 1 level changes for the three factors respectively. The total number of factor level changes then equals 11. Table 10.4 compares all 8! arrangements of the eight runs of the 2^3

Table 10.4 Classification of the run orders of the full 2^3 factorial design in terms of the number of level changes and the maximum time count

Number of level changes	Number of run orders	Number of run orders with maximum time count of 0	2	4	6	8	10	12	14	16
7	144	0	0	0	0	48	0	0	0	96
8	624	0	0	0	0	0	96	96	48	384
9	2832	0	48	0	0	768	432	624	192	768
10	4464	0	0	96	96	624	816	1344	528	960
11	8736	48	288	816	384	1872	1776	1728	1056	768
12	7584	0	0	192	528	1728	2304	1344	1104	384
13	8352	48	480	1296	1872	2016	1440	624	480	96
14	3552	0	0	384	1200	672	768	480	48	0
15	2640	48	144	336	816	576	432	288	0	0
16	1008	0	96	240	192	432	48	0	0	0
17	336	0	96	144	48	48	0	0	0	0
18	48	0	0	48	0	0	0	0	0	0

design in terms of the number of level changes and the maximum absolute time counts. The table shows that the cheap run orders, for instance the run orders with less than or equal to 10 level changes, have almost no small time counts. This means that the cheap run orders have a weak balance for time trends. Contrary to the cheap run orders, the run orders with more than 15 level changes possess rather small maximum time counts. Hence, the expensive run orders perform well in the presence of a linear time trend. The difficulty is thus to strike a balance between cheap but ineffective designs and costly designs with good trend resistance. The next section deals with the construction of cost-efficient and trend-robust regression designs in the presence of either fixed or random block effects.

10.8 Optimal Run Orders for Blocked Experiments

The approaches described in the literature finding trend-free block design are mainly restricted to time trends of first order, fixed block effects, equal block sizes and equidistant time points. Moreover, the number of time points is assumed to be equal to the number of observations and the approaches focus on the construction of trend-resistant run orders for a given set of design points. Consequently, the results that can be found in the literature are applicable only to a small class of practical design problems. In this section a generic approach will be presented to the construction of cost-efficient and trend-resistant run orders for arbitrary design problems.

10.8.1 Model and estimation

If a polynomial time trend of order q is present, the statistical model (10.1) for an experiment with either fixed or random block effects γ_i can be written as

$$y_{ij} = \mathbf{f}'(\mathbf{x}_{ij})\boldsymbol{\beta} + \mathbf{g}'(t_{ij})\boldsymbol{\tau} + \gamma_i + \varepsilon_{ij}, \quad i = 1, 2, \ldots, b; j = 1, 2, \ldots, k_i, \tag{10.24}$$

where $\mathbf{g}(t)$ is a q-dimensional vector representing the polynomial expansion for the postulated time trend, t_{ij} is the time point at which the jth observation within block i is taken and $\boldsymbol{\tau}$ is the vector of q parameters of the time trend. In matrix notation, model (10.24) becomes

$$\mathbf{y} = \beta_0 \mathbf{1}_n + \mathbf{X}\boldsymbol{\beta} + \mathbf{G}\boldsymbol{\tau} + \mathbf{Z}\boldsymbol{\gamma} + \boldsymbol{\varepsilon}, \tag{10.25}$$

where $\mathbf{G}$ represents the $n \times q$ design matrix for the time trend, if the blocks are assumed random. If the blocks are assumed fixed, then the intercept has to be dropped in order to obtain an estimable model. In the rest of the chapter, we will, however, assume that the block effects are random. As in the absence of trend effects, the case of fixed block effects can be seen as a special case.

If the block effects are treated as random, the fixed effects β_0, $\boldsymbol{\beta}$ and $\boldsymbol{\tau}$ can be estimated using generalized least squares:

$$\begin{aligned}
\begin{bmatrix} \hat{\beta}_{0,\mathrm{GLS}} \\ \hat{\boldsymbol{\beta}}_{\mathrm{GLS}} \\ \hat{\boldsymbol{\tau}}_{\mathrm{GLS}} \end{bmatrix} &= \left(\begin{bmatrix} \mathbf{1}'_n \\ \mathbf{X}' \\ \mathbf{G}' \end{bmatrix} \mathbf{V}^{-1} [\mathbf{1}_n \, \mathbf{X} \, \mathbf{G}] \right)^{-1} \begin{bmatrix} \mathbf{1}'_n \\ \mathbf{X}' \\ \mathbf{G}' \end{bmatrix} \mathbf{V}^{-1}\mathbf{y}, \\
&= \left(\begin{bmatrix} \mathbf{W}' \\ \mathbf{G}' \end{bmatrix} \mathbf{V}^{-1} [\mathbf{W} \, \mathbf{G}] \right)^{-1} \begin{bmatrix} \mathbf{W}' \\ \mathbf{G}' \end{bmatrix} \mathbf{V}^{-1}\mathbf{y},
\end{aligned} \tag{10.26}$$

where the variance-covariance matrix $\mathbf{V}$ is given in (10.16) and $\mathbf{W} = [\mathbf{1}_n \, \mathbf{X}]$. The variance-covariance matrix of these estimators is

$$\text{var}\begin{bmatrix} \hat{\beta}_{0,\text{GLS}} \\ \hat{\boldsymbol{\beta}}_{\text{GLS}} \\ \hat{\boldsymbol{\tau}}_{\text{GLS}} \end{bmatrix} = \left(\begin{bmatrix} \mathbf{W}' \\ \mathbf{G}' \end{bmatrix} \mathbf{V}^{-1} [\mathbf{W}\,\mathbf{G}] \right)^{-1}. \qquad (10.27)$$

The primary interest is of course in β_0 and $\boldsymbol{\beta}$. Therefore, the q parameters in $\boldsymbol{\tau}$ are usually treated as nuisance parameters and the attention is focused on the upper left-hand part of the variance-covariance matrix (10.27). Using Harville's (1997) Theorem 13.3.8, it can be shown that the deteminant of this submatrix can be optimized by minimizing

$$\mathcal{D}_t = \frac{|\mathbf{G}'\mathbf{V}^{-1}\mathbf{G}|}{\left| \begin{bmatrix} \mathbf{W}' \\ \mathbf{G}' \end{bmatrix} \mathbf{V}^{-1} [\mathbf{W}\,\mathbf{G}] \right|}. \qquad (10.28)$$

The resulting design and run order is called a $\mathcal{D}_t$-optimal run $\delta_{\mathcal{D}_t}$. The extent to which a run order is trend-free is expressed by the trend factor

$$\text{TF} = \left\{ \frac{|\mathbf{W}'\mathbf{V}^{-1}\mathbf{W} - \mathbf{W}'\mathbf{V}^{-1}\mathbf{G}(\mathbf{G}'\mathbf{V}^{-1}\mathbf{G})^{-1}\mathbf{G}'\mathbf{V}^{-1}\mathbf{W}|}{|\mathbf{W}'\mathbf{V}^{-1}\mathbf{W}|} \right\}^{1/(p+1)}, \qquad (10.29)$$

which lies between zero and one. The numerator of this expression represents the amount of information contained within the experiment if a time trend is present. The denominator represents the amount of information that would be obtained if no trend were active (see also Section 10.5.1). When the numerator and the denominator are equal, then the trend factor equals one and the design is said to be trend-free. It is shown in Tack and Vandebroek (2002) that this is the case if the orthogonality condition

$$\begin{bmatrix} \mathbf{X}' \\ \mathbf{Z}' \end{bmatrix} \mathbf{G} = 0 \qquad (10.30)$$

is satisfied. This result extends that of Bradley and Yeh (1980) who consider qualitative variables and fixed block effects only. If (10.30) is not satisfied, the trend factor will be less than one and the run order does not completely eliminate the effects of the postulated time trend on the results. A trend factor of 0.5 means that the experiment has to be replicated twice in order to obtain the same amount of information as in the absence of the time trend.

If costs are an issue, a trade-off between trend resistance and cost-efficiency can be found by computing a so-called $(\mathcal{D}_t, C)$-optimal run order $\delta_{(\mathcal{D}_t,C)}$ which maximizes the amount of information per unit cost. The corresponding criterion value equals $C\mathcal{D}_t^{-1/(p+1)}$, where C denotes the total cost of the experiment.

10.8.2 Computational results

Consider as an example an experiment set up to estimate a full quadratic model in two variables x_1 and x_2. Suppose that three blocks of five observations are available and that, within each block, the observations can be taken at five equidistant time points, which are coded to lie between -1 and $+1$. Using the points of the 3^2 factorial design, we have computed $\mathcal{D}_t$- and $(\mathcal{D}_t, C)$-optimal run orders for time trends of first, second and third order. We will denote these time trends by $\mathbf{g}_1(t)$, $\mathbf{g}_2(t)$ and $\mathbf{g}_3(t)$ respectively. For the degree of correlation η, we used 0.0, 0.4, 1.5 and 9.0. Finally, the transition cost for altering a factor level from $+1$ to 0, 0 to ± 1 and -1 to $+1$ or vice versa is twice as high as when the factor level remains the same.

The trend factors of the run orders computed are shown in Table 10.5. The table shows that the $\mathcal{D}_t$-optimal run orders with a postulated linear time trend are trend-

Table 10.5 Trend factors of optimal run orders

	$\mathcal{D}_t$-optimality			$(\mathcal{D}_t, C)$-optimality		
	$\mathbf{g}_1(t)$	$\mathbf{g}_2(t)$	$\mathbf{g}_3(t)$	$\mathbf{g}_1(t)$	$\mathbf{g}_2(t)$	$\mathbf{g}_3(t)$
$\eta = 0.0$	1.000	0.899	0.897	0.961	0.842	0.832
$\eta = 0.4$	1.000	0.977	0.977	0.903	0.961	0.938
$\eta = 1.5$	1.000	0.992	0.991	0.990	0.958	0.930
$\eta = 9.0$	1.000	0.988	0.988	0.964	0.958	0.876

free for all four variance ratios η. As an illustration, these run orders are displayed in Table 10.6. It can be verified that the main effects of x_1 and x_2 are indeed linear trend-free across the blocks if a common linear time trend is assumed for each block. The design points of the four run orders in Table 10.6 are displayed in Figure 10.7. It can be seen that the optimal design points are different for each of the η-values examined.

Table 10.5 also shows that the trend factors of the $\mathcal{D}_t$-optimal run orders decrease as the order of the time trend grows larger. This implies that the higher the order of the time trend, the more difficult it is to obtain an optimal balance for trend effects. Not unexpectedly, the $\mathcal{D}_t, C)$-optimal run orders have smaller trend factors than their $\mathcal{D}_t$-optimal counterparts.

Finally, the performance of the $\mathcal{D}_t$- and the $(\mathcal{D}_t, C)$-optimal run orders in terms of the amount of information obtained per unit cost is compared in Table 10.7. The $(\mathcal{D}_t, C)$-optimal run orders clearly outperform the $\mathcal{D}_t$-optimal ones. For the $(\mathcal{D}_t, C)$-optimal run orders, the cost of information increases as the order of the time trend grows larger.

Table 10.6 $\mathcal{D}_t$-optimal run orders for a linear time trend

		$\eta = 0.0$		$\eta = 0.4$		$\eta = 1.5$		$\eta = 9.0$	
Block	Time	x_1	x_2	x_1	x_2	x_1	x_2	x_1	x_2
1	−1	+1	−1	−1	−1	+1	+1	+1	+1
1	−0.5	+1	−1	−1	−1	+1	−1	0	0
1	0	+1	+1	+1	+1	−1	0	+1	−1
1	+0.5	−1	+1	−1	+1	0	+1	−1	+1
1	+1	−1	−1	+1	−1	−1	−1	−1	−1
2	−1	−1	+1	0	+1	0	−1	−1	−1
2	−0.5	0	+1	+1	−1	−1	+1	−1	+1
2	0	−1	−1	+1	0	+1	+1	+1	0
2	+0.5	+1	0	−1	+1	−1	0	0	+1
2	+1	+1	−1	−1	−1	+1	−1	+1	−1
3	−1	−1	−1	+1	−1	−1	−1	+1	−1
3	−0.5	−1	0	−1	+1	+1	0	−1	+1
3	0	+1	+1	−1	0	−1	+1	0	−1
3	+0.5	0	−1	+1	+1	0	−1	−1	0
3	+1	−1	+1	0	−1	+1	+1	+1	+1

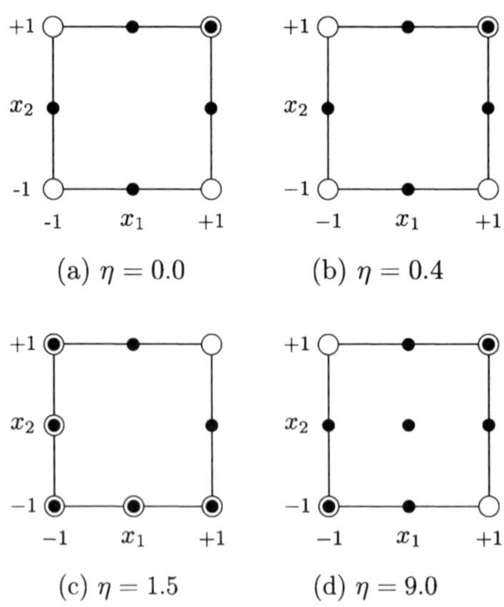

Figure 10.7 Projections of the $\mathcal{D}_t$-optimal run orders in Table 10.6 (• is a design point, ⦿ is a design point replicated twice, and ○ is a design point replicated three times)

Table 10.7 $(\mathcal{D}_t, C)$-criterion values of optimal run orders.

	$\mathcal{D}_t$-optimality			$(\mathcal{D}_t, C)$-optimality		
	$\mathbf{g}_1(t)$	$\mathbf{g}_2(t)$	$\mathbf{g}_3(t)$	$\mathbf{g}_1(t)$	$\mathbf{g}_2(t)$	$\mathbf{g}_3(t)$
$\eta = 0.0$	0.242	0.218	0.212	0.345	0.291	0.288
$\eta = 0.4$	0.265	0.259	0.265	0.314	0.306	0.298
$\eta = 1.5$	0.365	0.397	0.416	0.462	0.460	0.447
$\eta = 9.0$	1.328	1.395	1.405	1.593	1.541	1.487

10.9 A Time Trend in the Pastry Dough Mixing Experiment

In the pastry dough mixing experiment introduced in Section 10.2, the runs were conducted one at a time. This is because the laboratory equipment was used to test the different settings of the experimental variables. In many food processes, such as the baking process under study, a large variation can often be observed between different times of the day. Therefore, runs carried out early in the morning in many cases produce different results than those performed later in the day. One way to take this into account is by modelling a linear time trend. A $\mathcal{D}_t$-optimal run order for this model is given in the left panel of Table 10.8. This design is the best one found for η-values between 1 and 10. This shows once more that precise knowledge about η is often not needed. Although the design is not entirely trend-free, its trend factor amounts to 99.53% when $\eta = 1$ and to 99.46% when $\eta = 10$. The projection of the design obtained by ignoring the blocks is displayed in Figure 10.8(a). The projection is perfectly symmetric since it contains duplicates of all corner points and one replicate of the midpoints of the vertices of the design region. As a result, the design is slightly different from that in Figure 10.4b.

It can be verified that the $\mathcal{D}_t$-optimal design has a large number of changes in the factor levels. If changing the levels is cumbersome, this can be taken into account when designing the experiment. The $(\mathcal{D}_t, C)$-optimal design in the right panel of Table 10.8 is the best option if changing the level of a factor is twice as expensive as holding it fixed and $\eta = 1$. This design contains a much smaller number of factor-level changes, so that the cost of running the experiment is 25% smaller. The $\mathcal{D}_t$-criterion value of this cost-efficient design is, however, considerably worse. In terms of $\mathcal{D}_t$-efficiency, the $(\mathcal{D}_t, C)$-optimal design, the projection of which is given in Figure 10.8(b), performs 20% worse than the $\mathcal{D}_t$-optimal design. The variances of the parameter estimates are also considerably larger. This is due to the nonorthogonality with respect to the blocks of the experiment. The $(\mathcal{D}_t, C)$-optimal design displayed in Table 10.8 is no longer optimal when $\eta = 10$. For such large values of η, the $\mathcal{D}_t$-optimal design is also optimal with respect to the $(\mathcal{D}_t, C)$-criterion.

Table 10.8 $\mathcal{D}_t$- and $(\mathcal{D}_t, C)$-optimal designs for the pastry dough experiment

		$\mathcal{D}_t$-optimal design			$(\mathcal{D}_t, C)$-optimal design		
Block	Time	x_1	x_2	x_3	x_1	x_2	x_3
1	−1	−1	0	+1	+1	+1	−1
	−0.33	+1	+1	+1	+1	+1	+1
	+0.33	0	+1	−1	+1	−1	+1
	+1	+1	−1	0	+1	−1	−1
2	−1	+1	−1	+1	−1	−1	−1
	−0.33	−1	−1	−1	−1	+1	−1
	+0.33	+1	+1	−1	+1	+1	−1
	+1	−1	+1	+1	+1	−1	−1
3	−1	−1	+1	+1	+1	0	+1
	−0.33	+1	−1	+1	+1	0	−1
	+0.33	−1	−1	−1	+1	+1	0
	+1	+1	+1	−1	−1	+1	0
4	−1	+1	+1	0	0	−1	−1
	−0.33	+1	−1	−1	0	−1	+1
	+0.33	0	−1	+1	−1	0	+1
	+1	−1	0	−1	−1	0	−1
5	−1	+1	0	−1	+1	−1	0
	−0.33	−1	+1	−1	+1	+1	0
	+0.33	−1	−1	0	0	+1	−1
	+1	0	+1	+1	0	+1	+1
6	−1	0	−1	−1	−1	−1	+1
	−0.33	−1	−1	+1	−1	+1	+1
	+0.33	−1	+1	0	+1	+1	+1
	+1	+1	0	+1	+1	−1	+1
7	−1	−1	+1	−1	−1	+1	+1
	−0.33	+1	+1	+1	−1	+1	−1
	+0.33	−1	−1	+1	−1	−1	−1
	+1	+1	−1	−1	−1	−1	+1

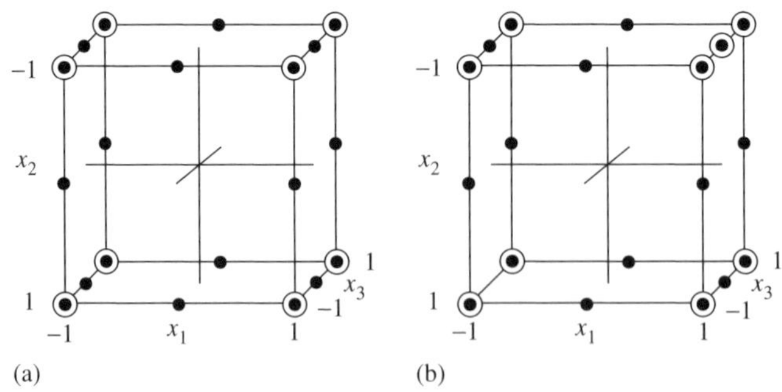

Figure 10.8 Projections of the designs from Table 10.4 obtained by ignoring the blocks: (a) projection of the $\mathcal{D}_t$-optimal design; (b) projection of the $(\mathcal{D}_t, C)$-optimal design (• is a design point and ⦿ is a design point replicated twice)

10.10 Summary

In this chapter, a number of issues that complicate the design of industrial experiments are discussed. Firstly, we focus on the problem of designing efficient experiments in the presence of fixed or random block effects. Next, it is shown how a blocked experiment can be designed if a time trend is present. Finally, it is demonstrated how cost considerations can affect the choice of a design. A pastry dough mixing experiment is used to illustrate the methods discussed.

Acknowledgement

The research that led to this chapter was carried out while the first author was a postdoctoral researcher of the Fund for Scientific Research Flanders.

Appendix: Design Construction Algorithms

As theoretical results do not exist for most practical situations, design construction algorithms are indispensable for the optimal design of practical experiments. The algorithms available in the literature for designing blocked experiments can be classified as exchange algorithms. Each of the algorithms requires the specification of a list of candidate points. A reasonable list of candidates consists of the points of a factorial design. When the model of interest is a main effects model, or a model containing mixed effects and interactions, the points of a 2^m factorial provide an adequate set of candidates. When the regression model contains quadratic effects in some of the factors, then the list of candidates should contain at least three levels for these factors. Of course, possible restrictions on the factor levels, as in mixture experiments, have to be taken into account when constructing the set of candidates. In order to find efficient designs, the set of candidate points should certainly include corner points and cover the entire design region. Apart from the list of candidate points, the input to the algorithms contains the number of observations n, the number of blocks b, the block sizes k_i $(i = 1, 2, \ldots, b)$, the number of fixed model parameters $p + 1$, the order of the model, the number of explanatory variables m and the structure of their polynomial expansion.

The oldest algorithms are the BLKL algorithm of Atkinson and Donev (1989) and the algorithm of Cook and Nachtsheim (1989). Both algorithms were developed for computing $\mathcal{D}$-optimal designs for estimating the fixed block effects model (10.2). A Fortran 77 code of the BLKL algorithm is given in Atkinson and Donev (1992). The BLKL algorithm first computes a feasible, non-singular n-point starting design. The starting design is then improved by repeatedly exchanging design points by points from a list of candidates. The algorithm stops when no further improvement in the $\mathcal{D}$-optimality criterion can be achieved. In order to avoid being stuck in a local optimum, more than one starting design is generated and the exchange procedure is repeated. Each of these repetitions is called a try. The number of tries is user-specified. Unlike Atkinson and Donev (1989), Cook and Nachtsheim (1989) only use one try. In order to obtain a starting design, they compute a $(p + 1)$-point design for model (10.6) and use this design to compute a non-singular design for the fixed block effects model (10.1). The resulting design is then improved by using two different strategies. Firstly, it is investigated whether exchanging design points with points from the candidate list improves the design. Next, the design is further improved by moving design points from one block to another and vice versa. This is done in a so-called interchange step.

An algorithm to compute $\mathcal{D}$-optimal designs for the random block effects model (10.15) was developed by Goos and Vandebroek (2001), who combined the algorithms of Atkinson and Donev (1989) and Cook and Nachtsheim (1989). As Atkinson and Donev, Goos and Vandebroek use several tries, i.e. they generate several starting designs, to avoid ending up in a local optimum. In order to improve this starting design, they use both an exchange procedure and an interchange procedure, as was also done by Cook and Nachtsheim. In order to evaluate the performance of the algorithm, Goos (2002) used the algorithm to compute $\mathcal{D}$-optimal designs for a number of situations in which the optimal design was known. It turns out that the optimal design is found in more than 98% of the tries. For large η, the designs produced by the algorithm of Goos and Vandebroek can also be compared to those obtained by applying the algorithms of Atkinson and Donev and Cook and Nachtsheim. This comparison is justified by the result from Section 10.5.2.1. In all problem instances examined, the algorithm of Goos and Vandebroek compares favourably to these algorithms. The input to the algorithm of Goos and Vandebroek is identical to that of the BLKL algorithm apart from the fact that a prior guess of the degree of correlation η has to be provided. Typically, information on η is available from prior experiments of a similar kind.

If trend effects are present and cost considerations have to be taken into account, the algorithm of Tack and Vandebroek (2002) to compute trend-resistant and cost-efficient run orders in the presence of fixed or random block effects can be used. The purpose of this algorithm is to generate optimal run orders by allocating n design points selected from a list of candidates to n out of a series of available time points. The algorithm can generate designs that minimize the $\mathcal{D}_t$- or $(\mathcal{D}_t, C)$-optimality criteria outlined in Section 10.8.1. Its input should contain the order of the assumed polynomial time trend $\mathbf{g}(t)$, a list of time points at which the observations can be taken, and cost information. The generic algorithm of Tack and Vandebroek (2002) is based on the design algorithm of Atkinson and Donev (1996) to compute trend-resistant run orders in the absence of blocking factors. In the algorithm, the starting designs are improved using three exchange strategies. Firstly, the effect is investigated of exchanging a design point by a point from the list of candidates. Secondly, the effect of moving a design point from one time point to a new time point within the same block is examined. Finally, the interchange of any design point $\mathbf{x}_i$ at time point t_k within block r with another design point $\mathbf{x}_j$ at time point t_l within block s is evaluated.

References

Anbari, F. T. and Lucas, J. M. (1994). Super-efficient designs: how to run your experiment for higher efficiency and lower cost, In: *Proceedings 48th Annual Quality Congress*, Las Vegas, pp. 852–863.

Anderson, R. L. and Bancroft, R. A. (1952). *Statistical Theory in Research*, New York: McGraw-Hill Book Co.

Atkins, J. E. and Cheng, C.-S. (1999). Optimal regression designs in the presence of random block effects, *Journal of Statistical Planning and Inference*, **77**, pp. 321–335.

Atkinson, A. C. and Donev, A. N. (1989). The construction of exact $\mathcal{D}$-optimum experimental designs with application to blocking response surface designs, *Biometrika*, **76**, pp. 515–526.

Atkinson, A. C. and Donev, A. N. (1992). *Optimum Experimental Designs*, Oxford: Clarendon Press.

Atkinson, A. C. and Donev, A. N. (1996). Experimental designs optimally balanced for trend, *Technometrics*, **38**, pp. 333–334.

Box, G. E. P. and Hunter, J. S. (1957). Multi-factor experimental designs for exploring response surfaces, *Annals of Mathematical Statistics*, **28**, pp. 195–241.

Bradley, R. A. and Yeh, C.-M. (1980). Trend-free block designs: Theory, *The Annals of Statistics*, **8**, pp. 883–893.

Cheng, C.-S. (1985). Run orders of factorial designs, In: L. LeCam, and R. A. Olshen, (eds), *Proceedings Berkeley Conference in Honor of Jerzy Neyman and Jack Kiefer*, Wadsworth, pp. 619–633.

Cheng, C.-S. (1990). Construction of run orders of factorial designs, In: S. Ghosh, (ed.), *Statistical Design and Analysis of Industrial Experiments*, New York and Basel: Marcel Dekker, Inc.

Cheng, C.-S. (1995). Optimal regression designs under random block-effects models, *Statistica Sinica*, **5**, pp. 485–497.

Cheng, C.-S. and Jacroux. M. (1988). The construction of trend-free run orders of two-level factorial designs, *Journal of the American Statistical Association*, **83**, pp. 1152–1158.

Cheng, S.-W. and Wu, C. F. J. (2002). Choice of optimal blocking schemes in two-level and three-level designs, *Technometrics*, **44**, pp. 269–277.

Cook, R. D. and Nachtsheim, C. J. (1989). Computer-aided blocking of factorial and response-surface designs, *Technometrics*, **31**, pp. 339–346.

Coster, D. C. (1993). Tables of minimum cost, linear trend-free run sequences for two- and three-level fractional factorial designs, *Computational Statistics and Data Analysis*, **16**, pp. 325–336.

Coster, D. C. and Cheng, C.-S. (1988). Minimum cost trend-free run orders of fractional factorial designs, *The Annals of Statistics*, **16**, pp. 1188–1205.

Cox, D. R. (1951). Some systematic experimental designs, *Biometrika*, **38**, pp. 312–323.

Cox, D. R. (1958). *Planning of Experiments*, New York: John Wiley & Sons, Ltd.

Daniel, C. and Wilcoxon, F. (1966). Fractional 2^{p-q} plans robust against linear and quadratic trends, *Technometrics*, **8**, pp. 259–278.

Dickinson, A. W. (1974). Some run orders requiring a minimum number of factor level changes for 2^4 and 2^5 main effects plans, *Technometrics*, **16**, pp. 31–37.

Draper, N. R. and Stoneman, D. M. (1968). Factor changes and linear trends in eight-run two-level ractorial designs, *Technometrics*, **10**, pp. 301–311.

Fries, A. and Hunter, W. G. (1980). Minimum aberration 2^{k-p} designs, *Technometrics*, **22**, pp. 601–608.

Ganju, J. (2000). On choosing between fixed and random block effects in some no-interaction models, *Journal of Statistical Planning and Inference*, **90**, pp. 323–334.

Gilmour, S. G. and Trinca, L. A. (2000). Some practical advice on polynomial regression analysis from blocked response surface designs, *Communications in Statistics: Theory and Methods*, **29**, pp. 2157–2180.

Goos, P. (2002). *The Optimal Design of Blocked and Split-plot Experiments*, New York: Springer.

Goos, P. and Vandebroek, M. (2001). D-optimal response surface designs in the presence of random block effects, *Computational Statistics and Data Analysis*, **37**, pp. 433–453.

Goos, P. and Vandebroek, M. (2004). Estimating the intercept in an orthogonally blocked experiment with random block effects, *Communications in Statistics: Theory and Methods*, **33**, pp. 873–890.

Harville, D. A. (1997). *Matrix Algebra from a Statistician's Perspective*, New York: Springer.

Jacroux, M. (1990). Methods for constructing trend-resistant run orders of 2-level factorial experiments, In: S. Ghosh, (ed.), *Statistical Design and Analysis of Industrial Experiments*, New York: Marcel Dekker, pp. 441–457.

Jacroux, M. (1998). On the determination and construction of $\mathcal{E}$-optimal block designs in the presence of linear trends, *Journal of Statistical Planning and Inference*, **67**, pp. 331–342.

Jacroux, M., Majumdar, D. and Shah, K. R. (1995). Efficient block designs in the presence of trends, *Statistica Sinica*, **5**, pp. 605–615.

Jacroux, M., Majumdar, D. and Shah, K. R. (1997). On the determination and construction of optimal block designs in the presence of linear trends, *Journal of the American Statistical Association*, **92**, pp. 375–382.

John, J. A. and Williams, E. R. (1995). *Cyclic and Computer Generated Designs*, London: Chapman & Hall.

Joiner, B. L. and Campbell, C. (1976). Designing experiments when run order is important, *Technometrics*, **18**, pp. 249–259.

Ju, H. L. (1992). Split plotting and randomization in industrial experiments, Ph.D. dissertation, University of Delaware.

Khuri, A. I. (1992). Response surface models with random block effects, *Technometrics*, **34**, pp. 26–37.

Khuri, A. I. (1994). Effect of blocking on the estimation of a response surface, *Journal of Applied Statistics*, **21**, pp. 305–316.

Khuri, A. I. and Cornell, J. A. (1987). *Response Surfaces: Designs and Analyses*, New York: Marcel Dekker.

Kiefer, J. (1959). Optimum experimental designs, *Journal of the Royal Statistical Society B*, **21**, pp. 272–319.

Kurotschka, V. G. (1981), A General approach to optimum design of experiments with qualitative and quantitative factors, *Proceedings of the Indian Statistical Institute Golden Jubilee International Conference on Statistics: Applications and New Directions*, Calcutta, edited by J. K. Ghosh and J. Roy, pp. 353–368.

Lawler, E. and Bell, M. (1966). A method for solving discrete optimization problems, *Operations Research*, **14**, pp. 1098–1112.

Lin, M. and Dean, A. M. (1991). Trend-free block designs for varietal and factorial experiments, *The Annals of Statistics*, **19**, pp. 1582–1596.

Mukerjee, R., A. Dey and Chatterjee, K. (2002). Optimal main effect plans with non-orthogonal blocking, *Biometrika*, **89**, pp. 225–229.

Myers, R. H. and Montgomery, D. C. (2002). *Response Surface Methodology: Process and Product Optimization Using Designed Experiments*, New York: John Wiley & Sons, Ltd.

Neuhardt, J. B. and Bradley, H. E. (1971). On the selection of multi-factor experimental arrangements with resource constraints, *Journal of the American Statistical Association*, **66**, pp. 618– 621.

Pignatiello, J. J. (1985). A Minimum cost approach for finding fractional factorials, *IIE Transactions*, **17**, pp. 212–218.

Sitter, R. R., Chen, J. and Feder, M. (1997). Fractional resolution and minimum aberration in blocked 2^{n-k} designs, *Technometrics*, **39**, pp. 382–390.

Sun, D. X., Wu, C. F. J. and Chen, Y. (1997). Optimal blocking schemes for 2^n and 2^{n-p} designs, *Technometrics*, **39**, pp. 298–307.

Tack, L. and Vandebroek, M. (2001). $(\mathcal{D}_t, C)$-Optimal run orders, *Journal of Statistical Planning and Inference*, **98**, pp. 293–310.

Tack, L. and Vandebroek, M. (2002). Trend-resistant and cost-efficient block designs with fixed or random block effects, *Journal of Quality Technology*, **34**, pp. 422–436.

Trinca, L. A. and Gilmour, S. G. (2000). An algorithm for arranging response surface designs in small blocks, *Computational Statistics and Data Analysis*, **33**, pp. 25–43.

Wu, C. F. J. and Hamada, M. (2000). *Experiments: Planning, Analysis, and Parameter Design Optimization*, New York: John Wiley & Sons, Ltd.

Yeh, C.-M., Bradley, R. A. and Notz, W. I. (1985). Nearly trend-free block designs, *Journal of American Statistical Association*, **80**, pp. 985–992.

Yen, V. (1985). Cost optimal saturated regression designs, In: *Proceedings ASA Business and Economic Statistics Section*, Las Vegas, pp. 286–288.

Index

Applied Optimal Designs Edited by M.P.F. Berger and W.K. Wong
 ISBN: 0-470-85697-1 (HB)